Lecture Notes in Computational Science and Engineering

52

Editors

Timothy J. Barth
Michael Griebel
David E. Keyes
Risto M. Nieminen
Dirk Roose
Tamar Schlick

Karl Heinz Hoffmann   Arnd Meyer   (Eds.)

Parallel Algorithms and Cluster Computing

Implementations, Algorithms and Applications

With 187 Figures and 18 Tables

Springer

Editors

Karl Heinz Hoffmann
Institute of Physics – Computational Physics
Chemnitz University of Technology
09107 Chemnitz, Germany
email: hoffmann@physik.tu-chemnitz.de

Arnd Meyer
Faculty of Mathematics – Numerical Analysis
Chemnitz University of Technology
09107 Chemnitz, Germany
email: a.meyer@mathematik.tu-chemnitz.de

Library of Congress Control Number: 2006926211

Mathematics Subject Classification: I17001, I21025, I23001, M13003, M1400X, M27004, P19005, S14001

ISBN-10 3-540-33539-0 Springer Berlin Heidelberg New York
ISBN-13 978-3-540-33539-9 Springer Berlin Heidelberg New York

Springer is a part of Springer Science+Business Media
springer.com

Printed in The Netherlands

Typesetting: by the authors and techbooks using a Springer LaTeX macro package
Cover design: *design & production* GmbH, Heidelberg

Printed on acid-free paper SPIN: 11739067 46/techbooks 5 4 3 2 1 0

Acknowledgement

The editors and authors of this book worked together in the SFB 393 "Parallele Numerische Simulation für Physik und Kontinuumsmechanik" over a period of 10 years. They gratefully acknowledge the continued support from the German Science Foundation (DFG) which provided the basis for the intensive collaboration in this group as well as the funding of a large number of young researchers.

Preface

High performance computing has changed the way in which science progresses. During the last 20 years the increase in computing power, the development of effective algorithms, and the application of these tools in the area of physics and engineering has been decisive in the advancement of our technological world. These abilities have allowed to treat problems with a complexity which had been out of reach for analytical approaches. While the increase in performance of single processes has been immense the increase of massive parallel computing as well as the advent of cluster computers has opened up the possibilities to study realistic systems. This book presents major advances in high performance computing as well as major advances due to high performance computing. The progress made during the last decade rests on the achievements in three distinct science areas.

Open and pressing problems in physics and mechanical engineering are the driving force behind the development of new tools and new approaches in these science areas. The treatment of complex physical systems with frustration and disorder, the analysis of the elastic and non-elastic movement of solids as well as the analysis of coupled fluid systems, pose problems which are open to a numerical analysis only with state of the art computing power and algorithms. The desire of scientific accuracy and quantitative precision leads to an enormous demand in computing power. Asking the right questions in these areas lead to new insights which have not been available due to other means like experimental measurements.

The second area which is decisive for effective high performance computing is a realm of effective algorithms. Using the right mathematical approach to the solution of a science problem posed in the form of a mathematical model is as crucial as asking the proper science question. For instance in the area of fluid dynamics or mechanical engineering the appropriate approach by finite element methods has led to new developments like adaptive methods or wavelet techniques for boundary elements.

The third pillar on which high performance computing rests is computer science. Having asked the proper physics question and having developed an appropriate effective mathematical algorithm for its solution it is the implementation of that algorithm in an effective parallel fashion on appropriate hardware which then leads to the desired solutions. Effective parallel algorithms are the central key to achieving the necessary numerical performance which is needed to deal with the scientific questions asked. The adaptive load balancing which makes optimal use of the available hardware as well as the development of effective data transfer protocols and mechanisms have been developed and optimized.

This book gives a collection of papers in which the results achieved in the collaboration of colleagues from the three fields are presented. The collaboration took place within the Sonderforschungsbereich SFB 393 at the Chemnitz University of Technology. From the science problems to the mathematical algorithms and on to the effective implementation of these algorithms on massively parallel and cluster computers we present state of the art technology. We highlight the connections between the fields and different work packages which let to the results presented in the science papers.

Our presentation starts with the Implementation section. We begin with a view on the implementation characteristics of highly parallelized programs, go on to specifics of FEM and quantum mechanical codes and then turn to some general aspects of postprocessing, which is usually needed to analyse the obtained data further.

The second section is devoted to Algorithms. The main focus is on FEM algorithms, starting with a discussion on efficient preconditioners. Then the focus is on a central aspect of FEM codes, the aspect ratio, and on problems and solutions to non-matching meshes at domain boundaries. The Algorithm section ends with discussing adaptive FEM methods in the context of elastoplastic deformations and a view on wavelet methods for boundary value problems.

The Applications section starts with a focus on disordered systems, discussing phase transitions in classical as well as in quantum systems. We then turn to the realm of atomic organization for amorphous carbons and for heterophase interphases in Titanium-Silicon systems. Methods used in classical as well as in quantum mechanical systems are presented.We finish by a glance on fluid dynamics applications presenting an analysis of Lyapunov instabilities for Lenard-Jones fluids.

While the topics presented cover a wide range the common background is the need for and the progress made in high performance parallel and cluster computing.

Chemnitz
March 2006

Karl Heinz Hoffmann
Arnd Meyer

Contents

Part I

Implementions

Parallel Programming Models for Irregular Algorithms

Gudula Rünger

Technische Universität Chemnitz, Fakultät für Informatik
09107 Chemnitz, Germany
ruenger@informatik.tu-chemnitz.de

Applications from science and engineering disciplines make extensive use of computer simulations and the steady increase in size and detail leads to growing computational costs. Computational resources can be provided by modern parallel hardware platforms which nowadays are usually cluster systems. Effective exploitation of cluster systems requires load balancing and locality of reference in order to avoid extensive communication. But new sophisticated modeling techniques lead to application algorithms with varying computational effort in space and time, which may be input dependent or may evolve with the computation itself. Such applications are called irregular. Because of the characteristics of irregular algorithms, efficient parallel implementations are difficult to achieve since the distribution of work and data cannot be determined a priori. However, suitable parallel programming models and libraries for structuring, scheduling, load balancing, coordination, and communication can support the design of efficient and scalable parallel implementations.

1 Challenges for parallel irregular algorithms

Important issues for gaining efficient and scalable parallel programs are load balancing and communication. On parallel platforms with distributed memory and clusters, load balancing means spreading the calculations evenly across processors while minimizing communication. For algorithms with regular computational load known at compile time, load balancing can be achieved by suitable data distributions or mappings of task to processors. For irregular algorithms, static load balancing becomes more difficult because of dynamically changing computation load and data load.

The appropriate load balancing technique for regular and irregular algorithms depends on the specific algorithmic properties concerning the behavior of data and task:

- The algorithmic structure can be data oriented or task oriented. Accordingly, load balancing affects the distribution of data or the distribution of tasks.
- Input data of an algorithm can be regular or more irregular, like sparse matrices. For regular and some irregular input data, a suitable data distribution can be selected statically before runtime.
- Regular as well as irregular data structures can be static or can be dynamically growing and shrinking during runtime. Depending on the knowledge before runtime, suitable data distributions and dynamic redistributions are used to gain load balance.
- The computational effort of an algorithm can be static, input dependent or dynamically varying. For a static or input dependent computational load, the distribution of tasks can be planned in advance. For dynamically varying problems a migration of tasks might be required to achieve load balancing.

The communication behavior of a parallel program depends on the characteristics of the algorithm and the parallel implementation strategy but is also intertwined with the load balancing techniques. An important issue is the locality of data dependencies in an algorithm and the resulting communication pattern due to the distribution of data.

- Locality of data dependencies: In the algorithm, data structures are chosen according to the algorithmic needs. They may have local dependencies, e.g. to neighboring cells in a mesh, or they may have global dependencies to completely different parts of the same or other data structures. Both local and global data dependencies can be static, input dependent or dynamically changing.
- Locality of data references: For the parallel implementation of an algorithm, aggregate data structures, like arrays, meshes, trees or graphs, are usually distributed according to a data distribution which maps different parts of the data structure to different processors. Data dependencies between data on the same processor result in local data references. Data dependencies between data mapped to different processors cause remote data reference which requires communication. The same applies to task oriented algorithms where a distribution of tasks leads to remote references by the tasks to data in remote memory.
- Locality of communication pattern: Depending on the locality of data dependencies and the data distribution, locality of communication pattern occurs. Local data dependencies usually lead either to local data references or to remote data references which can be realized by communication with neighboring processors. This is often called locality of communication. Global data dependencies usually result in more complicated remote access and communication patterns.

Communication is also caused by load balancing when redistributing data or migrating tasks to other processors. Also, the newly created distribution of data or tasks create a new pattern of local and remote data references and thus cause new communication patterns after a load balancing step. Although the specific communication may change after redistribution, the locality of the communication pattern is often similar.

The static planning of load balance during the coding phase is difficult for irregular applications and there is a need for flexible, robust, and effective programming support. Parallel programming models and environments address the question how to express irregular applications and how to execute the application in parallel. It is also important to know what the best performance can be and how it can be obtained. The requirement of scalability is essential, i.e. the ability to perform efficiently the same code for larger applications on larger cluster systems. Another important aspect is the type of communication. Specific communication needs, like asynchronous or varying communication demands, have to be addressed by a programming environment and correctness as well as efficiency are crucial.

Due to diverse application characteristics not all irregular applications are best treated by the same parallel programming support. In the following, several programming models and environments are presented:

- Task pool programming for hierarchical algorithms,
- Data and communication management for adaptive algorithms,
- Library support for mixed task and data parallel algorithms,
- Communication optimization for structured algorithms.

The programming models range from task to data oriented modes for expressing the algorithm and from self-organizing task pool approaches to more data oriented flexible adaptive modes of execution.

2 Task pool programming for hierarchical algorithms

The programming model of *task pools* supports the parallel implementation of task oriented algorithms and is suitable for hierarchical algorithms with dynamically varying computational work and complex data dependencies.

The main concept is a decomposition of the computational work into tasks and a task pool which stores the tasks ready for execution. Processes or threads are responsible for the execution of tasks. They extract tasks from the task pool for execution and create new tasks which are inserted into the task pool for a later computation, possibly by another process or thread. Complex data dependencies between tasks are allowed and may lead to complex interaction between the tasks, forming a virtual task graph. Usually, task pools are provided as programming library for shared memory platforms. Library routines for the creation, insertion, and extraction of tasks are available. A fixed number of processes or threads is created at program start to execute an arbitrary number of tasks with arbitrary dependence structures.

Load balancing and mapping of tasks is achieved automatically since a process extracts a task whenever processor time is available. There are several possibilities for the internal realization of task pools, which affect load balancing. Often the tasks are kept in task queues, see also Fig. 1:

- Central task pools: All tasks of the algorithm are kept in one task queue from which all threads extract tasks for execution and into which the newly created tasks are inserted. Access conflicts are avoided by a lock mechanism for shared memory programming.
- Decentralized task pools: Each thread has its own task queue from which the thread extracts tasks and into which it inserts newly created tasks. No access conflicts can occur and so there is no need for a lock mechanism. But load imbalances can occur for irregularly growing computational work.
- Decentralized task pools with task stealing: This variant of the decentralized task pool offers a task stealing mechanism. Threads with an empty task queue can steal tasks from other queues. Load imbalance is avoided but task stealing needs a locking mechanism for correct functionality.

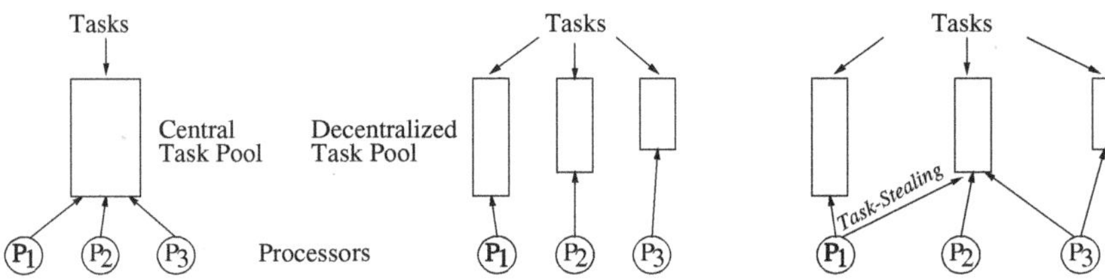

Fig. 1. Different types of task pool variants for shared memory

Due to the additional overhead of task pools it is suggested to use them only when required for highly irregular and dynamic algorithms. Examples are the hierarchical radiosity method from computer graphics and hierarchical n-body algorithms.

The hierarchical radiosity method

The radiosity algorithm is an observer-independent global illumination method from computer graphics to simulate diffuse light in three-dimensional scenes [10]. The method is based on the energy radiation between surfaces of objects and accounts for direct illumination and multiple reflections between surfaces within the environment. The radiosity method decomposes the surface of objects in the scene into small elements A_j, $j = 1, \ldots, n$, with almost constant radiation energy. For each element, the radiation energy is represented by a *radiosity value* B_j (of dimension [Watt/m^2]) describing the radiant energy per unit time and per unit area dA_j of A_j. The radiosity values of the elements

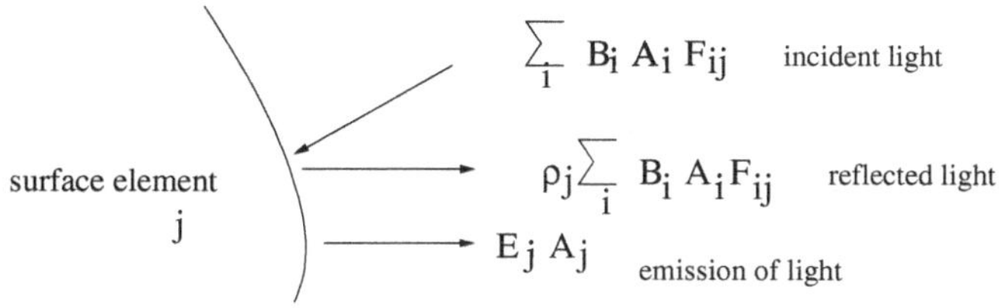

Fig. 2. Illustration of the radiosity equation

are determined by solving a linear equation system relating the different radiation energies of the scene using configuration factors (which express the geometrical positioning of the elements), see also Fig. 2:

$$B_jA_j = E_jA_j + \rho_j \sum_{i=1}^{n} F_{ij}B_iA_i, \quad j = 1, \ldots, n. \tag{1}$$

The element's emission energy is E_j. The factor ρ_j describes the diffuse reflectivity property of A_j. The dimensionless factors F_{ij} (called *configuration factors* or *form factors*)

$$F_{ij} = \frac{1}{A_i} \int_{A_i} \int_{A_j} \frac{cos(\delta_i)cos(\delta_j)}{\pi r^2} \mathrm{d}A_j \mathrm{d}A_i \tag{2}$$

describe the portions of radiance $\Phi_j = B_jA_j$ (of dimension [Watt]) incident on A_j, see also Fig. 3. Using the symmetry relation $F_{ij}A_i = F_{ji}A_j$ yields the linear system of equations for the radiosity values B_j

$$B_j = E_j + \rho_j \sum_{i=1}^{n} F_{ji}B_i, \quad j = 1, \ldots, n. \tag{3}$$

The computation of configuration factors as well as the solution of the linear system can be performed by different numerical methods (see [13] and its references). A variety of methods have been proposed to reduce the computational costs, including the hierarchical radiosity method [12] which realizes an efficient computational technique for solving the transport equations that

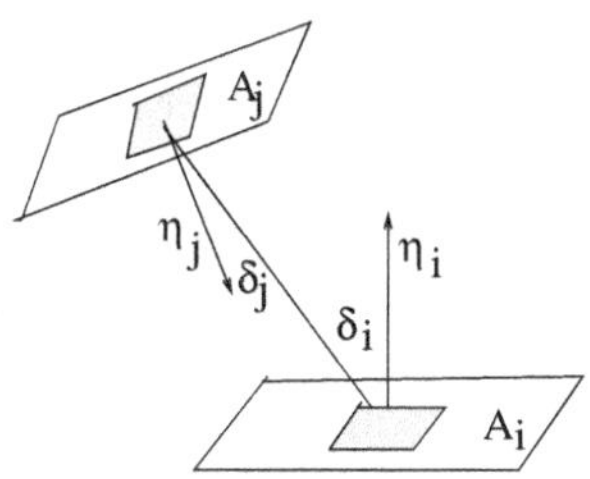

Fig. 3. Illustration of the form factors

specify the radiosity values of surface patches in complex scenes. The mutual illumination of surfaces is computed more precisely for nearby surfaces and less precisely for distant surfaces.

A common implementation strategy uses quadtrees to store the surfaces and interaction lists to store data dependencies due to mutual illumination. For each input polygon the subdivision into a hierarchy of smaller portions of the surface is organized in a quadtree, see Figure 4. The patches or elements attached to the four children of a node represent a partition of the patch attached to the parent's node. For each patch q of each input polygon, an interaction list is maintained containing patches of other input polygons, which are visible from the patch q, i.e. which can reflect light to q.

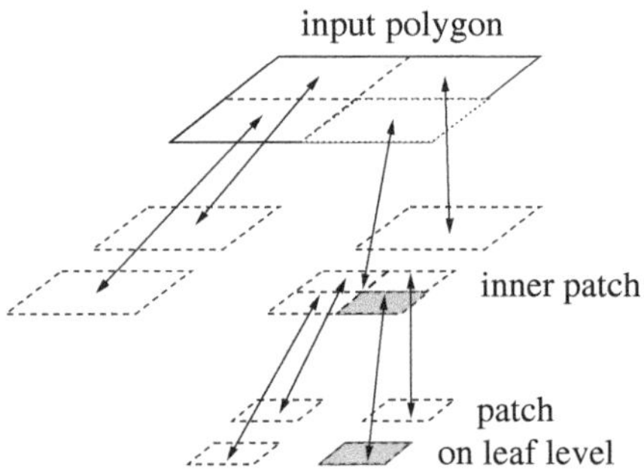

Fig. 4. Illustration of the quadtree data structure for the hierarchical radiosity method

During the computation of the form factors, the quadtrees are built up adaptively in order to guarantee the computations to be of sufficient precision. Therefore, the quadtrees need not be balanced. The method computes the energy transport (i.e. the configuration factor) between two patches or elements only if it is not too large; otherwise the patches are subdivided, see Fig. 5. Thus, each patch or element has its individual set of interaction elements or patches for which the configuration factors have to be computed. The hierarchical method alternates iteration steps of the Jacobi method to solve the energy system (3) with a re-computation of the quadtree and the interaction sets based on $F_{ji}B_i$ with radiosity values B_i from the last iteration step.

The computational structure of the hierarchical radiosity method is task oriented, input dependent, and dynamically varying during runtime. The data structures are a set of quadtrees with multiple dynamically changing interactions stored in interaction sets. Thus, an efficient parallel implementation of the hierarchical radiosity method requires a highly dynamic parallel programming model with load balancing during runtime, which is provided by task pools.

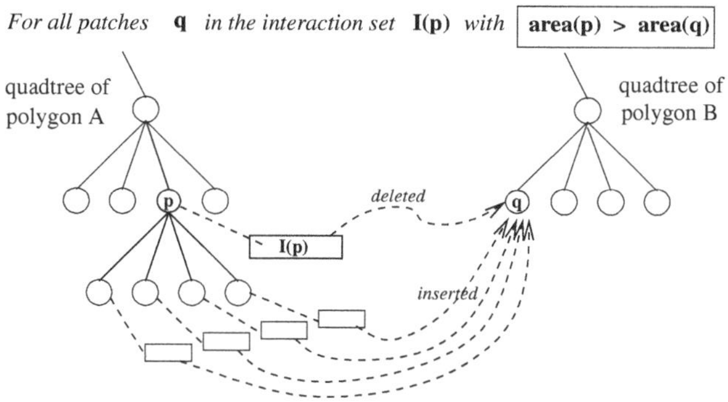

Fig. 5. Energy-based subdivision of a patch p and corresponding changes of interaction lists $I(p)$ in the hierarchical radiosity method

Extensive work concerning task pool implementations of hierarchical algorithms has been presented in [44]. How to realize a large potential of parallelism with the task pool approach in order to employ a large number of processes is investigated in [29]. In [27] additional task pool variants are presented and their performance impact is investigated for different irregular applications.

Task pool teams

For use on parallel platforms with distributed memory or on clusters, the idea of task pools has to be extended in order to include communication. One approach are *task pool teams* which combine task pools running on single cluster nodes with explicit communication [15, 19]. To support complex task and dependence structures in a dynamically growing task graph, a powerful communication mechanism with asynchronous and dynamic communication between tasks is needed. Asynchronous communication is required since it is not known in advance when and where a communication partner may be ready for execution. Dynamic communication is required since it is impossible to know in advance that a communication, e.g. for providing data, is requested by another task.

Task pool teams for an SMP (symmetric multiprocessor) cluster combines thread based task pools on SMP nodes with specific communication protocols handling the communication requirements mentioned above. The realization uses Pthreads for SMPs and MPI for communication between SMP nodes [11]. A number of worker threads and one specific communication thread run on each SMP node; the communication protocol for task pool teams supports asynchronous communication between SMP nodes exploiting the communication thread. For the application programmer, the programming support provides library routines for explicitly creating and extracting tasks; communication patterns are also explicitly inserted in the parallel program.

The hierarchical radiosity method is one of the most challenging problems concerning irregularity of data access and dynamic behavior of load and communication. Communication is required for remote data access as well as for load balancing between nodes. Task pool teams provide an appropriate tool for handling load imbalances on SMP nodes as well as on entire cluster platforms. Load balance between SMP nodes is realized explicitly offering additional possibilities for optimizing the expensive redistribution of data or migration of tasks.

The programming model of task pool teams also has been used to realize different parallel variants of the simulation of diffusion processes using random Sierpinski Carpets [9,20]. In this application the task pool team approach is especially suitable to efficiently realize the boundary update phase of the algorithm which is necessary after time steps. The first variant has a synchronous parallel update phase which exploits the specific communication thread provided by task pool teams; the exchange of boundary information is started by collective communication of the worker threads and the communication thread is responsible for the processing of the data received. The second implementation variant realizes an asynchronous boundary-update using only the task pool team's communication mechanism. Measurements of the execution time on a Xeon-Cluster with SCI network and a Beowulf cluster with Fast-Ethernet show that the synchronous approach is slightly better.

The central aim of using task pool teams is to support communication protocols that are suitable for dynamic and asynchronous communication, but which do not rely on specific attributes of the MPI library such as thread safety [17]. Thus, the implementation provides great flexibility concerning the underlying communication libraries, the parallel platform used, and specific application algorithms.

3 Data and communication management for adaptive algorithms

Adaptive algorithms adjust their behavior according to the specified properties of the result to be computed. This includes an adaption of computation and/or data and is usual guided by a required precision of a solution. A typical example is the adaptive finite element method for solving partial differential equations numerically.

The finite element method uses a discretization of the physical space into a mesh of finite elements. The numerical solution of the given problem is then computed iteratively on the mesh. The resulting numerical algorithm has mainly local dependencies since only values stored for neighboring mesh cells are used for the computation. Because of these local dependencies, a useful parallel implementation strategy exploits a decomposition of the mesh into parts of equal size in order to obtain load balance [3]. Communication is minimized by using a decomposition into blocks with small boundaries such that a small amount of data is exchanged with the neighboring processors.

The adaptive finite element method starts with a coarse mesh and performs a stepwise refinement of the mesh according to the requested precision of the approximation solution. From the mathematical point of view, error estimation is an important point to guide the refinement. From the implementation point of view, the appropriate data structures implementing the mesh and an effective realization of the refinement is crucial. Treelike data structures for storing the adaptive mesh and its refinement structures are one option.

For a parallel implementation, one has to deal with dynamically growing or shrinking data structures during runtime and varying communication needs. However, there is a difference compared to the hierarchical algorithm introduced in Sect. 2. The data dependencies of the finite element method are local in the sense that data exchange is required only with neighboring mesh cells. This property still holds for the adaptive method. Neighboring mesh cells might be refined into several new mesh cells but the new neighbors for communication are immediately known. No further unknown interaction occurs. Thus, an appropriate parallel programming model for the adaptive finite element method is a dynamic use of load balancing methods by graph partitioning methods known from the non-adaptive case. Graph partitioning methods can be used during runtime to achieve load balance.

Graph partitioning

The decomposition of data meshes is more complicated in the irregular case and is related to the NP-hard graph partitioning problem which partitions a given graph into subgraphs of almost equal size while cutting a minimal number of edges. Graph partitioning algorithms use one- and multi-dimensional partitioning or recursive bisection [7, 30]. Recursive bisection includes the partitioning according to coordinate values which is especially suitable for sparse matrices [46], recursive graph bisection exploiting local properties of the graph [26], or recursive spectral bisection using eigenvalues of the adjacency matrix as global property [31]. Multi-level algorithms for graph partitioning use multiple levels of the graph with different refinement which are produced in sequence of consecutive steps [14, 23, 24]. Programming support for the partitioning of unstructured graphs or reordering of sparse matrices is provided by the METIS System [22, 25].

The execution time for graph partitioning algorithms adds an additional overhead to the parallel execution time of the application problem to be solved, since the graph partitioning and repartitioning have to be done at runtime. The resulting communication time can be very high such that the incorporation of repartitioning may result in a more expensive parallel program. Thus, irregular algorithms with dynamically varying data structures usually require an additional mechanism for an efficient implementation.

Data and communication management for adaptive finite element methods

The efficient parallel implementation of a hexahedral adaptive finite element method has been presented in [16]. In this approach, the communication is encapsulated in order to provide a general mechanism for repartitioning in graph based algorithms.

The main characteristics inducing irregularity to adaptive hexahedral FEM are adaptively refined data structures and hanging nodes. A hanging node is a vertex of an element that can be a mid node of a face or an edge. Such nodes can occur when hexahedral elements are refined irregularly, i.e. when neighboring volumes have different levels of refinement. For a correct numerical computation, hanging nodes require projections onto nodes of other refinement levels during the solution process. The adaptive refinement process with computations on different refinement levels creates hierarchies of data structures and requires the explicit storage of these structures including their relations. These characteristics lead to irregular communication behavior and load imbalances in a parallel program.

The task pool team approach from Sect. 2 provides a concept which allows any kind of data references or communication patterns, including strong irregular communication. For adaptive FEM, however, the locality of references and communication is slightly different than described for the algorithms presented in the last section.

- The adaptive FEM has a data oriented program structure with dynamically growing (or shrinking) data structures. Based on the input data a suitable initial distribution of data can be chosen at program start.
- The computational effort varies dynamically according to the dynamic behavior of the data. The process is guided by the FEM algorithm with appropriate error estimation methods.
- The data dependencies are of local character, i.e. there are dependencies to neighboring cells in the mesh or graph data structures. This leads to local communication patterns in a parallel program with only a few neighboring processors.
- For efficient parallel execution, load redistribution is required during runtime. But load increases resulting from adaptive refinement are usually concentrated on a few processors. Appropriate load redistribution may affect all processors but are of local nature between neighboring processors.
- Communication occurs for remote data access when data are needed for calculation. Also shared data structures have to be maintained consistently during program run. In most cases the exact time of communication is not known in advance. Message sizes or communication partners vary and depend on the input. However, the communication partners work synchronously.

A parallel FEM implementation can exploit the locality of references and local communication pattern. A suitable data management and communication layer for adaptive three-dimensional hexahedral FEM is presented in [18].

The data management assumes a distribution of volumes across the processors where neighboring volumes share faces, edges, or nodes of the mesh data structure. The mapping of neighboring volumes to different processors requires an appropriate storage of those shared data such that the data management guarantees correct storage and a correct access to the data. The following data structures have been proposed:

- Shared data are stored as duplicates in all processors holding parts of the corresponding volumes. The solution vector is distributed correspondingly. Specific restrictions guarantee the correct computation on those data.
- The data structure and its distribution are refined consistently in two steps: a local refinement step for data which are only kept in the memory of one specific processor and a remote refinement step for data with duplicates in the memories of other processors.
- Coherence lists store the information about the distribution of data to support remote refinement and fast identification of communication partners. Due to the locality properties of adaptive FEM the remote refinement applies to neighboring processors.

The communication layer provides support for different communication situations that can occur in adaptive FEM:

- Synchronization and exchange of results between neighboring processors during the computation step.
- Accumulation of local subtotals which yield the total result after a computation step.
- Exchange of administration information after a refinement of volumes, including remote refinement, creation or update of duplicate lists, and identification of hanging nodes.

The second and third communication situations are of irregular type and are handled with a specific protocol. This protocol deals with asynchronous communication since the exact moment of communication is unknown and can take place with varying communication partners and message sizes. A collection phase gathers information about remote duplicates and its owners needed for the communication. Additionally, each processor asynchronously provides information about duplicate data requested by other processors in the collection phase.

A suitable method for repartitioning and load balancing is the use of a graph partitioning tools like ParMetis. The communication protocol presented in [18] can be combined with graph partitioning and guarantees the correct behavior after each repartitioning. The protocol has been used for the parallelization of the program version SPC-PM3AdH which realizes a 3-dimensional adaptive hexahedral FEM method suitable to solve elliptic partial differential equations [2]. The parallelization method can be applied to all adaptive algorithms which provide similar locality properties and similar communication patterns as adaptive FEM codes.

4 Library support for mixed task and data parallel algorithms

A large variety of application problems from scientific computing and related areas have an inherent modular structure of cooperating subtasks calling each other. Examples include environmental models combining atmospheric, surface water, and ground water models, as well as aircraft simulations combining models for fluid dynamics, structural mechanics, and surface heating, see [1] for an overview. But modular structures can also be found within numerical algorithms built up from submethods. For the efficient parallel implementation of those applications an appropriate parallel programming model is needed which can express a (hierarchical) modular structure.

The SPMD (single program multiple data) model proposed in [5] was from its inception more general than a data parallel model and did allow a hierarchical expression of parallelism, however most implementations exploit only a data parallel form. But many research groups have proposed models for mixed task and data parallel executions with the goal of obtaining parallel programs with faster execution time and better scalability properties, see [1, 45] for an overview of systems and programming approaches and see [4] for a detailed investigation of the benefits of combining task and data parallel executions. Several models support the programmer in writing efficient parallel programs for modular applications without dealing too much with the underlying communication and coordination details of a specific parallel machine. Language approaches include Braid, Fortran M, Fx, Opus, and Orca, see [1]. Fortran M [8] allows the creation of processes which can communicate with each other by predefined channels and which can be combined with HPF Fortran for a mixed task and data parallel execution.

The TwoL (Two Level) model uses a stepwise transformation approach for creating an efficient parallel program from a module specification of the algorithm (upper level) with calls to basic modules (lower level) [34, 36]. The transformation approach can be realized by a compiler tool and is suitable for statically known module structures. Different recent realizations of the specification mechanism have been proposed in [28] and [41]. For dynamically growing and varying modular structures, however, support at runtime is required which is provided by the runtime library Tlib for multiprocessor task programming.

Multiprocessor task programming

For the implementation on distributed memory machines or clusters, the modular structure can be captured by using tasks which incorporate the modules of the application. Those tasks are often called multiprocessor tasks (M-tasks) since each task can be executed on an arbitrary number of processors, concurrently with other M-tasks of the same program executed on disjoint processor groups. Internally an M-task can have a data-parallel or SPMD structure but may also have an internal hierarchical and modular structure. The entire M-task program consists of a set of cooperating, hierarchically structured M-tasks which mimic the module dependence structure of the algorithm. M-task programming can be used to design parallel programs with mixed task and data parallelism where a coarse structure of tasks form a coarse-grained hierarchical M-task graph, see Fig. 6. The execution mode is group-SPMD where at each point in execution time disjoint processor groups execute separate SPMD modules of the algorithm.

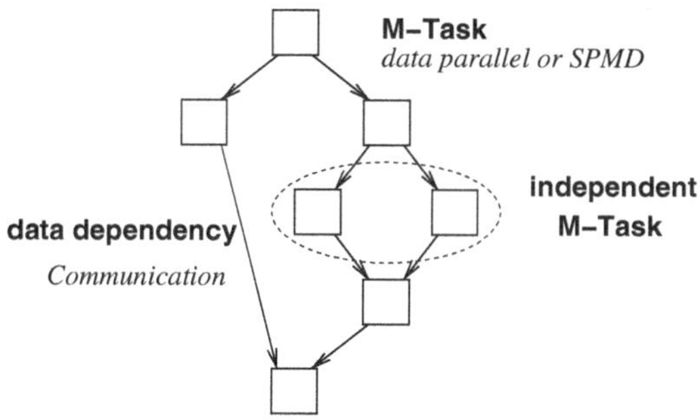

Fig. 6. Illustration of an M-task graph and its potential parallelism. Nodes denote M-tasks and arrows denote data dependencies between tasks which might result in communication

The advantage of the described form of mixed task and data parallelism is a potential for reducing the communication overhead and for improving scalability, especially if collective communication operations are used. Collective communication operations performed on smaller processor groups lead to smaller execution times due to the logarithmic or linear dependence of the communication times on the number of processors [38, 48]. As a consequence, the concurrent execution of independent tasks on disjoint processor subsets of appropriate size can result in smaller parallel execution times than the consecutive execution of the same M-tasks one after another on the entire set of processors.

Modular structures can be found in many numerical algorithms of multi-level form [21]. As an example we describe the potential of M-task parallelism in solution methods for ordinary differential equations.

Modular structures of Runge-Kutta methods

Numerical methods for solving systems of ordinary differential equations exhibit a nested or hierarchical structure which makes them amenable to a mixed task and data parallel realization. An example is the diagonal-implicitly iterated Runge-Kutta (DIIRK) method which is an implicit solution method with integrated step-size and error control for ordinary differential equations

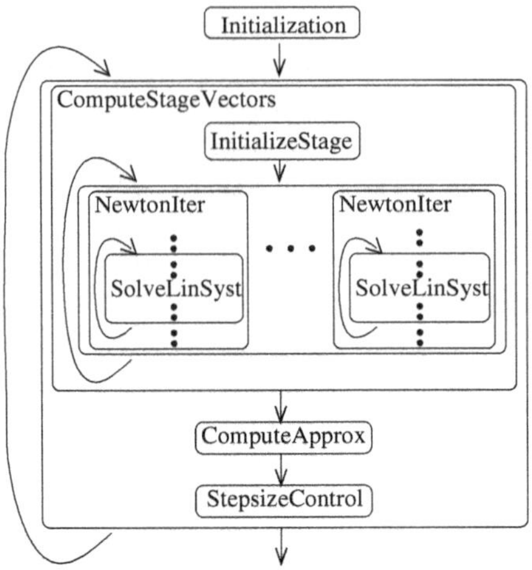

Fig. 7. Illustration of the modular structure of the diagonal-implicitly iterated Runge-Kutta method

(ODEs) arising, e.g., when solving time-dependent partial differential equations with the method of lines [47].

The modular structure of this solver is given in Figure 7 where boxes denote M-tasks and back arrows denote loop structures. In each time step, the solver computes a fixed number of s stage vectors (ComputeStageVectors) which are then combined to the final approximation (ComputeApprox). The method offers a potential of parallel execution of M-tasks since the computation of the different stage vectors are independent of each other [35]. Each single computation of a stage vector requires the solution of a non-linear equation system whose size is determined by the system size of the ordinary differential equations. The non-linear systems are solved with a modified Newton method (NewtonIter) requiring the solution of a linear equation system (SolveLinSyst) in each iteration step. Depending on the characteristics of the linear system and the solution method used, a further internal mixed task and data parallel execution can be used, leading to another level in the task hierarchy. A parallel implementation can exploit the modular structure in several different ways but can also execute the entire method in a pure SPMD fashion. The different implementations differ in the coordination of the M-tasks and usually differ in the resulting parallel execution time.

Runtime library Tlib

The runtime library Tlib supports M-task programming with varying, hierarchically structured, recursive M-tasks cooperating according to a given coordination program [39]. The coordination program contains activations of coordination operations of the Tlib library and user-defined M-task functions.

The parallel programming model for Tlib programs is a group-SPMD model at each point in the execution time. This results in a multilevel group-SPMD model with several different hierarchies of processor groups due to the dynamic change of processor groups during runtime and nested M-task calls.

Tlib mainly provides two kinds of operations:

- a family of split operations to structure the given set of processors and
- a family of mapping operations to assign specific M-tasks to specific processor groups for execution.

M-tasks cooperate through parameters which can include composed data structures so that Tlib programs have to deal with data placement, data distribution, and data redistribution. The specific challenge for selecting a data distribution lies in the dynamic character of M-task programs in Tlib, since the actual M-task structure and the processor layout are not necessarily known in advance. The advantage of this dynamic behavior is that arbitrary, hierarchically structured and recursive M-task programs can be coded easily, providing an easy way to express divide-and-conquer algorithms or irregular algorithms. For a data distribution and a correct cooperation of arbitrary M-tasks a specific data format is needed which fits the dynamic needs of the model [40].

5 Communication optimization for structured algorithms

The locality of dependencies can also be exploited to optimize the communication for parallel algorithms with a more structured data or task dependence graph. The parallel programming model of orthogonal processor groups provides a group-SPMD model for applications with a two- or higher-dimensional data or task grid where dependencies are mainly aligned in the dimensions of this grid. The advantage is a reduction of the communication to smaller groups of processors which leads to a reduction of the communication overhead as already mentioned in Sect. 4. The entire set of processors executing an application program is organized in a virtual two- or higher-dimensional processor grid and a fixed number of different decompositions of this set into disjoint processor subsets is defined. The subsets of a decomposition into disjoint processor groups correspond to the lower-dimensional hyperplanes of the virtual processor grid which are geometrically orthogonal to each other.

An application algorithm is mapped onto the processor grid according to its natural grid based data or task structure. The program executes a sequence of phases in which a processor alternatively executes the tasks assigned to it in an SPMD or group-SPMD way. In a group-SPMD phase the application program uses exactly one of the decompositions into processor hyperplanes and a processor within a hyperplane performs an SPMD computation together with other processors in the same group, including group internal collective communication operations. At each time of the execution only one partition can be active.

For programming with virtual processor grids, a suitable coding of processor grids and grid based data or task structures is needed. Parameterized data or task distributions can be used as a flexible technique for the mapping of data or tasks to processor subgroups.

Parameterized data or task mapping

Parameterized data distributions describe the data distribution for data grids of arbitrary dimension [6,37]. A task is assigned to exactly one processor, but each processor might have several tasks assigned to it.

A parameterized cyclic mapping of a d-dimensional task grid to a virtual processor grid of size $p_1 \times \cdots \times p_d$ is specified by a block-size b_l, $l = 1, \ldots, d$, in each dimension l. The block-size b_l determines the number of consecutive elements that each processor obtains from each cyclic block in this dimension. For a total number of p processors, the distribution of the data or task grid $\mathcal{T}$ of size $n_1 \times \cdots \times n_d$ is described by an assignment vector of the form:

$$((p_1, b_1), \ldots, (p_d, b_d)) \tag{4}$$

with $p = \prod_{l=1}^{d} p_l$ and $1 \leq b_l \leq n_l$ for $l = 1, \ldots, d$. For simplicity we assume $n_l/(p_l \cdot b_l) \in \mathbb{N}$.

The set of processors is divided into disjoint subsets of processors due to the virtual d-dimensional processor number in a d-dimensional processor grid. For the two-dimensional processor grid of size $p_1 \times p_2$, there are two different decompositions into a disjoint set of p_2 row groups R_q, $1 \leq q \leq p_2$, and into a disjoint set of p_1 column groups C_q, $1 \leq q \leq p_1$:

$$R_q = \{(r, q) \mid r = 1, \ldots, p_1\} \quad , \qquad C_q = \{(q, r) \mid r = 1, \ldots, p_2\} \tag{5}$$

with $|R_q| = p_1$ and $|C_q| = p_2$. The row and column groups build separate orthogonal partitions of the set of processors P, i.e.,

$$\bigcup_{q=1}^{p_2} R_q = \bigcup_{q=1}^{p_1} C_q = P \text{ and } R_q \cap R_{q'} = \emptyset = C_q \cap C_{q'} \text{ for } q \neq q'.$$

The row groups R_q and the column groups C_q are orthogonal processor groups.

A mapping of the task or data grid to orthogonal processor groups uses the data distribution vector (4) and the decomposition of the set of processors (5), see Fig. 8. Row i of the task grid $\mathcal{T}$ is assigned to the processors of a single row group $Ro(i) = R_k$ with $k = \left\lfloor \frac{i-1}{b_2} \right\rfloor \bmod p_2 + 1$. Similarly, column j of the task grid T is assigned to the processors of a single column group $Co(j) = C_k$ with $k = \left\lfloor \frac{j-1}{b_1} \right\rfloor \bmod p_1 + 1$. A program mapped to orthogonal processor groups can use the row and column groups $Ro(i)$ and $Co(i)$, i.e. the program can use the original task indices. Thus, after the mapping the task structure is still visible and the orthogonal processor groups according to the given mapping are known implicitly.

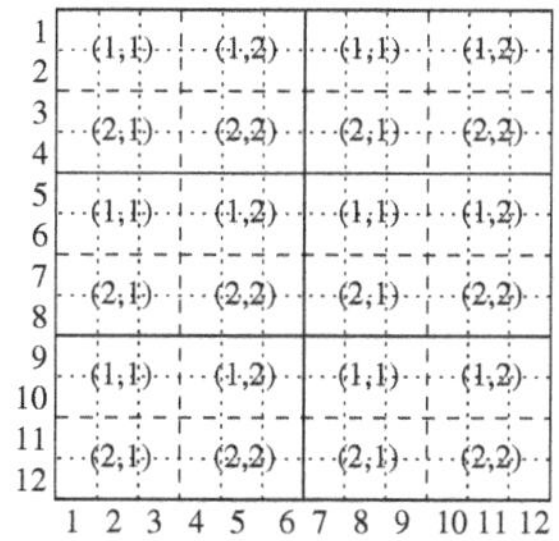

Fig. 8. Illustration of a task or data grid of size 12×12 and a mapping to a processor grid of size 2×2 with block-sizes $b_1 = 3, b_2 = 2$

Programming orthogonal structures

A runtime library ORT has been implemented to support parallel programming in the group-SPMD model with orthogonal processor groups. The library provides functions to build processor partitions and functions to assign tasks to the processor groups. The application programmer starts with a specification of the message-passing program using the library functions within a C and MPI program. Processor partitions are built up internally by the library when calling some starting routines. Tasks are mapped to processor groups with library calls giving tasks and processor groups as parameters. The advantage of a library support is to have a comfortable specification mechanism for group-SPMD programs and an executable implementation at the same time. The library is implemented on top of MPI so that the specification program is executable on message-passing machines. This programming style allows the application programmer to specify the program phases and their organization in a clear and readable program code. The execution model for programs in the ORT programming model has the following characteristics:

- Processors are organized in a two- or multi-dimensional grid structure depending on the specific application and the mapping.
- Processor groups are supported in each hyperplane of the processor grid, i.e. if processors are organized in an s-dimensional grid, hyperplanes of dimensions $1, \ldots, s-1$ can be used to build subgroups.
- Multiple program execution controls, one for each group, can exist, i.e., one or all groups of a single partition can be active, see Fig. 9.
- Interactions between concurrently working groups are possible.
- In the program with a two-dimensional grid, row and column groups can be identified using corresponding function calls. Generalized functions for more than two dimensions work analogously.
- The entire program is executed in an SPMD style but the specific `section` statements pick subgroups to work and communicate together within a part of the program. For those parts, the execution model is a group-SPMD model.

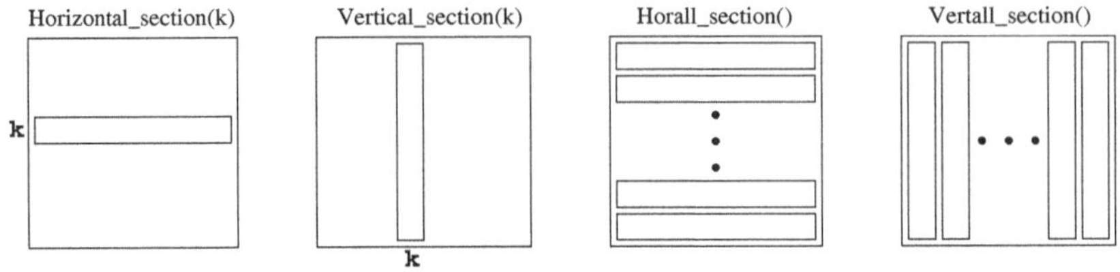

Fig. 9. The gray parts show active processor groups for different vertical and horizontal `section` commands in the two-dimensional case

Flexible composition of component based algorithms

Block-cyclic data distributions and virtual processor grids have also been used to parallelize efficiently the calculation of Lyapunov characteristics of many-particle systems [32]. The simulation algorithm consists of a large number of time steps calculating Lyapunov exponents and vectors, which have to be re-orthogonalized periodically [33]. For a large number of particles an use of parallel platforms is needed to reduce the computation time.

The challenge of the parallel simulation lies in the parallel implementation of the re-orthogonalization module and the flexible coupling of the re-orthogonalization with the molecular dynamics integration routine. The flexible composition is achieved with an interface combining a module for the parallel re-orthogonalization and a module for the parallel molecular dynamics integration routine. Both modules exploit a two-dimensional virtual processor grid and a block-cyclic data distribution in parameterized vector form given in Formula (4). Thus, many different data distributions can be used. The interface is responsible for the correct cooperation, especially concerning the data distribution, which may require an automatic redistribution.

The module for the parallel re-orthogonalization can be chosen from a set of different parallel modules realizing different algorithms. For the Gram-Schmidt orthogonalization and QR factorization based on blockwise Householder reflection several parallel variants with different versions of a block-cyclic data distribution have been implemented and tested [43]. Investigations for the parallel modified Gram-Schmidt algorithm have been presented in [42]. Depending on the molecular dynamics system and the specific parallel hardware different parallel re-orthogonalization modules show the best performance. The flexible program environment guarantees that the best parallel orthogonalization can be included and works correctly. Thus, the parallel program calculating Lyapunov characteristics combines the parallel programming model of orthogonal virtual processor groups introduced in this section and the parallel programming model for M-task programming with redistribution between modules from Sect. 4.

References

1. H. Bal and M. Haines. Approaches for Integrating Task and Data Parallelism. *IEEE Concurrency*, 6(3):74–84, July-August 1998.
2. S. Beuchler and A. Meyer. SPC-PM3AdH v1.0, Programmer's Manual. Technical Report SFB393/01-08, Chemnitz University of Technology, 2001.
3. W. J. Camp, S. J. Plimpton, B. A. Hendrickson, and R. W. Leland. Massively Parallel Methods for Engineering and Science Problems. *Comm. of the ACM*, 37:31–41, 1994.
4. S. Chakrabarti, J. Demmel, and K. Yelick. Modeling the benefits of mixed data and task parallelism. In *Symposium on Parallel Algorithms and Architecture (SPAA)*, pages 74–83, 1995.
5. F. Darema, D. A. George, V. A. Norton, and G. F. Pfister. A single-program-multiple-data computational mode for EPEX/FORTRAN. *Parallel Comput.*, 7(1):11–24, 1988.
6. A. Dierstein, R. Hayer, and T. Rauber. The ADDAP system on the ipsc/860: Automatic data distribution and parallelization. *J. Parallel Distrib. Comput.*, 32(1):1–10, 1996.
7. U. Elsner. Graph Partitioning. Technical Report SFB 393/97_27, TU Chemnitz, Chemnitz, Germany, 1997.
8. I. Foster and K.M. Chandy. Fortran M: A Language for Modular Parallel Programming. *J. Parallel Distrib.Comput.*, 25(1):24–35, April 1995.
9. A. Franz, C. Schulzky, S. Seeger, and K.H. Hoffmann. An efficient implementation of the exact enumeration method for random walks on sierpinski carpets. *Fractals*, 8(2):155–161, 2000.
10. C.M. Goral, K.E. Torrance, D.P. Greenberg, and B. Battaile. Modeling the interaction of light between diffuse surfaces. *Computer Graphics*, 18(3):212–222, 1984. Proceedings of the SIGGRAPH '84.
11. W. Gropp, E. Lusk, and A. Skjellum. *Using MPI: Portable Parallel Programming with the Message-Passing Interface.* MIT Press, 1999.
12. P. Hanrahan, D. Salzman, and L.Aupperle. A rapid hierarchical radiosity algorithm. *Computer Graphics*, 25(4):197–206, 1991. Proceedings of SIGGRAPH '91.
13. P.S. Heckbert. *Simulating Global Illumination using Adaptive Meshing.* PhD thesis, University of California, Berkeley, 1991.
14. B. Hendrickson and R. Leland. A multi-level algorithm for partitioning graphs. In *Proc. of the ACM/IEEE Conf. on Supercomputing*, San Diego, USA, 1995.
15. J. Hippold. Dezentrale Taskpools auf Rechnern mit verteiltem Speicher. Diplomarbeit, TU Chemnitz, Fakultät für Informatik, Dezember 2001.
16. J. Hippold, A. Meyer, and G. Rünger. A Parallelization Module for Adaptive 3-Dimensional FEM on Distributed memory. In *Proc. of the Int. Conference of Computational Science (ICCS-2004)*, LNCS 3037, pages 146–154. Springer, 2004.
17. J. Hippold and G. Rünger. A Communication API for Implementing Irregular Algorithms on SMP Clusters. In J. Dongarra, D. Lafarenza, and S. Orlando, editors, *Proc. of the 10th EuroPVM/MPI*, volume 2840 of *LNCS*, pages 455–463, Venice, Italy, 2003. Springer.
18. J. Hippold and G. Rünger. A Data Management and Communication Layer for Adaptive, Hexahedral FEM. In M. Danelutto, D. Laforenza, and M. Vanneschi,

editors, *Proc. of Euro-Par 2004*, volume 3149 of *LNCS*, pages 718–725, Pisa, Italy, 2004. Springer.
19. J. Hippold and G. Rünger. Task Pool Teams: A Hybrid Programming Environment for Irregular Algorithms on SMP Clusters. *To appear: Concurrency and Computation: Practice and Experience*, 2006.
20. M. Hofmann. Verwendung von Task Pool Teams Konzepten zur parallelen Implementierung von Diffusionsprozessen auf Fraktalen. Studienarbeit, TU Chemnitz, Fakultät für Informatik, Februar 2005.
21. S. Hunold, T. Rauber, and G. Rünger. Multilevel Hierarchical Matrix Multiplication on Clusters. In *Proc. of the 18th Annual ACM International Conference on Supercomputing, ICS'04*, pages 136–145, June 2004.
22. G. Karypis and V. Kumar. METIS Unstructured Graph Partitioning and Sparse Matrix Ordering System. Technical Report http://www.cs.umn.edu/metis, Department of Computer Science, University of Minnesota, Minneapolis, MN, USA, 1995.
23. G. Karypis and V. Kumar. Multilevel k-way Partitioning Scheme for Irregular Graphs. *J. Parallel Distrib. Comput.*, 48:96–129, 1998.
24. G. Karypis and V. Kumar. A Fast and Highly Quality Multilevel Scheme for Partitioning Irregular Graphs. *SIAM Journ. of Scientific Computing*, 20(1):359–392, 1999.
25. G. Karypis, K. Schloegel, and V. Kumar. ParMETIS Parallel Graph Partitioning and Sparse Matrix Ordering Library, Version 3.0. Technical report, University of Minnesota, Department of Computer Science and Engineering, Army HPC Research Center, Minneapolis, MN, USA, 2002.
26. B. Kernighan and S. Lin. An Efficient Heuristic Procedure for Partitioning Graphs. *Bell System Technical Journ.*, 29:291–307, 1970.
27. M. Korch and T. Rauber. A Comparison of Task Pools for Dynamic Load Balancing of Irregular Algorithms. *Concurrency and Computation: Practice and Experience*, 16:1–47, 2004.
28. J. O'Donnell, T. Rauber, and G. Rünger. Functional realization of coordination environments for mixed parallelism. In *Proc. of the IPDPS04 workshop on Advances in Parallel and Distributed Computational Models, CD-ROM*, Santa Fe, New Mexico, USA, 2004. IEEE.
29. A. Podehl, T. Rauber, and G. Rünger. A Shared-Memory Implementation of the Hierarchical Radiosity Method. *Theoretical Computer Science*, 196(1-2):215–240, 1998.
30. A. Pothen. Graph Partitioning Algorithms with Applications to Scientific Computing. In D. F. Keyws, A. H. Sameh, and V. Venkatakrishnan, editors, *Parallel Numerical Algorithms*. Kluwer, 1996.
31. A. Pothen, H. D. Simon, and K. P. Liou. Partitioning Sparse Matrices with Eigenvectors of Graphs. *SIAM Journ. on Matrix Analysis and Applications*, 11:430–452, 1990.
32. G. Radons, G. Rünger, M. Schwind, and G. Yang. Parallel algorithms for the determination of Lyapunov characteristics of large nonlinear dynamical systems. In *To appear: Proc. of PARA04 Workshop on State-of-the-Art in Scientific Computing 2004*. Denmark, 2006.
33. G. Radons and H. L. Yang. Static and Dynamic Correlations in Many-Particle Lyapunov Vectors. nlin.CD/0404028, and references therein.

34. T. Rauber and G. Rünger. The Compiler TwoL for the Design of Parallel Implementations. In *Proc. 4th Int. Conf. on Parallel Architectures and Compilation Techniques (PACT'96)*, IEEE, pages 292–301, 1996.
35. T. Rauber and G. Rünger. Diagonal-Implicitly Iterated Runge-Kutta Methods on Distributed Memory Machines. *Int. Journal of High Speed Computing*, 10(2):185–207, 1999.
36. T. Rauber and G. Rünger. A Transformation Approach to Derive Efficient Parallel Implementations. *IEEE Transactions on Software Engineering*, 26(4):315–339, 2000.
37. T. Rauber and G. Rünger. Deriving array distributions by optimization techniques. *Journal of Supercomputing*, 15:271–293, 2000.
38. T. Rauber and G. Rünger. Modelling the runtime of scientific programs on parallel computers. In Y. Pan and L.T. Yang, editors, *Parallel and Distributed Scientific and Engineering Computing*, volume 15 of *Adv. in Computations: Theory and Practice*, pages 51–65. Nova Science Publ., 2004.
39. T. Rauber and G. Rünger. Tlib - A Library to Support Programming with Hierarchical Multi-Processor Tasks. *J. Parallel Distrib. Comput.*, 65:347 – 360, 2005.
40. T. Rauber and G. Rünger. A data re-distribution library for multi-processor task programming. *To appear: International Journal of Foundations of Conmputer Science*, 2006.
41. R. Reilein-Ruß. Eine komponentenbasierte Realisierung der TwoL-Spracharchitektur. Dissertation, TU Chemnitz, Fakultät für Informatik, 2005.
42. G. Rünger and M. Schwind. Comparison of different parallel modified gram-schmidt algorithm. In *Proc. of Euro-Par 2005*, LNCS, Lisboa, Portugal, 2005. Springer.
43. M. Schwind. Implementierung und Laufzeitevaluierung paralleler Algorithmen zur Gram-Schmidt Orthogonalisierung und zur QR-Zerlegung. Diplomarbeit, TU Chemnitz, Fakultät für Informatik, Dezember 2004.
44. J.P. Singh, C. Holt, T. Totsuka, A. Gupta, and J. Hennessy. Load balancing and data locality in adaptive hierarchical N-body methods: Barnes-hut, fast multipole, and radiosity. *J. Parallel Distrib. Comput.*, 27:118–141, 1995.
45. D. Skillicorn and D. Talia. Models and languages for parallel computation. *ACM Computing Surveys*, 30(2):123–169, 1998.
46. M. Ujaldon, E. L. Zapata, S. D. Sharma, and J. Saltz. Parallelization Techniques for Sparse Matrix Applications. *J. Parallel Distrib. Comput.*, 38(2):256–266, 1996.
47. P.J. van der Houwen, B.P. Sommeijer, and W. Couzy. Embedded Diagonally Implicit Runge–Kutta Algorithms on Parallel Computers. *Mathematics of Computation*, 58(197):135–159, January 1992.
48. Z. Xu and K. Hwang. Early Prediction of MPP Performance: SP2, T3D and Paragon Experiences. *Parallel Comput.*, 22:917–942, 1996.

Basic Approach to Parallel Finite Element Computations: The DD Data Splitting

Arnd Meyer

Technische Universität Chemnitz, Fakultät für Mathematik
09107 Chemnitz, Germany
a.meyer@mathematik.tu-chemnitz.de

1 Introduction

From Amdahl's Law we know: the efficient use of parallel computers can not mean a parallelization of some single steps of a larger calculation, if in the same time a relatively large amount of sequential work remains or if special convenient data structures for such a step have to be produced with the help of expensive communications between the processors. From this reason, our basic work on parallel solving partial differential equations was directed to investigating and developing a natural fully parallel run of a finite element computation – from parallel distribution and generating the mesh – over parallel generating and assembling step – to parallel solution of the resulting large linear systems of equation and post–processing.

So, we will define a suitable data partitioning of all large finite element (F.E.) data that permits a parallel use within all steps of the calculation.

This is given in detail in the following Sect. 2. Considering a typical iteration method for solving a linear finite element system of equations, as is done in Sect. 3, we conclude that the only relevant communication technique has to be introduced within the preconditioning step. All other parts of the computation show a purely local use of private data. This is important for both message passing systems (local memory) and shared memory computers as well. The first environment clearly uses the advantage of having as less interprocessor communication as possible. But even in the shared memory environment we obtain advantages from our data distribution. Here, the use of private data within nearly all computational steps does not require any of the well–known expensive semaphore–like mechanisms in order to secure writing conflicts. The same concept as in the distributed memory case permits the use of the same code for both very different architectures.

2 Finite element computation and data splitting

Let

$$a(u, v) = \langle f, v\rangle \tag{1}$$

be the underlying bilinear form belonging to a partial differential equation (p.d.e.) $\mathcal{L}u = f$ in Ω with boundary conditions as usual. Here, $u \in H^1(\Omega)$ with prescribed values on parts Γ_D of the boundary $\partial\Omega$ is the unknown solution, so (1) holds for all $v \in H_0^1(\Omega)$ (with zero values on Γ_D). The Finite Element Method defines an approximation u_h of piecewise polynomial functions depending on a given fine triangulation of Ω.

Let $\mathbb{V}_h$ denote this finite dimensional subspace of finite element functions and $\mathbb{V}_{h0} = \mathbb{V}_h \cap H_0^1(\Omega)$. So,

$$a(u_h, v) = \langle f, v\rangle \quad \forall v \in \mathbb{V}_{h0} \tag{2}$$

is the underlying F. E. equation for defining $u_h \in \mathbb{V}_h$ (with prescribed values on Γ_D). In more complicated situations such as linear elasticity u is a vector function.

With the help of the finite element nodal base functions

$$\Phi = (\varphi_1, \cdots, \varphi_N)$$

we map u_h to the N-vector $\underline{u}$ by

$$u_h = \Phi \underline{u} \tag{3}$$

Often $\varphi_i(\mathbf{x}_j) = \delta_{ij}$ for the nodes $\mathbf{x}_j$ of the triangulation, so $\underline{u}$ contains the function values of $u_h(\mathbf{x}_j)$ at the j-th position, but it is basically the vector of coefficients of the expansion of u_h with respect to the nodal base Φ.

With (3) (for u_h and for arbitrary $v = \Phi\underline{v}$) (2) is equivalent to the linear system

$$K\underline{u} = \underline{b} \tag{4}$$

with

$$\begin{array}{lll} K = (k_{ij}) & k_{ij} = a(\varphi_j, \varphi_i) & \\ \underline{b} = (b_i) & b_i = \langle f, \varphi_i\rangle & i, j = 1, \cdots, N. \end{array}$$

So, from the definition, we obtain 2 kinds of data:

I: large vectors containing "nodal values" (such as $\underline{u}$)

II: large vectors and matrices containing functional values such as $\underline{b}$ and K.

From the fact that these functional values are integrals over Ω, the type-II-data is splitted over some processors as partial sums, when the parallelization idea starts with domain decomposition.

That is, let

$$\overline{\Omega} = \bigcup_{s=1}^{p} \overline{\Omega}_s \quad , \quad (\Omega_s \cap \Omega_{s'} = \emptyset, \forall s \neq s')$$

be a non-overlapping subdivision of Ω . Then, the values of a local vector

$$\underline{b}_s = (b_i)_{\mathbf{x}_i \in \overline{\Omega}_s} \in \mathbb{R}^{N_s}$$

are calculated from the processor P_s running on Ω_s-data independently of all other processors and the true right hand side satisfies

$$\underline{b} = \sum_{s=1}^{p} H_s^T \underline{b}_s \tag{5}$$

with a special (only theoretically existent) $(N_s \times N)$ -Boolean-connectivity matrix H_s. If the i-th node $\mathbf{x}_i$ in the global count has node number j locally in $\overline{\Omega}_s$ then $(H_s)_{ji} = 1$ (otherwise zero).

The formula (5) is typical for the distribution of type-II-data, for the matrix we have

$$K = \sum_{s=1}^{p} H_s^T K_s H_s \ , \tag{6}$$

where K_s is the local stiffness matrix belonging to $\overline{\Omega}_s$, calculated within the usual generate/assembly step in processor P_s independently of all other processors. Note that the code running in all processors at the same time in generating and assembling K_s is the same code as within a usual Finite Element package on a sequential one processor machine. This is an enormous advantage that relatively large amount of operations included in the element by element computation runs ideally in parallel. Even on a shared memory system, the matrices K_s are pure private data on the processor P_s and the assembly step requires no security mechanisms.

The data of type I does not fulfill such a summation formula as (5), here we have

$$\underline{u}_s = H_s \underline{u} \tag{7}$$

which means the processor P_s stores that part of $\underline{u}$ as private data that belongs to nodes of $\overline{\Omega}_s$.

Note that some identical values belonging to "coupling nodes" $\mathbf{x}_j \in \overline{\Omega}_s \cap \overline{\Omega}_{s'}$ are stored in more than one processor. If not given beforehand such a compatibility has to be guaranteed for type-I-data. This is the main difference to a F.E. implementation in [10], where the nodes are distributed exclusively over the processors. But from the fact that we have all boundary information of the local subdomain available in P_s , the introduction of modern hierarchical techniques (see Sect. 4) is much cheaper.

Another advantage of this distinguishing of the two data types is found in using iterative solvers for the linear system (4) in paying attention to (5), (6) and (7). Here we watch that vectors of same type are updated by vectors of same type, so again this requires no data communication over the processors. Moreover, all iterative solvers need at least one matrix–vector–multiply per step of the iteration. This is nothing else than the calculation of a vector of functional values, so it changes a type-I into a type-II-vector without any data transfer again:

Let $u_h = \Phi \underline{u}$ an arbitrary function in $\mathbb{V}_h$, so $\underline{u} \in \mathbb{R}^N$ an arbitrary vector, then $\underline{v} = K\underline{u}$ contains the functional values

$$v_i = a(u_h, \varphi_i) \ \ i = 1, \cdots, N,$$

and

$$\underline{v} = \left(\sum H_s^T K_s H_s\right) \underline{u} = \sum H_s^T K_s \underline{u}_s = \sum H_s^T \underline{v}_s,$$

whenever $\underline{v}_s := K_s \underline{u}_s$ is done locally in processor P_s. From the same reason, the residual vector $\underline{r} = K\underline{u} - \underline{b}$ of the given linear system is calculated locally as type-II-data.

3 Data flow within the conjugate gradient method

The preconditioned conjugate gradient method (PCGM) has found to be the appropriate solver for large sparse linear systems, if a good preconditioner can be introduced. Let this preconditioner signed with C, the modern ideas for constructing C and the results are given in the next chapters. Then PCGM for solving $K\underline{u} = \underline{b}$ is the following algorithm.

PCGM

Start: define start vector $\underline{u}$

$$\underline{r} := K\underline{u} - \underline{b}, \quad \underline{q} := \underline{w} := C^{-1}\underline{r}, \quad \gamma_o := \gamma := \underline{r}^T \underline{w}$$

Iteration: until stopping criterion fulfilled do

(1) $\underline{v} := K\underline{q}$

(2) $\delta := \underline{v}^T \underline{q}, \quad \alpha := -\gamma/\delta$

(3) $\underline{u} := \underline{u} + \alpha \underline{q}$

(4) $\underline{r} := \underline{r} + \alpha \underline{v}$

(5) $\underline{w} := C^{-1}\underline{r}$

(6) $\hat{\gamma} := \underline{r}^T \underline{w}, \quad \beta := \hat{\gamma}/\gamma, \ \gamma := \hat{\gamma}$

(7) $\underline{q} := \underline{w} + \beta \underline{q}$

Remark 1: The stopping criterion is often :

$$\gamma < \gamma_o \cdot tol^2 \Rightarrow stop.$$

Here, the quantity

$$\underline{r}^T C^{-1} \underline{r} = \underline{z}^T K C^{-1} K \underline{z}$$

with the actual error $\underline{z} = \underline{u} - \underline{u}^*$ is decreased by tol^2, so the $KC^{-1}K$–Norm of $\underline{z}$ is decreased by tol.

Remark 2: The convergence is guaranteed if K and C are symmetric, positive definite. The rate of convergence is linear depending on the convergence quotient

$$\eta = \frac{1 - \sqrt{\xi}}{1 + \sqrt{\xi}} \quad \text{with} \quad \xi = \lambda_{\min}(C^{-1}K)/\lambda_{\max}(C^{-1}K).$$

For the parallel use of this method, we define

	$\underline{u}, \underline{w}, \underline{q}$	to be type-I-data
and	$\underline{b}, \underline{r}, \underline{v}$	to be type-II-data
		(from the above discussions) .

So the steps (1), (3), (4) and (7) do not require any data communication and are pure arithmetical work with private data. The both inner products for δ and γ in step (2) and (6) are simple sums of local inner products over all processors:

$$\gamma = \underline{r}^T \underline{w} = \left(\sum H_s^T \underline{r}_s\right)^T \underline{w} = \sum \underline{r}_s^T H_s \underline{w} = \sum_{s=1}^{p} \underline{r}_s^T \underline{w}_s .$$

So the parallel preconditioned conjugate gradient method is the following algorithm (running locally in each processor P_s):

PPCGM

Start:

Choose $\underline{u}$, set $\underline{u}_s = H_s \underline{u}$ in P_s

$\underline{r}_s := K_s \underline{u}_s - \underline{b}_s$, $\underline{w} := C^{-1}\underline{r}$ (with $\underline{r} = \sum H_s^T \underline{r}_s$)

set $\underline{w}_s = H_s \underline{w}$ in P_s

$\gamma_s := \underline{r}_s^T \underline{w}_s$ $\gamma := \gamma_o := \sum_{s=1}^{p} \gamma_s$

Iteration: until stopping criterion fulfilled do

(1) $\underline{v}_s := K_s \underline{q}_s$

(2) $\delta_s := \underline{v}_s^T \underline{q}_s \qquad \delta := \sum_{s=1}^{p} \delta_s, \quad \alpha := -\gamma/\delta$

(3) $\underline{u}_s := \underline{u}_s + \alpha \underline{q}_s$

(4) $\underline{r}_s := \underline{r}_s + \alpha \underline{v}_s$

(5) $\underline{w} := C^{-1} \underline{r}$ (with $\underline{r} = \sum H_s^T \underline{r}_s$)

set $\underline{w}_s = H_s \underline{w}$

(6) $\gamma_s := \underline{r}_s^T \underline{w}_s, \quad \hat{\gamma} := \sum_{s=1}^{p} \gamma_s, \quad \beta := \hat{\gamma}/\gamma, \quad \gamma := \hat{\gamma}$

(7) $\underline{q}_s := \underline{w}_s + \beta \underline{q}_s$

Remark 3: The connection between the subdomains Ω_s is included in step (5) only, all other steps are pure local calculations or the sum of one number over all processors.

A proper definition of the preconditioner C fulfills three requirements:

(A) The arithmetical operations for step (5) are cheap (proportionally to the number of unknowns)
(B) The condition number $\kappa(C^{-1}K) = \xi^{-1}$ is small, independent of the discretization parameter h (mesh spacing) or only slightly growing for $h \to 0$, such as $\mathcal{O}(|\ln h|)$.
(C) The number of data exchanges between the processors for realizing step (5) is as small as possible (best: exactly one data exchange of values belonging to the coupling nodes).

Remark 4: For no preconditioning at all ($C = I$) or for the simple diagonal preconditioner ($C = diag(K)$), (A) and (C) are perfectly fulfilled, but (B) not. Here we have $\kappa(C^{-1}K) = \mathcal{O}(h^{-2})$. So the number of iterations would grow with h^{-1} not optimally.

The modern preconditioning techniques such as

- the domain decomposition preconditioner (i.e. local preconditioners for interior degrees of freedom combined with Schur-complement preconditioners on the coupling boundaries, see [1,3–7])
- hierarchical preconditioners for 2D problems due to Yserentant [9]
- Bramble-Pasciak-Xu-preconditioners (and related ones see [1,8]) for hierarchical meshes in 2D and 3D

and others have this famous properties. Here (A) and (B) are given from the construction and from the analysis. The property (C) is surprisingly fulfilled perfectly. Nearly the same is true, when Multigrid–methods are used as preconditioner within PPCGM, but from the inherent recursive work on the coarser meshes, we cannot achieve exactly one data exchange over the coupling boundaries per step but L times for an L level grid.

Remark 5: All these modern preconditioners can be found as special cases of the Multiplicative or Additive Schwarz Method (MSM/ASM [8]) depending on various splittings of the F.E. subspace $\mathbb{V}_h$ into special chosen subspaces.

4 An example for a parallel preconditioner

The most simple but efficient example of a preconditioner fulfilling (A), (B), (C) is YSERENTANT's hierarchical one [9]. Here we have generated the fine mesh from L levels subdivision of a given coarse mesh. One level means subdividing all triangles into 4 smaller ones of equal size (in the most simple case). Then, additionally to the nodal basis Φ of the fine grid, we can define the so called hierarchical basis $\Psi = (\psi_1, \cdots, \psi_N)$ spanning the same space $\mathbb{V}2_h$. So there exists an $(N \times N)-$matrix Q transforming Φ into Ψ:

$$\Psi = \Phi Q.$$

From the fact that a stiffness matrix defined with the base Ψ

$$K_H = (a(\psi_j, \psi_i))_{i,j=1}^N$$

would be much better conditioned but nearly dense, we obtain from $K_H = Q^T K Q$ the matrix $C = (QQ^T)^{-1}$ as a good preconditioner:

$$\kappa(K_H) = \kappa((QQ^T)K) = \kappa(C^{-1}K).$$

From [9] the multiplying $\underline{w} := QQ^T\underline{r}$ can be very cheaply implemented if the level by level mesh generation is stored within a special list. Surprisingly, this multiply is perfectly parallel, if the lists from the mesh subdivision are stored locally in P_s (mesh in $\overline{\Omega}_s$).

Let $\underline{w} = Q\underline{y}$, $\underline{y} = Q^T\underline{r}$, then the multiply $\underline{y} = Q^T\underline{r}$ is nothing else than transforming the functional values of a "residual functional" $r_i = \langle r, \varphi_i \rangle$ with respect to the nodal base functions into functional values with respect to the hierarchical base functions: $y_i = \langle r, \psi_i \rangle$.
So this part of the preconditioner transforms type-II-data into type-II-data without communication and $\underline{y} = \sum H_s^T \underline{y}_s$.

Then the type-II-vector $\underline{y}$ is assembled into type-I

$$\underline{\tilde{y}} = \underline{y}, \quad \underline{\tilde{y}}_s = \underline{y}_s + \sum_{j \neq s} \underbrace{H_s H_j^T \underline{y}_j}_{\textit{from other processors}}$$

containing now nodal data, but values belonging to the hierarchical base. So the function

$$w = \Phi \underline{w} = \Psi \underline{\tilde{y}}$$

is represented by $\underline{w}$ after back transforming

$$\underline{w} = Q\underline{\tilde{y}}$$

This is again a transformation of type-I-data into itself, so the preconditioner requires exactly one data exchange of values belonging to coupling boundary nodes per step of the PPCGM iteration.

Remark 6: For better convergence, a coarse mesh solver is introduced. Additionally use of Jacobi-preconditioning is a worthy idea for beating jumping coefficients and varying mesh spacings etc., so

$$C^{-1} = J^{-1/2} Q \begin{pmatrix} C_o^{-1} & \mathbb{O} \\ \mathbb{O} & I \end{pmatrix} Q^T J^{-1/2}$$

with $J = \text{diag}(K)$.

Remark 7: The better modern BPX-Preconditioners are implemented similarly fulfilling (A), (B), (C) in the same way.

5 Examples in linear elasticity

Let us demonstrate the power of this parallel finite element code at a 2D benchmark example.

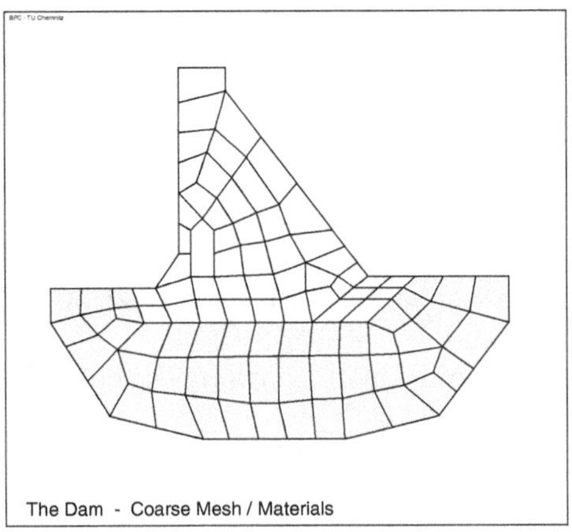

Fig. 1. Numerical example

We have used the GC Power-Plus-128 parallel computer at TU Chemnitz having up to 128 processors PC601 for computation from the years before 2000. The maximal arithmetical speed of 80 MFlops/s is never achieved for our typical application with unstructured meshes. Here, the matrix vector multiply $\underline{v}_s := K_s \underline{q}_s$ (locally at the same time) dominates the computational time. Note that K_s is a list of non-zero elements together with column index for efficient storing this sparse matrix. The use of the next generation parallel machine CLIC with up to 524 processors of Pentium-III type divides both the computation time as well as the communication by a factor of 10, so the same parallel efficiency is achieved.

We consider the elastic deformation of a dam. Figure 1 shows the 1–level–mesh, so the coarse mesh contained 93 quadrilaterals. From distributing over $p = 2^m$ processors, we obtain subdomains with the following number of coarse quadrilaterals.

Table 1. Number of quadrilaterals per processor

p	# quadrilaterals	max.speed–up
p=1	93	–
p=2	47	1.97
p=4	24	3.87
p=8	12	7.75
p=16	6	15.5
p=32	3	31
p=64	2	46.5
p=128	1	93

It is typical for *real life* problems that the mesh distribution cannot be optimal for a larger number of processors. In the following table we present some total running times and the measured percentage of the time for communication for finer subdivisions of the above mesh until 6 levels.

The most interesting question of scalability of the total algorithm is hard to analyze from this rare time measurements of this table. If we look at the two last diagonal entries, the quotient 85.5/64.1 is 1.3. From the table before and the growth of the iteration numbers a ratio of $(4 * 15.5/46.5) * (144/134) = 1.4$ would be expected, so we obtained a realistic scale–up of near one for this example. The percentage of communication tends to zero for finer discretizations

Table 2. Times and percentage of communication

L	N	It	p=1	p=4	p=16	p=64
1	3,168	83	2.1"/0%	1.6"/75%	2.4"/90%	
2	12,288	100	9.2"/0%	3.7"/40%	3.3"/70%	
3	48,384	111	39.8"/0%	11.7"/20%	6.1"/50%	
4	192,000	123	–/–	46.2"/10%	16.9"/25%	
5	764,928	134	–/–	–/–	64.1"/10%	19.4"/25%
6	2,085,248	144	–/–	–/–	–/–	85.5"/9%

and constant number of processors. Much more problematic is the comparison of two non–equal processor numbers, such as 16 and 64 in the table. Certainly, the larger number of processors requires more communication start–up's in the dot–products (log_2p). Within the subdomain communication the start-up's can be equal but need not. This depends on the resulting shapes of the subdomains within each processor, so the decrease from 10% to 9% in the above table is typical but very dependent on the example and the distribution process.

Whereas such 2D examples give scale–up values of near 90% for fine enough discretizations, much smaller values of the scale–up are achieved in 3D. The reason is the more complicate connection between the subdomains. Here, we have the crosspoints, the coupling faces belonging to 2 processors as in 2D but additionally coupling edges with an unstructured rich relationship between the subdomains. So the data exchange within the preconditioning step (5) of PPCGM is much more expensive and 50% communication time is typical for our parallel computer.

References

1. J. H. Bramble, J. E. Pasciak, A. H. Schatz (1986-89): The Construction of Preconditioners for Elliptic Problems by Substructuring I – IV, *Mathematics of Computation*, 47:103–134, 1986, 49:1–16, 1987, 51:415–430, 1988, 53:1–24, 1989.
2. I. H. Bramble, J. E. Pasciak, J. Xu, Parallel Multilevel Preconditioners, *Math. Comp.*, 55:191, 1-22, 1990.
3. G. Haase, U. Langer, A. Meyer, The Approximate Dirichlet Domain Decomposition Method,
   Part I: An Algebraic Approach,
   Part II: Application to 2nd-order Elliptic B.V.P.s,
   Computing, 47:137-151/153-167, 1991.
4. G. Haase, U. Langer, A. Meyer, Domain Decomposition Preconditioners with Inexact Subdomain Solvers,
   J. Num. Lin. Alg. with Appl., 1:27-42, 1992.
5. G. Haase, U. Langer, A. Meyer, Parallelisierung und Vorkonditionierung des CG-Verfahrens durch Gebietszerlegung,
   in: G. Bader, et al, eds., *Numerische Algorithmen auf Transputer-Systemen, Teubner Skripten zur Numerik*, B. G. Teubner, Stuttgart 1993.
6. G. Haase, U. Langer, A. Meyer, S.V.Nepommnyaschikh Hierarchical Extension Operators and Local Multigrid Methods in Domain Decomposition Preconditioners, *East-West J. Numer. Math.*, 2:173-193, 1994.
7. A. Meyer, A Parallel Preconditioned Conjugate Gradient Method Using Domain Decomposition and Inexact Solvers on Each Subdomain,
   Computing, 45:217-234, 1990.
8. P.Oswald, Multilevel Finite Element Approximation: Theory and Applications, Teubner Skripten zur Numerik, B.G.Teubner Stuttgart 1994.
9. H. Yserentant, Two Preconditioners Based on the Multilevel Splitting of Finite Element Spaces, *Numer. Math.*, 58:163-184, 1990.

10. L.Gross, C.Roll, W.Schönauer, Nonlinear Finite Element Problems on Parallel Computers,
In:Dongarra, Wasnievski, eds.,
Parallel Scientific Computing, First Int. Workshop PARA '94, 247-261, 1994.

A Performance Analysis of ABINIT on a Cluster System

Torsten Hoefler[1], Rebecca Janisch[2], and Wolfgang Rehm[1]

[1] Technische Universität Chemnitz, Fakultät für Informatik
09107 Chemnitz, Germany
htor@cs.tu-chemnitz.de Prof.Rehm@informatik.tu-chemnitz.de
[2] Technische Universität Chemnitz
Fakultät für Elektrotechnik und Informationstechnik
09107 Chemnitz, Germany
rebecca.janisch@etit.tu-chemnitz.de

1 Introduction

1.1 Electronic structure calculations

In solid state physics, bonding and electronic structure of a material can be investigated by solving the quantum mechanical (time-independent) Schrödinger equation,

$$\widehat{H}_{\text{tot}}\Phi = E_{\text{tot}}\Phi \quad , \tag{1}$$

in which the Hamilton operator $\widehat{H}_{\text{tot}}$ describes all interactions within the system. The solution Φ, the wavefunction of the system, describes the state of all N electrons and M atomic nuclei, and E_{tot} is the total energy of this state.

Usually, the problem is split by separating the electronic from the ionic part by making use of the Born-Oppenheimer approximation [1]. Next we consider the electrons as independent particles, represented by one-electron wavefunctions ϕ_i. Density functional theory (DFT), based on the work of Hohenberg and Kohn [2] and Kohn and Sham [3], then enables us to represent the total electronic energy of the system by a functional of the electron density $n(\mathbf{r})$:

$$n(\mathbf{r}) = \sum_i |\phi_i|^2 \tag{2}$$

$$\begin{aligned} \rightarrow E = E\left[n(r)\right] &= F[n] + \int V_{\text{ext}}(\mathbf{r})n(\mathbf{r})d\mathbf{r} \\ &= E_{\text{kin}}[n] + E_{\text{H}}[n] + E_{\text{xc}}[n] + \int V_{\text{ext}}(\mathbf{r})n(\mathbf{r})d\mathbf{r} \quad . \end{aligned} \tag{3}$$

Thus the many-body problem is projected onto an effective one-particle problem, resulting in a reduction of the degrees of freedom from $3N$ to 3. The one-particle Hamiltonian $\hat{H}$ now describes electron i, moving in the effective potential V_{eff} of all other electrons and the nuclei.

$$\begin{aligned} \hat{H}\ \phi_i &= \epsilon_i\ \phi_i \\ \left\{-\tfrac{\hbar^2 \Delta}{2m} + V_{\text{eff}}[n(\mathbf{r})]\right\} \phi_i(\mathbf{r}) &= \epsilon_i\ \phi_i(\mathbf{r}), \\ \text{where}\quad V_{\text{eff}}[n(\mathbf{r})] \quad = \quad V_{\text{eff}}(\mathbf{r}) &= V_{\text{H}}(\mathbf{r}) + V_{\text{xc}}(\mathbf{r}) + V_{\text{ext}}(\mathbf{r})\ . \end{aligned} \tag{4}$$

In (4), which are part of the so-called Kohn-Sham equations, $-\frac{\hbar^2 \Delta}{2m}$ is the operator of the kinetic energy, V_{H} is the Hartree and V_{xc} the exchange-correlation potential. V_{ext} is the external potential, given by the lattice of atomic nuclei. For a more detailed explanation of the different terms see e.g. [4]. The self-consistent solution of the Kohn-Sham equations determines the set of wavefunctions ϕ_i that minimize the energy functional (3). In order to obtain it, a starting density n_{in} is chosen from which the initial potential is constructed. The eigenfunctions of this Hamiltonian are then calculated, and from these a new density n_{out} is obtained. The density for the next step is usually a combination of input and output density. This process is repeated until input and output agree within the limits of the specified convergence criteria.
There are different ways to represent the wavefunction and to model the electron-ion interaction. In this paper we focus on pseudopotential+planewave methods.
If the wavefunction is expanded in plane waves,

$$\phi_i = \sum_{\mathbf{G}} c_{i,\mathbf{k}+\mathbf{G}} e^{i(\mathbf{k}+\mathbf{G})\mathbf{r}} \tag{5}$$

the Kohn-Sham equations assume the form [5]

$$\sum_{\mathbf{G}'} H_{\mathbf{k}+\mathbf{G},\mathbf{k}+\mathbf{G}'} \times c_{i,\mathbf{k}+\mathbf{G}'} = \epsilon_{i,\mathbf{k}} c_{i,\mathbf{k}+\mathbf{G}}\ , \tag{6}$$

with the matrix elements

$$\begin{aligned} H_{\mathbf{k}+\mathbf{G},\mathbf{k}+\mathbf{G}'} = &\ \frac{\hbar^2}{2m}|\mathbf{k}+\mathbf{G}|^2 \delta_{\mathbf{G}\mathbf{G}'} \\ &+ V_{\text{H}}(\mathbf{G}-\mathbf{G}') + V_{\text{xc}}(\mathbf{G}-\mathbf{G}') + V_{\text{ext}}(\mathbf{G}-\mathbf{G}')\ . \end{aligned} \tag{7}$$

In this form the matrix of the kinetic energy is diagonal, and the different potentials can be described in terms of their Fourier transforms. Equation (6) can be solved independently for each k-point on the mesh that samples the first Brillouin zone. In principle this can be done by conventional matrix diagonalization techniques. However, the cost of these methods increases with the third power of the number of basis states, and the memory required to store the Hamiltonian matrix increases as the square of the same number. The number of plane waves in the basis is determined by the choice of the cutoff energy $E_{\text{cut}} = \hbar^2/2m|\mathbf{k}+\mathbf{G}_{\text{cut}}|^2$ and is typically of the order of 100 per atom, if norm-conserving pseudopotentials are used. Therefore alternative techniques have been developed to minimize the Kohn-Sham energy functional (3), e.g. by conjugate gradient (CG) methods (for an introduction to this method see e.g. [6]). In a band-by-band CG scheme one eigenvalue (band) $\epsilon_{i,\mathbf{k}}$ is obtained at a time, and the corresponding eigenvector is orthogonalized with respect to the previously obtained ones.

1.2 The ABINIT code

ABINIT [7] is an open source code for *ab initio* electronic structure calculations based on the DFT described in Sect. 1.1. The code is the object of an ongoing open software project of the Université Catholique de Louvain, Corning Incorporated, and other contributors [8] .

ABINIT mostly aims at solid state research, in the sense that periodic boundary conditions are applied and the majority of the integrals that have to be calculated are represented in reciprocal space (k-space). It currently features the calculation of various electronic ground state properties (total energy, bandstructure, density of states,..), and several structural optimization routines. Furthermore it enables the investigation of electric and magnetic polarization and electronic excitations. Originally a pseudopotential+planewave code, ABINIT for a short time (since version 4.2.x) also features the projector-augmented wave method, but this is still under developement. In the following we refer to the planewave method.
To begin the self-consistency cycle, a starting density is constructed, and a starting potential derived. The eigenvalues and eigenvectors are determined by a band-by-band CG scheme [5], during which the density (i.e. the potential) is kept fixed until the whole set of functions has been obtained. The alternative of updating the density with each new band has been abandoned, to make a simple parallelization of the calculation over the k-points possible. Only at the end of one CG loop is the density updated by the scheme of choice (e.g. simple mixing, or Anderson mixing). For a comparison of different schemes see e.g. [9]). A more detailed description of the DFT implementation in general is given in [10].

Different levels of parallelization are implemented. The most efficient parallelization is the distribution of the k-points that are used to sample the Brillouin zone on different processors. Unfortunately the necessary number of k-points decreases with increasing system size, so the scaling with the number of atoms is rather unfavourable. One can partially make up for this by distributing the work related to different states (or bands) within a given k-point. Since the number of states increases with the system size, the overall

scaling with number of atoms improves. For example a blocked conjugate gradient algorithm can be used to optimize the wavefunctions, which provides the possibility to parallelize over the states within one block. Instead of a single eigenstate, as in the band-by-band scheme, **nbdblock** states are determined at the same time, where **nbdblock** is the number of bands in one block. Of course this leads to a small increase in the time that is needed to orthogonalize the eigenvectors with respect to those obtained previously. Furthermore, to guarantee convergence, a too high value for **nbdblock** should not be chosen. The ABINIT manual advises **nbdblock** ≤ 4 as a meaningful choice.
Both methods, k-point parallelization and parallelization over bands, are implemented using the MPI library. A third possibility of parallelization is given by the distribution of the work related to different wavefunction coefficients, which is realized with OpenMP compiler directives. It is used for example in the parallelization of the FFT algorithm, but this feature is still under develepement.

These parallelization methods, which are based on the underlying physics of the calculation, are useful only for a finite number of CPUs (a fact, that is not a special property of ABINIT, but common to all electronic structure codes). In a practical calculation, the required number of k-points, **nkpt**, for a specific geometry is determined by convergence tests. To decrease the computational effort, the k-point parallelization is then the first method of choice. The best speedup is achieved if the number of k-points is an integer multiple of the number of CPUs:

$$\texttt{nkpt} = n \times N_{\mathrm{CPU}} \quad \text{with} \quad n \in \mathbb{N}. \tag{8}$$

Ideally, $n = 1$.
If the number of available CPUs is larger than the number of k-points needed for the calculation, the speedup saturates. In this case, the additional parallelization over bands can improve the performance of ABINIT, if

$$N_{\mathrm{CPU}} = \texttt{nbdblock} \times \texttt{nkpt} \quad . \tag{9}$$

In principle the parallelization scheme also works for $N_{\mathrm{CPU}} =$ **nbdblock**, which results in a parallelization over bands only. However, this is rather inefficient, as will be seen below.

1.3 Related work

The biggest challenge after programming a parallel application is to optimize it according to a given parallel architecture. The first step of each optimization process is the performance and scalability measurement which is often called benchmarking. There are methods based on theoretical simulation [11,12] and methods based on benchmarking [13,14]. There are also studies which try to explain bottlenecks for scalability [15] or studies which compare different parallel systems [16]. In the following we analyze parallel efficiency and scalability of the application ABINIT on a cluster system and describe the results with the knowledge gained about the application. We also present several simple ideas to improve the scaling and performance of the parallel application.

2 Benchmark methodology

The parallel benchmark runs have been conducted on two different cluster systems. The first one is a local cluster at the University of Chemnitz which consists of 8 Dual Xeon 2.4GHz systems with 2GB of main memory per CPU. The nodes are interconnected with Fast Ethernet. We used MPICH2 1.0.2p1 [17] with the ch_p4 (TCP) device as the MPI communication library. The source code of ABINIT 4.5.2 was compiled with the Intel Fortran Compiler 8.1 (Build 20050520Z). The relevant entries of the `makefile_macros` are shown in the following:

```
FC=ifort
COMMON_FFLAGS=-FR -w -tpp7 -axW -ip -cpp
FFLAGS=$(COMMON_FFLAGS) -O3
FFLAGS_Src_2psp =$(COMMON_FFLAGS) -O0
FFLAGS_Src_3iovars =$(COMMON_FFLAGS) -O0
FFLAGS_Src_9drive =$(COMMON_FFLAGS) -O0
FFLAGS_LIBS=-O3 -w
FLINK=-static
```

Listing 1. Relevant `makefile_macros` entries for the Intel Compiler

The -O3 optimization had to be disabled for several directories, because of endless compiling. For the serial runs we also compiled ABINIT with the open source g95 Fortran compiler, with the following relevant flags:

```
FC=g95
FFLAGS=-O3 -march=pentium4 -mfpmath=sse -mmmx \
          -msse -msse2
FLINK=-static
```

Listing 2. Relevant `makefile_macros` entries for the g95 Compiler

The second system is a Cray Opteron Cluster (strider) of the High Performance Computing Center Stuttgart (HLRS), consisting of 256 2 GHz AMD Opteron CPUs with 2GB of main memory per CPU. The nodes are interconnected with a Myrinet 2000 network. ABINIT has been compiled with the

64-bit PGI Fortran compiler (version 5.0). On strider the MPI is implemented as a port of MPICH (version 1.2.6) over GM (version 2.0.8).

```
FC=pgf90
FFLAGS=-tp=k8-64 -Mextend -Mfree -04
FFLAGS_LIBS = -04
LDFLAGS=-Bstatic -aarchive
```

Listing 3. Relevant `makefile_macros` entries for the pgi Compiler

2.1 The input file

All sequential and parallelized benchmarks have been executed with an essentially identical input file which defines a (hexagonal) unit-cell of Si_3N_4 (two formula units). We used 56 bands and a planewave energy cut-off of 30 Hartree (resulting in $\approx$ 7700 planewaves). A Monkhorst-Pack k-point mesh [18] was used to sample the first Brillouin zone. The number of k-points along the axes of the mesh was changed with the **ngkpt** parameter as shown in Table 1.
To use parallelization over bands, we switched from the default band-by-band wavefunction optimisation algorithm to the blocked conjugate gradient algorithm (**wfoptalg** was changed to 1) and chose numbers of bands per block > 1 (**nbdblock**) according to (9) in Sect. 1.2.

Table 1. Number of k-points along the axes of the Monkhorst-Pack mesh, **ngkpt**, and resulting total number of k-points in the calculation, **nkpt**

nkpt	ngkpt
2 2 2	2
4 2 2	4
4 4 2	8
4 4 4	16

3 Benchmark results

3.1 Sequential analysis

Due to the fact that a calculation which is done on a single processor in the k-point parallelization is exactly the same as in the sequential case, a sequential analysis can be used to analyze the behaviour of the calculation itself.

Call graph

The call graph of a program shows all functions which are called during a program run. We used the `gprof` utility from the gcc toolchain and the program `cgprof` to generate a call graph. ABINIT called 300 different functions during our calculation. Thus, the full callgraph is much too complicated and we present only a short extract with all functions that use more than 4% of the total application runtime (Fig. 1).

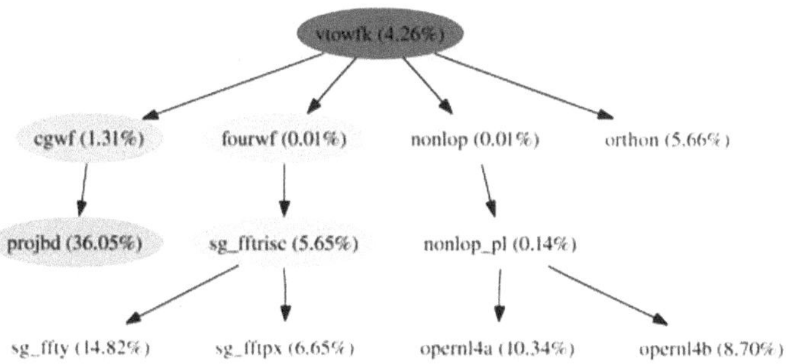

Fig. 1. The partial callgraph

The percentage of the runtime of the functions is given in the diagram, and the darkness of the nodes indicates the percentage of the subtree of these nodes (please keep in mind that not all functions are plotted). About 97% of the runtime of the application is spent in a subtree of `vtowfk` which computes the density at a given k-point with a conjugate gradient minimization method. The 8 most time consuming functions need more than 92% of the application runtime (they are called more than once). The most time demanding function is `projbd` which orthogonalizes a state vector against the ones obtained previously in the band-by-band optimization procedure.

Impact of the compiler

Compilation of the source files can be done with various compilers. We compared the open-source g95 compiler with the commercial Intel Fortran Compiler 8.1 (abbreviated with ifort). The Intel compiler is able to auto-parallelize the code (cmp. OpenMP), this feature has also been tested on our dual Xeon processors. The benchmarks have been conducted three times and the mean value is displayed. The results of our calculations are shown in Table 2.

This shows clearly that the Intel Compiler generates much faster code than the g95. However, the g95 compiler is currently under development and there is a lot of potential for optimizations. The auto-parallelization feature of the ifort is also not beneficial, this could be due to the thread spawning overhead at small loops.

Table 2. Comparison of different compilers

Compiler/Features	Runtime (s)
ifort	625.07
ifort -parallel	643.19
g95	847.73

Impact of the BLAS library

Mathematical libraries such as the BLAS Library are used to provide an abstraction of different algebraic operations. These operations are implemented in so called math libraries which are often architecture specific. Many of them are highly optimized and can accelerate the code by a significant factor (as compared to "normal programming"). ABINIT offers the possibility to exchange the internal math implementations with architecture-optimized variants. A comparison in runtime (all libraries compiled with the g95 compiler) is shown in Table 3.

Table 3. Comparison of different mathematical libraries

BLAS Library	Runtime (s)
internal	847.73
Intel-MKL	845.62
AMD-ACML	840.56
goto BLAS	860.60
Atlas	844.67

The speedup due to an exchange of the math library is negligible. One reason could be that the mathematical libraries are not efficiently implemented and do not offer a significant improvement in comparison to the reference implementation (internal). However, this seems unlikely since all the libraries are highly optimized and several were tested. A more likely explanation is that the libraries are not used very often in the code (calls and execution do not consume much time compared to the total runtime). To investigate this we analyzed the callgraph with respect to calls to math-library functions, and found that indeed all calls to math libraries such as `zaxpy`, `zswap`, `zscal`, ... make less than 2% of the application runtime (with the internal math library). Thus, the speedup of the whole application cannot exceed 2% even if the math libraries are improved.

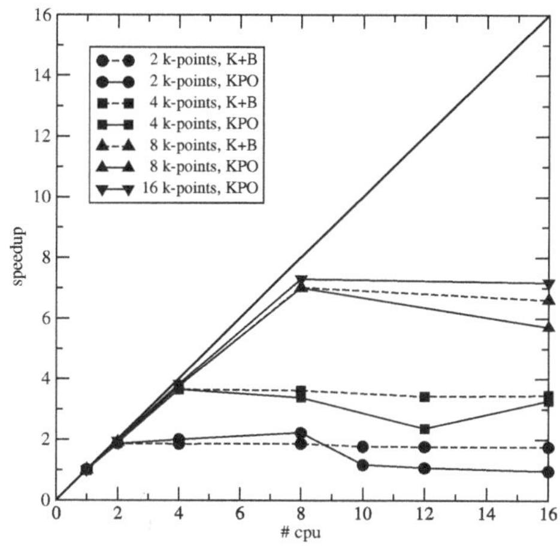

Fig. 2. Speedup vs. number of CPUs on the Xeon Cluster. Parallelization over k-points only (KPO - filled symbols): Except for the case of 16 k-points the scaling is almost ideal as long as $N_{CPU} \leq$ **nkpt** . Saturation is observed for $N_{CPU} >$ **nkpt**, only for 16 k-points this occurs already for $N_{CPU} = 8$. Parallelization over k-points and bands (K+B - open symbols): The number of bands per block at the different data points equals $N_{CPU}/$**nkpt**. Only negligible additional speedup is observed

3.2 Parallel analysis

Speedup analysis

Fig. 2 shows the speedup versus the number of CPUs on the Xeon Cluster. In all cases except the case of 16 k-points the scaling is almost ideal as long as $N_{CPU} \leq$ **nkpt**. Saturation is observed for $N_{CPU} >$ **nkpt**, only for 16 k-points this occurs already for $N_{CPU} = 8$, due to overheads from MPI barrier synchronizations in combination with process skew on the Xeon Cluster (see Sect. 3.2). The parallelization over bands, which needs intense communication, only leads to negligible additional speedup on this Fast Ethernet network. Fig. 3 shows the speedup vs. the number of CPUs on the Cray Opteron Cluster. For the parallelization over k-points only, an almost ideal speedup is obtained for small numbers (≤ 8) of CPUs. The less than ideal behaviour for larger numbers can be explained by a communication overhead, see Sect. 3.2. For $N_{CPU} >$ **nkpt** the speedup saturates, as expected. In this regime the speedup can be considerably improved (up to 250% in the case of 4 k-points and 16 CPUs) by including the parallelization over bands, as long as the number of bands per block remains reasonably small (≤ 4).

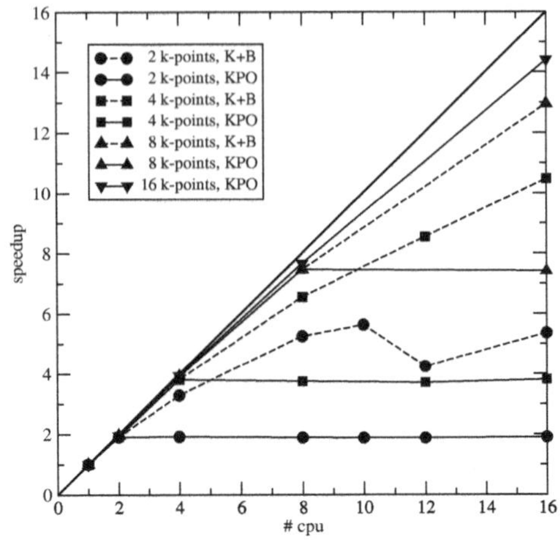

Fig. 3. Speedup vs. number of CPUs on the Cray Opteron Cluster. Parallelization over k-points only (KPO - filled symbols): For $N_{CPU} \leq$ **nkpt** the scaling is almost ideal. Saturation is observed for $N_{CPU} >$ **nkpt**. Parallelization over k-points and bands (K+B - open symbols): The number of bands per block at the different data points equals $N_{CPU}/$ **nkpt**. For reasonable numbers **nbdblock** ≤ 4) the speedup in the formerly saturated region improves considerably

Communication analysis I: Plain *k*-point parallelization

We used the MPE environment and Jumpshot [19] to perform a short analysis of the communication behaviour of ABINIT in the different working scenarios on the local Xeon Cluster. This parallelization method is investigated in two scenarios, the almost ideal speedup with 8 processors calculating 8 k-points and less than ideal speedup with 16 processors calculating 16 k-points. The MPI communication scheme of 8 processors calculating 8 k-points is shown in the following diagram. The processors are shown on the ordinate (rank 0-7), and the communication operations are shown for each of them. Each MPI operation corresponds to a different shade of gray. The processing is not depicted (black). The ideal communication diagram would show nothing but a black screen, every MPI operation delays the processing and increases the overhead.

Fig. 4 shows the duration of all calls to the MPI library. This gives a rough overview of the parallel performance of the application. The parallelization is very efficient, all processors are computing most of the time, some CPU time is lost during the barrier synchronization. The three self consistent field (SCF) steps can easily be recognized as the processing time (black) between the `MPI_Barrier` operations. The synchronization is done at the end of each step and afterwards a small amount of data is exchanged via `MPI_Allreduce`, but this does not add much overhead. The parallelization is very efficient and only a small fraction of the time is spent for communication. All processors

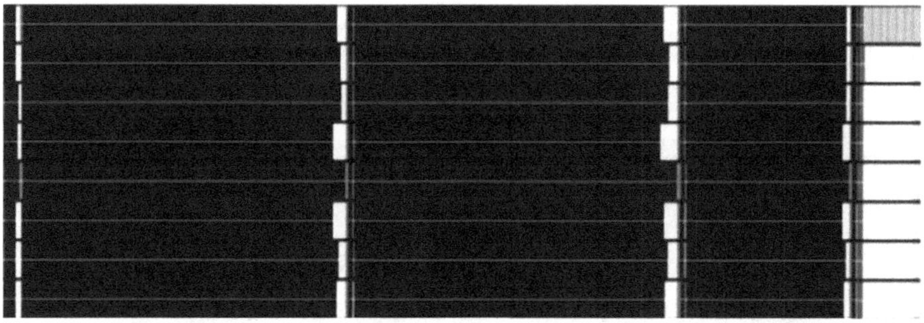

Fig. 4. Visualization of the MPI overhead for k-point parallelization over 8 k-points on 8 Processors. Each MPI operation corresponds to a different shade of gray: `MPI_Barrier` white, `MPI_Bcast` light gray, and `MPI_Allreduce` dark gray. The overhead is negligible and this case is efficiently parallelized

send their results to rank 0 in the last part, after the last SCF step and synchronization. This is done with many barrier operations and send-receive, and could significantly be enhanced with a single call to `MPI_Gather`.

The scheme for 16 processors calculating 16 k-points, shown in Fig. 5, is nearly identical, but the overhead resulting from barrier synchronization is much higher and decreases the performance. This is due to so called process skew, where the unpredictable and uncoordinated scheduling of operating system processes on the cluster system interrupts the application and introduces a skew between the processes. This skew adds up during the whole application runtime due to the synchronization at the end of each SCF step. Thus, the scaling of ABINIT is limited on our cluster system due to the operating system's service processes. Even if the problem is massively parallel the overhead is much bigger as for 8 processors due to the barrier synchronization.

Communication analysis II: Parallelization over bands

The communication diagram for 8 processors and a calculation with 4 k-points and `nbdblock=2` is shown in Fig. 6. The main MPI operations besides the `MPI_Barrier` and `MPI_Allreduce` are `MPI_Send` and `MPI_Recv` in this scenario. These operations are called frequently and show a master-slave principle where the block specific data is collected at a master for each block and is processed. The MPI-overhead is significantly higher than for the pure k-point parallelization and the overall performance is heavily dependent on the network performance. Thus the results for band parallelization are rather bad for the cluster equipped with Fast Ethernet, while the results with Myrinet are good.

The diagram for two k-points calculated on 8 processors is shown in Fig. 7. This shows that the communication overhead outweighs the calculation and the parallelization is rendered senseless. Nearly the whole application runtime

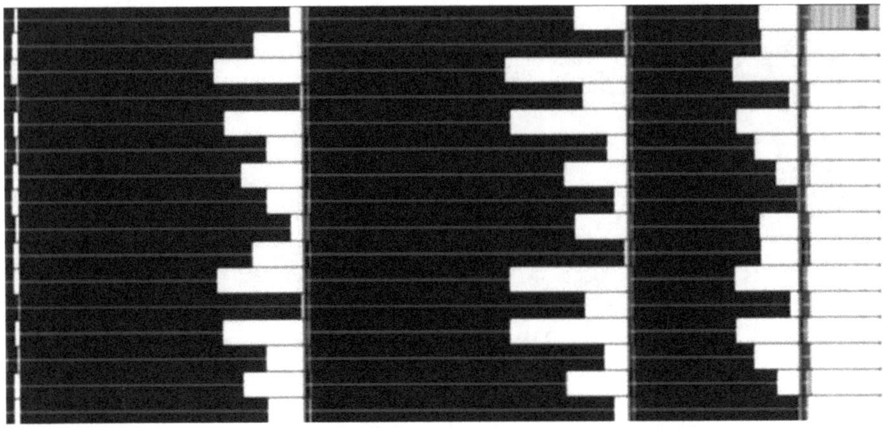

Fig. 5. Visualization of the MPI overhead for k-point parallelization over 16 k-points on 16 Processors. Each MPI operation corresponds to a different shade of gray: `MPI_Barrier` white, `MPI_Bcast` light gray, and `MPI_Allreduce` dark gray. The synchronization overhead is much bigger than in Fig. 4. This is due to the occuring process skew

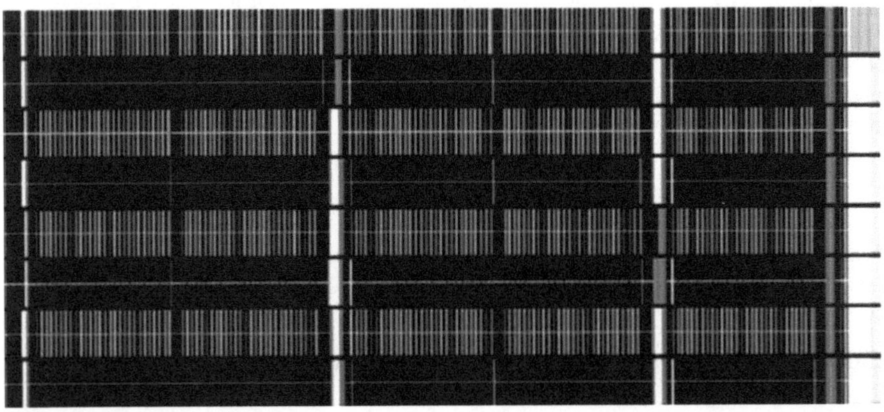

Fig. 6. Visualization of the MPI overhead for k-point and band parallelization over 4 k-points and 2 bands per block on 8 processors. Each MPI operation corresponds to a different shade of gray: `MPI_Barrier` white, `MPI_Bcast` light gray, and `MPI_Allreduce` dark gray. This figure shows the fine grained parallelism for band parallelisation. The overhead is visible but not dominationg the execution time

is overhead (all but black regions), mainly MPI_Bcast and MPI_Barrier. Thus, even allowing for the reasons of convergence, mentioned in Sect. 1.2, the band parallelization is effectively limited to cases with a reasonable MPI overhead, i.e. 4 bands per block on this system.

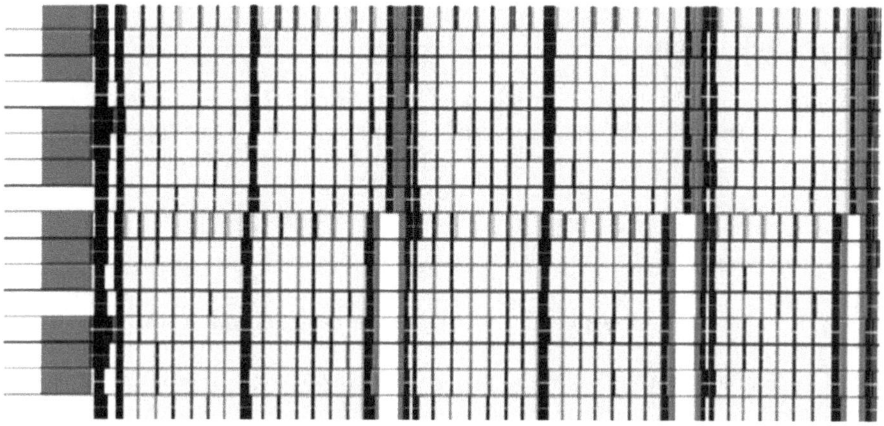

Fig. 7. Visualization of the MPI overhead for k-point and band parallelization over 2 k-points and 8 bands per block on 16 processors. Each MPI operation corresponds to a different shade of gray: MPI_Barrier white, MPI_Bcast light gray, and MPI_Allreduce dark gray. The overhead is clearly dominating the execution and the parallelization is rendered senseless

4 Conclusions

We have shown that the performance of the application ABINIT on a cluster system depends on different factors, such as compiler and communication network. Other factors which are usually crucial such as different implementations of mathematical functions are less important because the math libraries are rarely used in the critical path for our measurements. The choice of the compiler can decrease the runtime by almost 25%. Note that the promising feature of auto-parallelization is counterproductive. The different math libraries differ in less than 1% of the running time. The influence on the interconnect and parallelization technique is also significant. The embarassingly parallel k-point parallelization hardly needs any communication and is thus almost independent of the communication network. The scalability is limited to 8 on our cluster system due to operating system effects, which introduce process skew during each round. The scalability on the Opteron system is not limited. Thus, in principle this implementation in ABINIT is ideal for small systems demanding a lot of k-points, e.g. metals. For systems demanding large supercells, the communication wise more demanding band parallelization becomes attractive. However, the use of this implementation is only advantageous if a fast interconnect can be used for communication. Fast Ethernet is not suitable for this task.

Acknowledgements

The authors would like to thank X. Gonze for helpful comments. R.J. acknowledges the HLRS for a free trial account on the Cray Opteron Cluster strider.

References

1. M. Born and J.R. Oppenheimer. Zur Quantentheorie der Molekeln. *Ann. Physik*, 84:457, 1927.
2. P. Hohenberg and W. Kohn. Inhomogeneous electron gas. *Phys. Rev.*, 136:B864, 1964.
3. W. Kohn and L.J. Sham. Self-consistent equations including exchange and correlation effects. *Phys. Rev*, 140:A1133, 1965.
4. R. M. Martin. *Electronic Structure.* Cambridge University Press, Cambridge, UK, 2004.
5. M.C. Payne, M.P. Teter, D.C. Allan, T.A. Arias, and J.D. Joannopoulos. Iterative minimization techniques for ab initio total-energy calculations - molecular dynamics and conjugate gradients. *Rev. Mod. Phys.*, 64:1045, 1992.
6. J. Nocedal. Theory of algorithms for unconstrained optimization. *Acta Num.*, page 199, 1991.
7. X. Gonze, G.-M. Rignanese, M. Verstraete, J.-M. Beuken, Y. Pouillon, R. Caracas, F. Jollet, M. Torrent, G. Zerah, M. Mikami, P. Ghosez, M. Veithen, J.-Y. Raty, V. Olevano, F. Bruneval, L. Reining, R. Godby, G. Onida, D.R. Hamann, and D.C. Allan. A brief introduction to the ABINIT software package. *Z. Kristallogr.*, 220:558, 2005.
8. http://www.abinit.org/.
9. V. Eyert. A comparative study on methods for convergence acceleration of iterative vector sequences. *J.Comp.Phys.*, 124:271, 1995.
10. X. Gonze, J.-M. Beuken, R. Caracas, F. Detraux, M. Fuchs, G.-M. Rignanese, L. Sindic, M. Verstraete, G. Zerah, F. Jollet, M. Torrent, A. Roy, M. Mikami, Ph. Ghosez, J.-Y. Raty, and D.C. Allan. First-principles computation of material properties: the ABINIT software project. *Comp. Mat. Sci.*, 25:478, 2002.
11. A.D. Malony, V. Mertsiotakis, and A. Quick. Automatic scalability analysis of parallel programs based on modeling techniques. In *Proceedings of the 7th international conference on Computer performance evaluation : modelling techniques and tools*, pages 139–158, Secaucus, NJ, USA, 1994. Springer-Verlag New York, Inc.
12. R.J. Block, S. Sarukkai, and P. Mehra. Automated performance prediction of message-passing parallel programs. In *Proceedings of the 1995 ACM/IEEE conference on Supercomputing*, 1995.

13. M. Courson, A. Mink, G. Marcais, and B. Traverse. An automated benchmarking toolset. In *HPCN Europe*, pages 497–506, 2000.
14. X. Cai, A. M. Bruaset, H. P. Langtangen, G. T. Lines, K. Samuelsson, W. Shen, A. Tveito, and G. Zumbusch. Performance modeling of pde solvers. In H. P. Langtangen and A. Tveito, editors, *Advanced Topics in Computational Partial Differential Equations*, volume 33 of *Lecture Notes in Computational Science and Engineering*, chapter 9, pages 361–400. Springer, Berlin, Germany, 2003.
15. A. Grama, A. Gupta, E. Han, and V. Kumar. Parallel algorithm scalability issues in petaflops architectures, 2000.
16. G. Luecke, B. Raffin, and J. Coyle. Comparing the communication performance and scalability of a linux and an nt cluster of pcs.
17. W. Gropp, E. Lusk, N. Doss, and A. Skjellum. A high-performance, portable implementation of the mpi message passing interface standard. *Parallel Comput.*, 22(6):789–828, 1996.
18. H.J. Monkhorst and J.D. Pack. Special points for brillouin-zone integrations. *Phys. Rev. B*, 13:5188, 1976.
19. O. Zaki, E. Lusk, W. Gropp, and D. Swider. Toward scalable performance visualization with Jumpshot. *The International Journal of High Performance Computing Applications*, 13(3):277–288, Fall 1999.

Some Aspects of Parallel Postprocessing for Numerical Simulation

Matthias Pester

Technische Universität Chemnitz, Fakultät für Mathematik
09107 Chemnitz, Germany
pester@mathematik.tu-chemnitz.de

1 Introduction

The topics discussed in this paper are closely connected to the development of parallel finite element algorithms and software based on domain decomposition [1, 2]. Numerical simulation on parallel computers generally produces data in large quantities being kept in the distributed memory. Traditional methods of postprocessing by storing all data and processing the files with other special software in order to obtain nice pictures may easily fail due to the amount of memory and time required.

On the other hand, developing new and efficient parallel algorithms involves the necessity to evaluate the behavior of an algorithm immediately as an on-line response. Thus, we had to develop a set of visualization tools for parallel numerical simulation which is rather quick than perfect, but still expressive. The numerical data can be completely processed in parallel and only the resulting image is displayed on the user's desktop computer while the numerical simulation is still running on the parallel machine.

This software package for parallel postprocessing is supplemented by interfaces to external software running on high-performance graphic workstations, either storing files for an off-line postprocessing or using a TCP/IP stream connection for on-line data exchange.

2 Pre- and postprocessing interfaces in parallel computation

The main purpose of parallel computers in the field of numerical simulation is number crunching. The discretization of mathematical equations and the refinement of the meshes within the domain of interest lead to large arrays of numerical data which are stored in the distributed memory of a MIMD computer. The tasks of pre- and postprocessing, however, are not the primary and obvious fields of parallelization because of its mostly higher part of user interaction. Principally, a program has one input and one output of data and the parallel computer should act as a black box which accelerates the numerical simulation.

Thus, looking into this black box we have some internal interfaces adapting the sequential view for the user outside to the parallel behavior on the computer cluster inside [3]. The left column in Fig. 1 shows the typical processing sequence, and the right column specifies how this is split with respect to the location where it is processed.

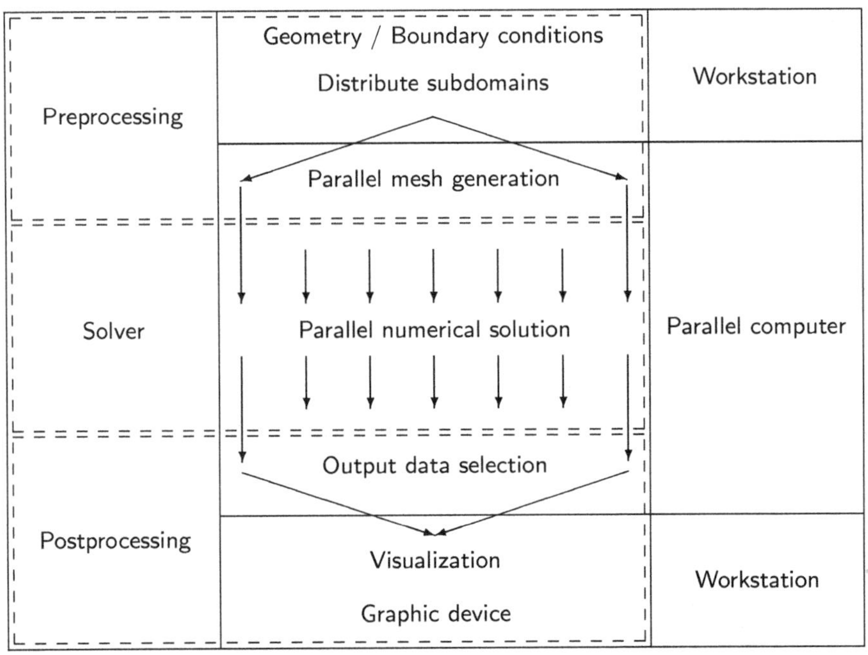

Fig. 1. Pre- and postprocessing interfaces on a parallel computer

In the preprocessing phase on a single workstation it is obviously useful to handle only coarse meshes and use them as input for a parallel program. More complex geometries may also be described separately and submitted to the program together with the coarse grid data. The mesh refinement can be performed with regard to this geometric requirements [4] (see Figs. 2,3).

In the postprocessing phase we have similar problems in the reverse way. The values which are computed in distributed memory must be summarized to some short convincing information on the desktop computer. If this infomation should express more than the total running time, it is mostly any kind of graphical output which can be quickly captured for evaluation.

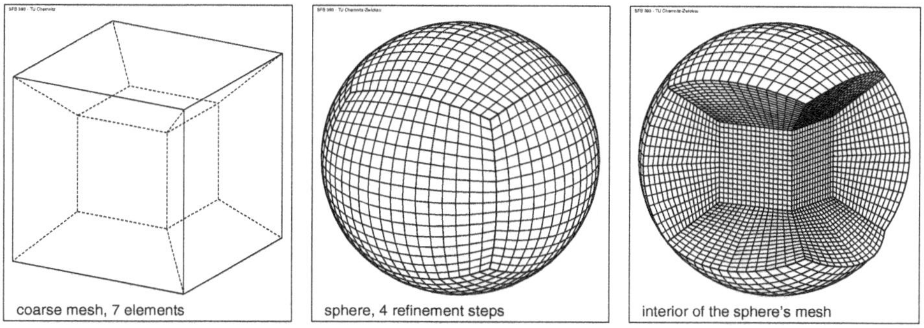

Fig. 2. Coarse mesh and refinement for a spherical surface

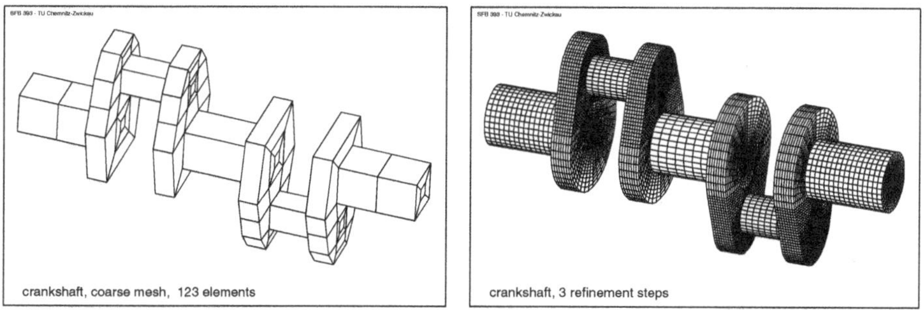

Fig. 3. Coarse mesh and refinement for a crankshaft

3 Implementation methods

3.1 Assumptions

The special circumstances arising from parallel numerical simulation allow two different policies of sharing the work for postprocessing among the parallel processors and the workstation:

1. Convert as much as possible from numerical data to graphical data (e.g. plot commands at the level of pixels and colors) on the parallel computer and send the result to the workstation which has only to display the images.
2. Send numerical data to the workstation and use high-end graphic tools to obtain any suitable visualization.

Beginning in the late 1980's, when a parallel computer was still an exotic equipment with special hardware and software, our aim was first of all a rather *quick and dirty* visualization of numerical results for low-cost desktop workstations. Hence, we preferred the first policy where most of the work is done on the parallel computer.

Therefore, our first implementation of graphical output from the parallel computer is based on a minimum of assumptions:

- The program runs on a MIMD computer with distributed memory and message passing. (Shared-memory machines can always simulate message passing.)
- A (virtual) hypercube topology is available for communication within the parallel computer. It may also be sufficient to have an efficient implementation of the global data exchange for any number of processors other than a hypercube.
- At least one processor (the *root* processor 0) is connected to the user's workstation directly or via ethernet.
- At least this *root* processor can use X11 library functions and contact the X server on the user's workstation.

Such a minimum of implementation requirements has been well-tried giving a maximum of flexibility and versatility on changing generations of parallel computers. The communication interface is not restricted to a certain standard library, it is rather an extension to PVM, MPI or any hardware specific libraries [5, 6]. Since the parallel graphics library uses only this interface as well as the basic X11 functions abstaining from special GUI's, it has not only been able to survive for the last two decades, but was also successfully applied in testing parallel algorithms and presenting their computational results.

3.2 Levels of implementation

According to the increasing requirements over a couple of years, our visualization library grew up in a few steps, each of them representing a certain level of usage (see Fig. 4).

- At the basic level there is a set of drawing primitives as an interface for Fortran and C programmers hiding the complex X11 data structures.
- The next step provides a simple interface for 2-dimensional finite element data structures to be displayed in a window. This includes the usual variety of display options including special information related to the subdomains or material data
- For 3-dimensional finite element data we only had to insert an interface for projecting 3D data to a 2D image plane (surface plot or sectional view). Then the complete functionality of the 2D graphics library is applicable for the projected data.
- Alternatively, it is possible to transfer the complete information about the 3D structure (as a file or a socket data stream) to a separate high-end graphic system with more features. Such a transfer was integrated in our graphics interface library as an add-on using for reference e. g. the IRIS Explorer [7] as an external 3D graphic system.

Details of those interfaces for programmers and users are given in [8].

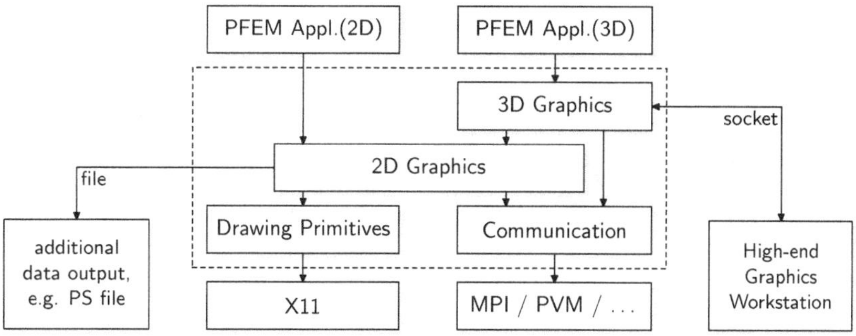

Fig. 4. Graphics library access for parallel finite element applications

3.3 Examples for 2D graphics

Because our requirements are content with a basic X-library support we have a quite simple interface for user interaction. The graphical menus are as good as "hand-made" (see Fig. 5). This is sufficient for a lot of options to select data and display styles as needed.

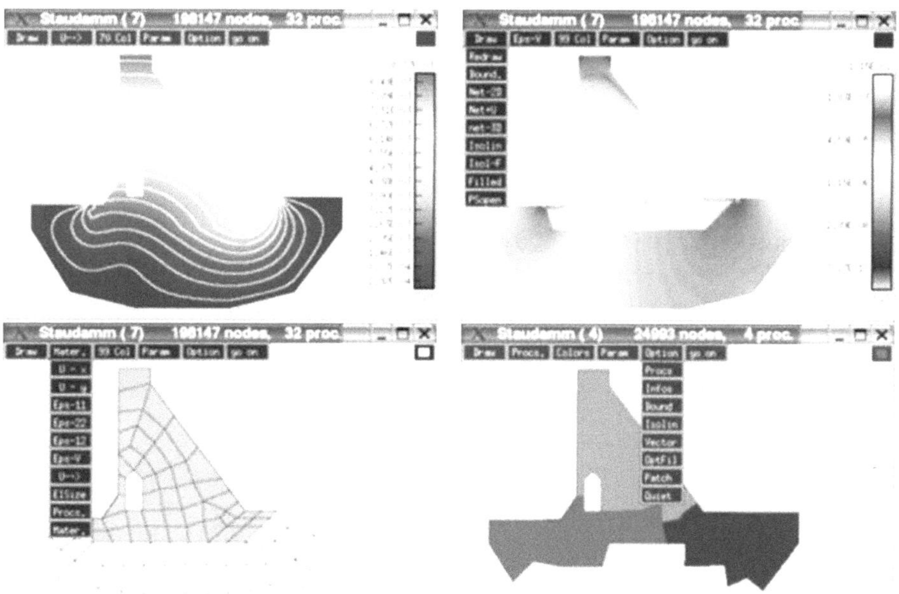

Fig. 5. Screenshots of the user interface with menus and various display options (deformation, stresses, material regions, coarse grid, subdomains)

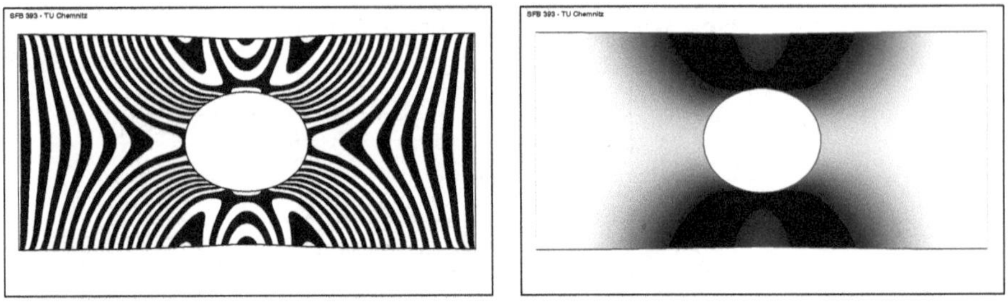

Fig. 6. Strains within a tensil-loaded workpiece in different color schemes

However, it is also provided that the programmer may pre-define a set of display options and show the same picture without interaction at running time, e. g. for presentations or frequently repeated testing.

Among others, the menu offers a lot of common drawing options, such as

- draw grid or boundary lines, isolines or colored areas;
- various scaling options, vector or tensor representation;
- zooming, colormaps, pick up details for any point within the domain.

In particular, the user may zoom into the 2D domain either by dragging a rectangular area with the mouse or by typing in the range of x- and y-coordinates. Different colormaps may give either a smooth display or one with higher contrast. For illustration, Fig. 6 shows one and the same example displayed once with a black and white, once with a grayscale coloring. The number of isolines can be explicitly redefined, leading to different densities of lines. In some cases a logarithmic scaling gives more information than a linear scale (Fig. 7).

If a solution of elastic or plastic deformation problems or in simulation of fluid dynamics has been computed, the first two components can be considered as the displacement or velocity vector. Such a vector field can be shown

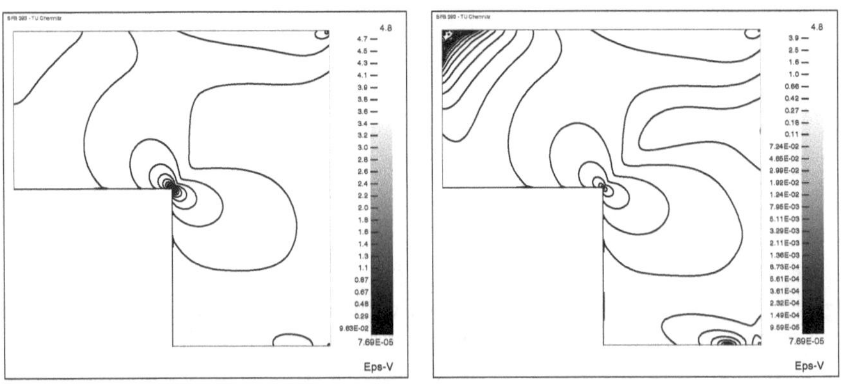

Fig. 7. Linear and logarithmic scale for isolines

by small arrows in grid points. A displacement (possibly scaled by a certain factor) may also be added to the coordinates to show the deformed mesh.

As mentioned above, the 2D graphics interface is used as base level for displaying certain views of 3D data.

3.4 Examples for 3D graphics

The primary aim of the 3D graphics support for our parallel finite element software was to have a low-cost implementation based on the already running 2D graphics. A static three-dimensional view which is recognizable for the user, may be either a surface plot or an intersecting plane:

1. In the case of intersecting a three-dimensional mesh by cutting each affected tetrahedron or hexahedron, the result is a mesh consisting of polygons with 3 to 6 edges. Inserting a new point (e.g. the barycenter of the polygon), this may easily be represented by a triangular finite element mesh (Fig. 8).

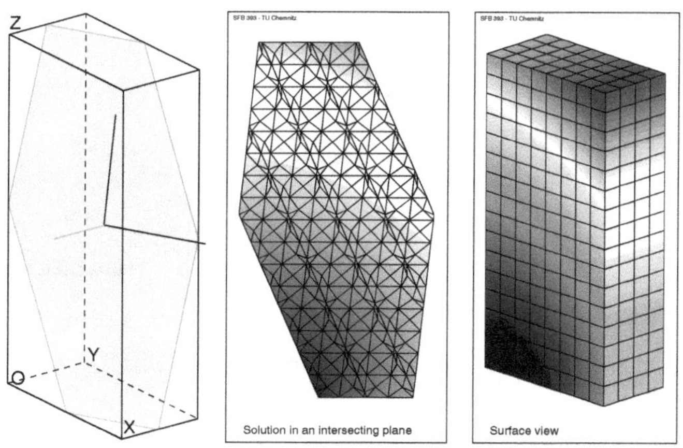

Fig. 8. Intersecting plane and surface view of a three-dimensional solid

2. The projection of any surface mesh of a tetrahedral or hexahedral finite element mesh to the image plane can be represented as a two-dimensional triangular or quadrangular finite element mesh. Of course, only such surface polygons are considered that point towards the viewer (Fig. 8, right). Anyhow, some of those faces may be hidden if the solid is not convex. In this case the mesh may be displayed as a wireframe, accepting some faulty display, or the display may be improved by a depth-sort algorithm (Fig. 9). A complete hidden-line handling was not the purpose of this graphics tool for parallel applications.

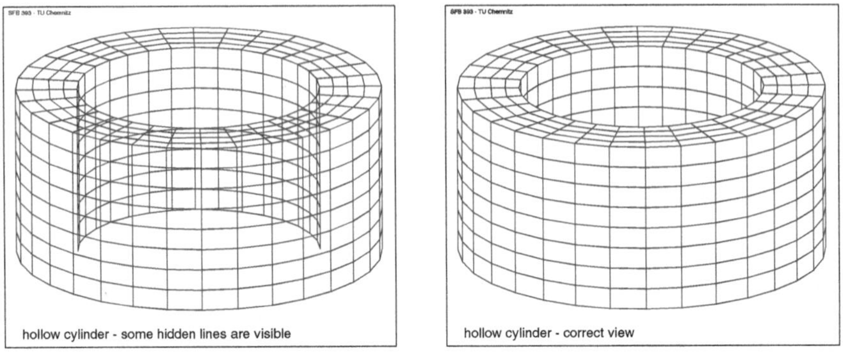

Fig. 9. Different displays for a non-convex solid

3. Another kind of "intersection" can be obtained by clipping all elements of a three-dimensional mesh above a given plane and show a surface view of the remaining body (Fig. 10).

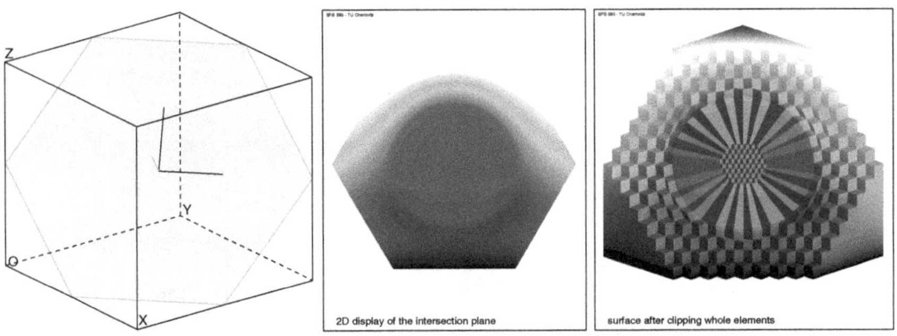

Fig. 10. Comparison of intersection and clipping with the same plane

4. Sometimes multiple clipping planes may give a better impression of the three-dimensional representation (Fig. 11,12).

In the case of an intersecting plane, the two-dimensional view will be just this plane. In the other cases there remains an additonal option to specify the view coordinate system for the projection of the surface. For that purpose, the user may define the viewer's position, i.e. the normal vector of the image plane, and for a perspective view also the distance from the plane. A bounding box with the position of the axes in the current view and the cut or clip planes is always displayed in a separate window as shown in the leftmost pictures of Figs. 8, 10, or 11.

Certainly, from a given view, any rotation of the coordinate system is supported to get another view.

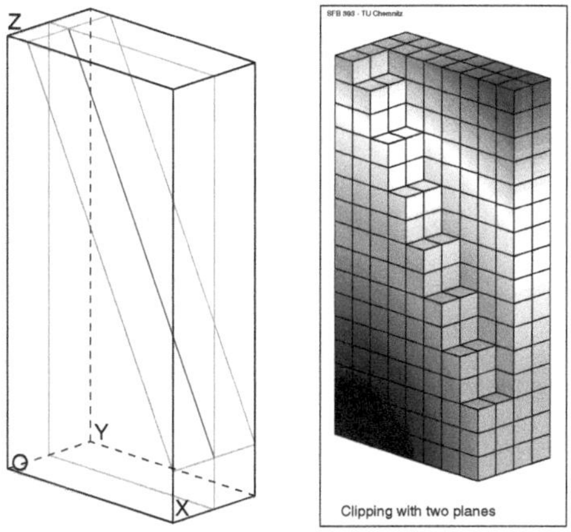

Fig. 11. Surface view of a three-dimensional solid after clipping at two planes

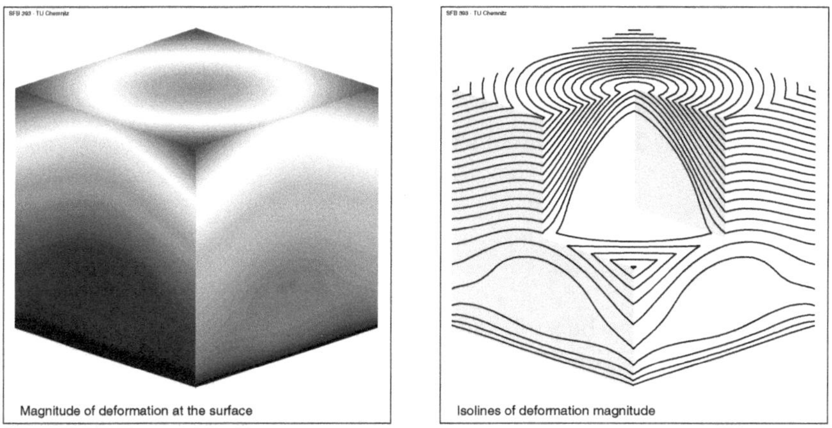

Fig. 12. Deformation of a solid that includes a sphere with different material, a view to the surface and the more interesting view into the interior, clipped by 3 planes

4 Additional tasks

Although the primary purpose of this visualization tool was to have an immediate display of results from numerical simulation with parallel finite element software, it is only a small step forward to use this software for some additional more or less general features.

4.1 Multiple views simultaneously

Since the programmer may switch off the menu and user interaction in the graphics window for the purpose of a presentation, frequently repeated tests, or time-dependend simulation as in Sect 4.3, it may be convenient to split the graphics window into two or more regions with separate views. Thus, e. g. multiple physical quantities or the mesh may be displayed simultaneously (see screenshot in Fig. 13).

Such settings may be made either by the user via the menu of the graphics window, or by the programmer via subroutines provided by the graphics library.

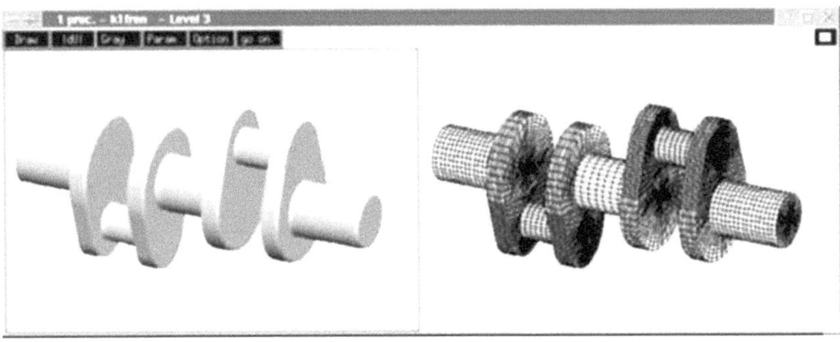

Fig. 13. The screenshot shows two displays in a split window

4.2 Postscript output

Here and there it may be sufficient to use a screen snapshot to get a printable image of the graphical output. But higher print quality can be obtained using real postscript drawing statements for scalable vector graphics instead of scaled pixel data. The procedure for this purpose is very simple. Each drawing primitive for the X11 interface has its counterpart to write the corresponding postscript commands. The user may switch on the postscript output and any subsequent drawings are simultaneously displayed on the screen and written to the postscript file until the postscript output is closed.

The format of the files is encapsulated postscript (EPS), best suited for including in LaTeX documents or to be converted to PDF format using `epstopdf`. The header of the postscript file contains some definitions and switches which may be adapted afterwards (see [8]).

4.3 Video sequences

Even though the performance of computers increases rapidly, there are always some tasks that cannot be simulated in real-time, e. g. in fluid dynamics. Hundreds of small time-steps have to be simulated one by one. In each of them, one or more linear or non-linear systems of equations must be solved in order to compute updates of quantities such as velocity, pressure, or density.

In each time-step our visualization tool may produce a single image of the current state on the screen.

At the end one is interested in the behavior of any physical quantity during the sequence of time-steps, i. e. to see an animation or a movie.

In order to simplify the simulation process, the visualization may be switched into a "batch mode", where the current view is updated automatically after each time-step, without user interaction. In most cases, however, the (parallel) numerical simulation together with the computation and display of graphical output will not be fast enough to see a "movie" on the screen in real-time.

Therefore, one of the additional tools is a separate program that captures the images from the screen (triggered by a synchronizing command of the simulation program) and saves them as a sequence of image files. After the simulation has finished the animated solution may be viewed with any suitable tool (xanim, mpegtools).

Refer to `http://www.tu-chemnitz.de/~pester/exmpls/` for a list of examples with animated solutions.

5 Remarks

This chapter was intended to give only a short overview on visualization software which was implemented for parallel computing already some years ago, but has also been enhanced more and more over the years. The tools are flexible with respect to the underlying hardware and parallel software. Within this scope such an overview cannot be complete. For more technical details and examples one may refer to the cited papers and Web addresses.

References

1. A. Meyer. A parallel preconditioned conjugate gradient method using domain decomposition and inexact solvers on each subdomain. *Computing*, 45 : 217–234, 1990.
2. G. Haase, U. Langer, and A. Meyer. Parallelisierung und Vorkonditionierung des CG-Verfahrens durch Gebietszerlegung. In G. Bader et al, editor, *Numerische Algorithmen auf Transputer-Systemen, Teubner Skripten zur Numerik*. B. G. Teubner, Stuttgart, 1993.
3. M. Pester. On-line visualization in parallel computations. In W. Borchers, G. Domick, D. Kröner, R. Rautmann, and D. Saupe, editors, *Visualization Methods in High Performance Computing and Flow Simulation*. VSP Utrecht and TEV Vilnius, 1996. Proceeedings of the International Workshop on Visualization, Paderborn, 18-21 January 1994.
4. M. Pester. Treating curved surfaces in a 3d finite element program for parallel computers. *Preprintreihe, TU Chemnitz-Zwickau*, SFB393/97-10, April 1997. Available from World Wide Web: `http://www.tu-chemnitz.de/sfb393/Files/PS/sfb97e10.ps.gz`.

5. G. Haase, T. Hommel, A. Meyer, and M. Pester. Bibliotheken zur Entwicklung paralleler Algorithmen. Preprint SPC 95_20, TU Chemnitz-Zwickau, Febr. 1996. Available from World Wide Web: `http://archiv.tu-chemnitz.de/pub/1998/0074/`.
6. M. Pester. Bibliotheken zur Entwicklung paralleler Algorithmen - Basisroutinen für Kommunikation und Grafik. *Preprintreihe, TU Chemnitz*, SFB393/02-01, Januar 2002. Available from World Wide Web: `http://www.tu-chemnitz.de/sfb393/Files/PDF/sfb02-01.pdf`.
7. NAG's IRIS Explorer – Documentation. Web documentation, The Numerical Algorithms Group Ltd, Oxford, 1995–2005. Available from World Wide Web: `http://www.nag.co.uk/visual/IE/iecbb/DOC/`.
8. M. Pester. Visualization tools for 2D and 3D finite element programs - user's manual. *Preprintreihe, TU Chemnitz*, SFB393/02-02, January 2002. Available from World Wide Web: `http://www.tu-chemnitz.de/sfb393/Files/PDF/sfb02-02.pdf`.

Part II

Algorithms

Efficient Preconditioners for Special Situations in Finite Element Computations

Arnd Meyer

Technische Universität Chemnitz, Fakultät für Mathematik
09107 Chemnitz, Germany
a.meyer@mathematik.tu-chemnitz.de

1 Introduction

From the very efficient use of hierarchical techniques for the quick solution of finite element equations in case of linear elements, we discuss the generalization of these preconditioners to higher order elements and to the problem of crack growth ,where the introduction to of the crack opening would destroy existing mesh hierarchies. In the first part of this paper, we deal with the higher order elements. Here, especially elements based on cubic polynomials require more complicate tasks such as the definition of ficticious spaces and the Ficticious Space Lemma. A numerical example demonstrates that iteration numbers similar to the linear case are obtained.

In the second part (Sect. 6 to 10) we present a preconditioner based on a change of the basis of the ansatz functions for efficient simulating the crack growth problem within an adaptive finite element code. Again, we are able to use a hierarchical preconditioner although the hierarchical structure of the mesh could be partially destroyed after the next crack opening.

2 Solving finite element equations by preconditioned conjugate gradients

We consider the usual weak formulation of a second order partial differential equation:

Find $u \in \mathbb{H}^1(\Omega)$ (fulfilling Dirichlet–type boundary conditions on $\Gamma_D \subset \partial\Omega$)

$$\text{with } a(u,v) = \langle f, v\rangle \quad \forall v \in \mathbb{H}^1_0(\Omega) = \{v \in \mathbb{H}^1(\Omega) : v = 0|_{\Gamma_D}\}. \qquad (1)$$

For simplicity let Ω be a polygonal domain in $\mathbb{R}^d$ ($d = 2$ or $d = 3$). So, using a fine triangulation (in the usual sense)

$\mathcal{T}_L = \{T \subset \Omega\}$, T are triangles (quadrilaterals) if $d = 2$,
T are tetrahedrons (pentahedrons/hexahedrons) if $d = 3$,

with the nodes a_i (represented by their numbers i), we define a finite dimensional subspace $\mathbb{V} \subset \mathbb{H}^1(\Omega)$ and define the finite element solution $u_h \in \mathbb{V}$ from

$$a(u_h, v) = \langle f, v \rangle \quad \forall v \in \mathbb{V} \cap \mathbb{H}_0^1(\Omega). \tag{2}$$

(The usual generalizations of approximating $a(\cdot,\cdot)$ or $\langle f, \cdot \rangle$ or the domain Ω from $\cup T$ are straightforward, but not considered here).

Let $\Phi = (\varphi_1(x), \ldots, \varphi_N(x))$ be the row vector of the finite element basis functions defined in $\mathbb{V}$, then we use the mapping $\mathbb{V} \ni u \longleftrightarrow \underline{u} \in \mathbb{R}^N$ by

$$u = \Phi \underline{u} \tag{3}$$

for each function $u \in \mathbb{V}$. With (3) the equation (2) is transformed into the vector space, equivalently to

$$K \underline{u}^{ex} = \underline{b}, \tag{4}$$

when

$$K = (a(\varphi_j, \varphi_i))_{i,j=1}^N, \quad \underline{b} = (\langle f, \varphi_i \rangle)_{i=1}^N \quad \text{and} \quad u_h = \Phi \underline{u}^{ex}.$$

So we have to solve the linear system (4), which is large but sparse. Whenever its dimension N exceeds about 10^3 the problems in using Gaussian elimination diverge, so we consider efficient iterative solvers, such as the conjugate gradient method with modern preconditioners. The preconditioner in the vector space is a symmetric positive definite matrix C^{-1} (constant over the iteration process) for which 3 properties should be fulfilled as best as possible:

P1: The action $\underline{w} := C^{-1}\underline{r}$ should be cheap ($\mathcal{O}(N)$ arithmetical operations). Here $\underline{r} = K\underline{u} - \underline{b}$ is the residual vector of an approximate solution $\underline{u} \approx \underline{u}^{ex}$ and $\underline{w}$, the preconditioned residual, has to be an approximation to the error $\underline{u} - \underline{u}^{ex}$.

P2: The condition number of $C^{-1}K$, i.e.

$$\kappa(C^{-1}K) = \lambda_{\max}(C^{-1}K)/\lambda_{\min}(C^{-1}K)$$

should be small, this results in a small number of $\sim \kappa(C^{-1}K)^{1/2}$ iterations for reducing a norm of $\underline{r}$ under a given tolerance ϵ.

P3: The action $\underline{w} = C^{-1}\underline{r}$ should work in parallel according to the data distribution of all large data (all vectors/matrices with $\mathcal{O}(N)$ storage) over the processors of a parallel computer.

In the past, preconditioning was a matrix–technique (compare: incomplete factorizations), nowadays the construction of efficient preconditioners uses the analytical knowledge of the finite element spaces. So, we transform the equation $\underline{w} = C^{-1}\underline{r}$ into the finite element space $\mathbb{V}$ for further investigation of more complicate higher order finite elements:

Lemma 1. *The preconditioning operation* $\underline{w} = C^{-1}\underline{r}$ *on* $\underline{r} = K\underline{u} - \underline{b}$ *in* $\mathbb{R}^N$ *is equivalent to the definition of a 'preconditioned function'* $w = \Phi\underline{w} \in \mathbb{V}$ *with*

$$w = \sum_{i=1}^{N} \psi_i \langle \mathfrak{r}, \psi_i \rangle$$

for a special basis Ψ *in* $\mathbb{V}$.

Proof: With $\underline{w} = C^{-1}\underline{r}$ we define $w = \Phi\underline{w}$. For a given $\underline{u} \in \mathbb{R}^N$, we have $u = \Phi\underline{u} \in \mathbb{V}$ and define the 'residual functional' $\mathfrak{r} \in \mathbb{H}^{-1}(\Omega)$ with

$$a(u, v) - \langle f, v \rangle = \langle \mathfrak{r}, v \rangle \quad \forall v \in \mathbb{H}_0^1(\Omega).$$

In the *PCG* algorithm, we have the values $\langle \mathfrak{r}, \varphi_i \rangle$ within our residual vector $\underline{r}$:

$$\underline{r} = K\underline{u} - \underline{b} = \Big(\sum a(\varphi_j, \varphi_i) u_j - \langle f, \varphi_i \rangle \Big)_{i=1}^{N}$$

$$= \big(a(u, \varphi_i) - \langle f, \varphi_i \rangle\big)_{i=1}^{N} = \big(\langle \mathfrak{r}, \varphi_i \rangle\big)_{i=1}^{N}$$

So, $w = \Phi\underline{w} = \Phi C^{-1}\underline{r}$ can be written with any factorization $C^{-1} = FF^T$ (square root of C^{-1} or Cholesky decomposition etc.) as

$$w = \Phi F F^T \underline{r} = \sum_{i=1}^{N} \psi_i \langle \mathfrak{r}, \psi_i \rangle,$$

whenever $\Psi = (\psi_1 \dots \psi_N) = \Phi F$ is another basis in $\mathbb{V}$, transformed with the regular $(N \times N)$–matrix F.

Remark 1: If no preconditioning is used, we have $C^{-1} = F = I$, the definition of w is

$$w = \sum_{i=1}^{N} \varphi_i \langle \mathfrak{r}, \varphi_i \rangle$$

with our nodal basis Φ.

Remark 2: The well–known hierarchical preconditioner due to [16] (see next section) constructs the matrix F directly from the basis transformation of nodal basis functions Φ into hierarchical base functions Ψ and the action $\underline{w} := C^{-1}\underline{r}$ is indeed

$$\underline{w} := FF^T\underline{r}.$$

3 Basic facts on hierarchical and BPX preconditioners for linear elements

Let the triangulation $\mathcal{T}_L$ be the result of L refinement steps starting from a given coarse mesh $\mathcal{T}_0$. For simplicity let each triangle T' in $\mathcal{T}_{l-1}$ be subdivided into 4 equal subtriangles of $\mathcal{T}_l$. Then the mesh history is stored within a list of nodal numbers, where each new node $a_i = \frac{1}{2}(a_j + a_k)$ from subdividing the edge (a_j, a_k) in $\mathcal{T}_{l-1}$ is stored together with his 'father' nodes:

$$(i, j, k) = (\text{Son, Father1, Father2}).$$

This list is ordered from coarse to fine due to the history. Note that in the quadrilateral case this list contains (Son, Father1, Father2) if an edge is subdivided, but additionally

$$(\text{Son, Father1, Father2, Father3, Father4})$$

with the 'Son' as interior node and all four vertices of a quadrilateral subdivided into 4 parts. With this definition we have the finite element spaces $\mathbb{V}_l$ belonging to the triangulation $\mathcal{T}_l$ equipped with the usual nodal basis Φ_l

$$\mathbb{V}_l = span\,\Phi_l,\ l = 0, \dots, L.$$

All functions in this basis are piecewise (bi–)linear with respect to $\mathcal{T}_l$. From

$$\mathbb{V}_0 \subset \mathbb{V}_1 \subset \cdots \subset \mathbb{V}_L, \tag{5}$$

we can define a hierarchical basis Ψ_L in $\mathbb{V}_L$ recursively:
Let $\Psi_0 = \Phi_0$ and Ψ_{l-1} the hierarchical basis in $\mathbb{V}_{l-1} = span\,\Psi_{l-1} = span\,\Phi_{l-1}$, then we define

$$\Psi_l = (\Psi_{l-1} \vdots \Phi_l^{new}) \tag{6}$$

where Φ_l^{new} contains all 'new' nodal basis functions of $\mathbb{V}_l$ belonging to nodes a_i that are new ('Sons') in $\mathcal{T}_l$ (not existent in $\mathcal{T}_{l-1}$).
For $l = L$, we have $\mathbb{V}_L = span\,\Psi_L = span\,\Phi_L$, so another basis additionally to Φ_L is defined and there exists a regular $(N \times N)$–Matrix F with $\Psi_L = \Phi_L F$ which is used in our preconditioning procedure as considered in Sect. 2.
From [16] the following facts are derived:

P2 is fulfilled with $\kappa(C^{-1}K) = \mathcal{O}(L^2) = \mathcal{O}(|\log h|^2)$ from the good condition of the 'hierarchical stiffness matrix' $K_H = F^T K F \quad (\kappa(K_H) = \kappa(C^{-1}K))$, which is a consequence of 'good' angles between the subspaces $\mathbb{V}_{l-1}$ and $(\mathbb{V}_l - \mathbb{V}_{l-1})$.

P1 is fulfilled from the recursive refinement formula: Consider the spaces $\mathbb{V}_{l-1}$ and $\mathbb{V}_l$ with the bases Φ_{l-1} and Φ_l ($\dim \mathbb{V}_l = N_l$). Then from $\mathbb{V}_{l-1} \subset \mathbb{V}_l$ there is an $(N_l \times N_{l-1})$–Matrix $\tilde{P}_{l-1}$ with

$$\Phi_{l-1} = \Phi_l \tilde{P}_{l-1}. \tag{7}$$

Here $\tilde{P}_{l-1}$ is explicitly known

$$\tilde{P}_{l-1} = \begin{pmatrix} I \\ \cdots \\ P_{l-1} \end{pmatrix} \tag{8}$$

from

$$\varphi_i^{(l-1)} = \varphi_i^{(l)} + \frac{1}{2} \sum_{j \in \mathcal{N}(i)} \varphi_j^{(l)}. \tag{9}$$

(The sum runs over all 'new' nodes j that are neighbors of i forming the set $\mathcal{N}(i)$.)
That is, P_{l-1} hat values $\frac{1}{2}$ at position (j, i) iff j is 'Son' of i or an edge (i, i') from $\mathcal{T}_{l-1}$ was subdivided into (i, j) and (i', j) in $\mathcal{T}_l$.
So, the value $\frac{1}{2}$ occurs exactly twice in each row of P_{l-1}.

From this definition for all $l = 1, \ldots, L$ follows that F is a product of transformations from level to level, from which the matrix vector multiply $\underline{w} := FF^T\underline{r}$ becomes very cheap according to the following two basic algorithms:

A1: $y := F^T\underline{r}$ is done by:
1. $y := \underline{r}$
2. for all entries within the list (backwards) do:

$$\begin{cases} y(\text{Father1}) := y(\text{Father1}) + \frac{1}{2}y(\text{Son}) \\ \\ y(\text{Father2}) := y(\text{Father2}) + \frac{1}{2}y(\text{Son}) \end{cases}$$

A2: $\underline{w} := Fy$ is done by:
1. $\underline{w} := y$
2. for all entries within the list do:

$$w(\text{Son}) := w(\text{Son}) + \frac{1}{2}(w(\text{Father1}) + w(\text{Father2}))$$

Note: In the quadrilateral case sometimes 4 fathers exist then $\frac{1}{2}$ is to be replaced by $\frac{1}{4}$ due to another refinement equation.

Remark 1: This preconditioner works perfectly in a couple of applications in $2D$. Basically it has been successfully used for simple potential problems, but a generalization to linear elasticity problems (plane stress or plane strain 2D) is simple. Here, we use this technique for each single component of the vector function $\mathbf{u} \in (\mathbb{H}^1(\Omega))^2$. The condition number $\kappa(C^{-1}K)$ is enlarged by the constant from Korn's inequality.

Remark 2: For 3D problems, a growing condition number $\kappa(C^{-1}K) = \mathcal{O}(h^{-1})$ would appear. To overcome this difficulty, the BPX preconditioner has to be used [3, 11]. According to (7) we have

$$\begin{aligned} \Phi_l &= \Phi_L Q_l \quad \forall l = 0, \dots, L \\ Q_l &= \tilde{P}_{L-1} \cdot \ldots \cdot \tilde{P}_l \text{ is } (N \times N_l) \text{ and } Q_L = I. \end{aligned} \tag{10}$$

Then the BPX preconditioner is defined as

$$w = \mathcal{C}^{-1}\mathfrak{r} = \sum_{l=0}^{L} \sum_{i=1}^{N_l} \varphi_i^{(l)} \langle \mathfrak{r}, \varphi_i^{(l)} \rangle \cdot d_i^l \tag{11}$$

which is (from Sect. 2) equivalent to:

$$\underline{w} = \sum_{l=0}^{L} Q_l D_l Q_l^T \underline{r}. \tag{12}$$

Here, $D_l = diag(d_1^l, \dots, d_{N_l}^l)$ are scale factors, which can be chosen as $2^{(d-2)l}$ or as inverse main diagonal entries of the stiffness matrices $K_l = Q_l^T K Q_l$ belonging to Φ_l.
For this preconditioner the fact $\kappa(C^{-1}K) \leq$ const (independent of h) can be proven and the algorithm for (12) is similar to (A1) and (A2).

4 Generalizing hierarchical techniques to higher order elements

From the famous properties of the preconditioning technique in Sect. 3, we should wish to construct similar preconditioners for higher order finite elements and especially for shell and plate elements.

We propose the same nested triangulation as in Sect. 3: $\mathcal{T}_0, \dots, \mathcal{T}_L$. Let n_l be the total number of nodes in $\mathcal{T}_l$, then in Sect. 3 we had $N_l = n_l$ (it was one degree of freedom per node). Now this is different, usually $N_l > n_l$ (at least for $l = L$, where the finite element space $\mathbb{V}_L = \mathbb{V}$ for approximating our bilinear form is defined).

For using hierarchical–like techniques we have 3 possibilities:

Technique 1: For some kind of higher order elements, we define the finite element spaces $\mathbb{V}_l$ on each level and obtain nested spaces

$$\mathbb{V}_0 \subset \mathbb{V}_1 \subset \ldots \subset \mathbb{V}_L.$$

In this case the same procedure as in Algorithms A1/A2 can be used, but due to a more complicate refinement formula (instead of (9)), the algorithms are to be adapted.

Example 1: Bogner–Fox–Schmidt–elements on quadrilaterals (with bicubic functions and 4 degrees of freedom per node), see [13, 14].

Technique 2: Usually we cannot guarantee that the spaces $\mathbb{V}_l$ are nested (i.e. $\mathbb{V}_{l-1} \not\subset \mathbb{V}_l$), but our finest space $\mathbb{V} = \mathbb{V}_L$ belonging to the triangulation $\mathcal{T}_L$ contains all piecewise linear functions on $\mathcal{T}_L$. So we have the nested spaces $\mathbb{V}_0^{(1)} \subset \ldots \subset \mathbb{V}_L^{(1)} \subset \mathbb{V}_L$, when $\mathbb{V}_l^{(1)}$ are defined of the piecewise linear functions on $\mathcal{T}_l$ (as in Sect. 3). Then we have to represent $\mathbb{V}_L = \mathbb{V}_L^{(1)} \dot{+} \mathbb{W}_L$ (direct sum) and prove that $\gamma = \cos\angle(\mathbb{V}_L^{(1)}, \mathbb{W}_L) < 1$. This angle is defined from the $a(\cdot,\cdot)$ energy–inner product

$$\gamma^2 = \sup \frac{a^2(u,v)}{a(u,u)a(v,v)}, \quad \text{where} \tag{13}$$

the supremum is taken over all $u \in \mathbb{V}_L^{(1)}$ and $v \in \mathbb{W}_L$. If $\gamma < 1$ (independent of h), the preconditioner works as in Sect. 3 for $l = 0, \ldots, L$ (and linear elements: A1/A2) and additionally there is one transformation from the nodal basis Φ_L of $\mathbb{V}_L$ into the hierarchical basis $(\Phi_L^{(1)} \vdots \Phi_L^{new})$ of $(\mathbb{V}_L^{(1)} \dot{+} \mathbb{W}_L)$ and back. Again we have to calculate a special refinement formula for this last step:

$$\Phi_L^{(1)} = \Phi_L \tilde{P}_L. \tag{14}$$

Here, $\tilde{P}_L$ has another structure as $\tilde{P}_l (l < L)$ from Sect. 3. The entries of P_L are defined from

$$\varphi_i^{(1)} = \varphi_i^{higher} + \sum_{j \in N(i)} \alpha_{ij} \varphi_j^{higher}, \tag{15}$$

where $\Phi_L^{(1)} = (\varphi_1^{(1)}, \ldots, \varphi_{n_L}^{(1)})$ are the piecewise linear basis functions and φ_j^{higher} the finite element basis functions that span $\mathbb{V}_L$ (for example piecewise polynomials of higher order).

Example 2: $\mathbb{V}_L = \mathbb{V}_L^{(2)}$ (piecewise quadratic polynomials on 6–node triangles). Here, $\alpha_{ij} = \frac{1}{2}$ iff i is vertex node of $\mathcal{T}_L$ and j the node on the midpoint of an edge $(a_i, a_{i'})$. From $\alpha_{ij} = \frac{1}{2}$ follows that Algorithm A1/A2 can be used without change (one level more, all edge nodes are 'sons' of the vertex nodes of this edges).

Example 3: $\mathbb{V}_L = \mathbb{V}_L^{(2)}$ (piecewise biquadratic polynomials on 9–node quadrilaterals). Here,

$$\alpha_{ij} = \begin{cases} \frac{1}{2} \; j \text{ midpoint of an edge } (i, i') \\ \frac{1}{4} \; j \text{ interior node} \end{cases}$$

Again Algorithm A1/A2 works without change, one level more.

Example 4: $\mathbb{V}_L = \mathbb{V}_L^{(2,red)}$ (reduced biquadratic polynomials on 8–node quadrilaterals).
Here, $\alpha_{ij} = \frac{1}{2}$, again use Algorithm A1/A2.

Example 5: $\mathbb{V}_L = \mathbb{V}_L^{(3,red)}$ (piecewise reduced cubic polynomials on triangles, the cubic bubble is removed such that all quadratic polynomials are included).
This example has to be considered, because this space occurs as fictitious space in Technique 3. Here, we define the following functions:
$u(x) \in \mathbb{V}_L$ is a reduced cubic polynomial on each $T \in \mathcal{T}_L$ (defined by 9 values on the vertices of T).
We choose $u_i = u(a_i)$ and $u_{i,j} = \frac{\partial u}{\partial \mathbf{s}_{ij}}|_{a_i}$, the tangential derivatives along the edges of T: $\mathbf{a}_{ij} = a_j - a_i$, $\mathbf{s}_{ij} = \mathbf{a}_{ij}/|\mathbf{a}_{ij}|$.
Globally, we have $|\mathcal{N}(i)| + 1$ degrees of freedom at each node a_i: $u_i = u(a_i)$ and $u_{i,j} = \frac{\partial u}{\partial \mathbf{s}_{ij}}|_{a_i}$ $\forall j \in \mathcal{N}(i)$. From this definition, we easily find

$$\varphi_i^{(1)}(x) = \varphi_i^{(3)}(x) + \sum_{j \in \mathcal{N}(i)} \frac{1}{|\mathbf{a}_{ij}|}(\varphi_{j,i}(x) - \varphi_{i,j}(x)). \tag{16}$$

Here, $\varphi_i^{(3)}(x)$ is the finite element basis function with

$$\varphi_i^{(3)}(a_j) = \delta_{ij} \text{ and } \nabla\varphi_i^{(3)}(a_j) = (0,0)^T \ \forall i,j \tag{17}$$

(with support of all T around a_i) and $\varphi_{i,j}(x)$ fulfills

$$\varphi_{i,j}(a_k) = 0 \,\forall i,j,k, \quad \frac{\partial}{\partial \mathbf{s}_{ij}}\varphi_{i,j}(a_i) = 1 \tag{18}$$

$$\frac{\partial}{\partial \mathbf{s}_{ik}}\varphi_{i,j}(a_i) = 0 \text{ for all edges } (i,k) \neq (i,j),$$

so the support are two triangles that share the edge (i,j).
In the splitting $\mathbb{V}_L^{(3,red)} = \mathbb{V}_L^{(1)} + \mathbb{W}_L$, $\mathbb{W}_L$ is spanned by all functions $\varphi_{i,j}(x)$.

Technique 3: There are examples, where neither the spaces $\mathbb{V}_l$ are nested, nor piecewise linear functions $\mathbb{V}_L^{(1)}$ are contained within the finest space $\mathbb{V}_L$. Here, we can use the Fictitious Space Lemma (see [9,11]) for the construction of a preconditioner which is written as

$$\mathcal{C}^{-1} = \mathcal{R}\tilde{\mathcal{C}}^{-1}\mathcal{R}^*, \tag{19}$$

if a fictitious space $\tilde{\mathbb{V}}$ exists (in general of higher dimension than $\mathbb{V}_L$) having a good preconditioner $\tilde{\mathcal{C}}^{-1}$. The key is the definition of a restriction operator

$$\mathcal{R} : \tilde{\mathbb{V}} \to \mathbb{V}_L$$

with small energy norm.

Example 6: For more complicate plate analysis, the space $\mathbb{V}^{HC}$ of Hermite–triangles is used (DKT–elements, HCT–elements). Here $u \in \mathbb{V}^{HC}$ is again a

reduced cubic polynomial on each $T \subset \mathcal{T}_L$, but we use globally 3 degrees of freedom per node: $u_i = u(a_i)$ and $\boldsymbol{\Theta}_i = \nabla u(a_i)$.
Here, ∇u is discontinuous on the edges, but continuous on each node a_i. From this property the spaces cannot be nested and the functions in $\mathbb{V}_L^{(1)}$ have discontinuous gradients on all vertices a_i, so $\mathbb{V}_L^{(1)} \not\subset \mathbb{V}^{HC}$.

From $\mathbb{V}^{HC} \subset \mathbb{V}^{(3,red)}$, we may define a restriction operator

$$\mathcal{R} : \mathbb{V}_L^{(3,red)} \to \mathbb{V}_L = \mathbb{V}^{HC}$$

and use the preconditioner for $\mathbb{V}_L^{(3,red)}$ from Example 5 for defining $\tilde{\mathcal{C}}^{-1}$, which leads to a good preconditioner for this space $\mathbb{V}^{HC}$. The definition of $\mathcal{R}$ is not unique, we use an easy choice, some kind of averaging of $\frac{\partial u}{\partial \mathbf{s}_{ij}}|_{a_i}$ to $\nabla u(a_i)$:

Let $\tilde{u} \in \mathbb{V}_L^{(3,red)}$ be represented by

$$u_i = \tilde{u}(a_i) \text{ and } u_{i,j} = \frac{\partial \tilde{u}}{\partial \mathbf{s}_{ij}}|_{a_i} \quad (\forall i, \forall j \in \mathcal{N}(i)\,)$$

then we define $u = \mathcal{R}\tilde{u} \in \mathbb{V}^{HC}$ with $u_i = u(a_i)$ and $\boldsymbol{\Theta}_i = \nabla u(a_i)$ from

$$\boldsymbol{\Theta}_i = \frac{1}{m_i} S_i\, \underline{u}_i \quad (\underline{u}_i = (u_{i,j})\ \forall j \in \mathcal{N}(i)\,).$$

The $(2 \times m_i)-$matrix S_i contains all normalized vectors $\mathbf{s}_{ij}$ for all edges meeting a_i and $m_i = |\mathcal{N}(i)|$. So,

$$\boldsymbol{\Theta}_i = \frac{1}{m_i} \sum_{j \in \mathcal{N}(i)} \mathbf{s}_{ij} \cdot \frac{\partial u}{\partial \mathbf{s}_{ij}}|_{a_i} = \frac{1}{m_i} \sum u_{i,j} \mathbf{s}_{ij} \tag{20}$$

5 Numerical examples to the above preconditioners

Let us demonstrate the preconditioners proposed in Sect. 4 at one example that allows some of the finite elements discussed in Example 1 to Example 6. We choose Ω as a rectangle with prescribed Dirichlet type boundary conditions and solve a simple Laplace equation

$$\left.\begin{aligned} -\Delta u &= 0 \text{ in } \Omega \\ u &= g \text{ on } \partial\Omega, \end{aligned}\right\}$$

hence, $\quad a(u,v) = \int_\Omega (\nabla u) \cdot (\nabla v) d\Omega \quad$ and we use the following discretization:

(a) piecewise linear functions ($\mathbb{V}_L = \mathbb{V}_L^{(1)}$)
on a triangular mesh (3–node–triangles)
(b) piecewise bilinear functions ($\mathbb{V}_L = \mathbb{V}_L^{(1)}$)
on a quadrilateral mesh (4–node–quadrilaterals)

(c) piecewise quadratic functions ($\mathbb{V}_L = \mathbb{V}_L^{(2)}$)
on a triangular mesh (6–node–triangles)
(d) piecewise biquadratic functions ($\mathbb{V}_L = \mathbb{V}_L^{(2)}$)
on quadrilaterals (9–node–quadrilaterals)
(e) piecewise reduced quadratic functions ($\mathbb{V}_L = \mathbb{V}_L^{(2,red)}$)
on quadrilaterals (8–node–quadrilaterals)
(f) piecewise reduced cubic functions ($\mathbb{V}_L = \mathbb{V}^{HC}$)
on a triangular mesh (Hermite cubic triangles)

The preconditioners for cases (a) to (e) were simply described in Examples 1 to 4 in Sect. 4. We will give the matrix representation for the preconditioner of case (f) from combining Examples 5 and 6:
For $\mathbb{V}_L = \mathbb{V}^{HC}$, we have to solve

$$a(u,v) = \langle f, v\rangle \ \forall v \in \mathbb{V}_L \cap \mathbb{H}_0^1(\Omega)$$

in solving

$$K\underline{u}^{ex} = \underline{b}.$$

Here $u = \Phi\underline{u}$ is represented by $u_i = u(a_i)$ and $\mathbf{\Theta}_i = \nabla u(a_i)$, so $\underline{u} \in \mathbb{R}^{3n}$. The preconditioner $\mathcal{C}^{-1}$ from Sect. 2 is an operator which maps the residual functional $\mathfrak{r}(u)$ to the preconditioned function w. According to the fictitious space lemma, we set

$$\mathcal{C}^{-1} = \mathcal{R}\tilde{\mathcal{C}}^{-1}\mathcal{R}^*$$

with $\mathcal{R} : \tilde{\mathbb{V}} = \mathbb{V}_L^{(3,red)} \to \mathbb{V}_L = \mathbb{V}^{HC}$ as in Example 6. In the fictitious space $\tilde{\mathbb{V}}$, we use the preconditioner of Example 5. From Example 5 the matrix representation of this preconditioner is

$$\tilde{C}^{-1} = Q_L \begin{pmatrix} FF^T & \mathbb{O} \\ \mathbb{O} & I \end{pmatrix} Q_L^T \tag{21}$$

with F from Sect. 2. Here, Q_L transforms the basis $(\Phi^{(3)} \vdots \Phi^{(edge)})$ of the cubic functions in $\mathbb{V}^{(3,red)}$ into the hierarchical basis $(\Phi^{(1)} \vdots \Phi^{(edge)})$.

$\Phi^{(1)} = (\varphi_1^{(1)}, \ldots, \varphi_n^{(1)})$		piecewise linear functions
$\Phi^{(3)} = (\varphi_1^{(3)}, \ldots, \varphi_n^{(3)})$		piecewise reduced cubic with property (17)
$\Phi^{(edge)} = (\varphi_{i,j})$	$\forall$ edges (i,j) at node i	piecewise reduced cubic with property (18)

So,

$$Q_L = \begin{pmatrix} I & \vdots & \mathbb{O} \\ P_L & \vdots & I \end{pmatrix} \tag{22}$$

and entries of P_L are found in (16). Combining this with Example 6, we have

$$C^{-1} = RQ_L \begin{pmatrix} FF^T & \mathbb{O} \\ \mathbb{O} & I \end{pmatrix} Q_L^T R^T, \tag{23}$$

when R is the matrix representation of $\mathcal{R}$. The implementation of C^{-1} is the following algorithm (note that RQ_L is done at once for saving storage, this is a $(3n \times 3n)$–matrix).
The preconditioner C^{-1} acts on a residual vector $\underline{r} \in \mathbb{R}^{3n}$ $(n = n_L)$ with the entries

$$r_i = \langle \mathfrak{r}, \varphi_{i,0} \rangle \quad \text{and} \quad \tilde{\Theta}_i = \begin{pmatrix} \langle \mathfrak{r}, \varphi_{i,1} \rangle \\ \langle \mathfrak{r}, \varphi_{i,2} \rangle \end{pmatrix}$$

defined with the basis functions $(\varphi_{i,\alpha})$ in $\mathbb{V}^{HC}$:

$$\begin{array}{rlrl} \varphi_{i,0}(a_j) &= \delta_{ij} & \varphi_{i,1}(a_j) &= \varphi_{i,2}(a_j) = 0 \\ \nabla\varphi_{i,0}(a_j) &= (0,0)^T & \nabla\varphi_{i,1}(a_j) &= \delta_{ij}(1,0)^T \\ & & \nabla\varphi_{i,2}(a_j) &= \delta_{ij}(0,1)^T. \end{array}$$

Analogously, the result $\underline{w} = C^{-1}\underline{r}$ contains entries w_i and $\mathbf{\Theta}_i$ (for $w(a_i)$ and $\nabla w(a_i)$). From the definition

$$C^{-1} = R\tilde{C}^{-1}R^T = RQ_L \begin{pmatrix} FF^T & \mathbb{O} \\ \mathbb{O} & I \end{pmatrix} Q_L^T R^T$$

we have to implement the matrix-vector-multiply

$$(RQ_L)^T \underline{r}$$

and $\underline{w} = RQ_L y$, which is done at once for saving storage, i.e. there is no need to store the (approximately 7n) values $u_{i,j} = s_{ij}^T \mathbf{\Theta}_i$. This is contained in Algorithms B1 (for $y := (RQ_L)^T \underline{r}$) and B2 (for $\underline{w} := RQ_L \underline{y}$).

<u>Algorithm B1:</u> for each edge (i, j) do

$$\left| \begin{array}{l} y_i := r_i - \mathbf{a}_{ij}^T(\tilde{\Theta}_i + \tilde{\Theta}_j)/|\mathbf{a}_{ij}|^2 \\ y_j := r_j + \mathbf{a}_{ij}^T(\tilde{\Theta}_i + \tilde{\Theta}_j)/|\mathbf{a}_{ij}|^2 \end{array} \right.$$

<u>Algorithm B2:</u> for each edge (i, j) do

$$\left| \begin{array}{l} \mathbf{\Theta}_i := \mathbf{\Theta}_i + \mathbf{a}_{ij}(\mathbf{a}_{ij}^T\tilde{\Theta}_i + w_j - w_i)/|\mathbf{a}_{ij}|^2 \\ \mathbf{\Theta}_j := \mathbf{\Theta}_j + \mathbf{a}_{ij}(\mathbf{a}_{ij}^T\tilde{\Theta}_j + w_j - w_i)/|\mathbf{a}_{ij}|^2 \end{array} \right.$$

$\mathbf{\Theta}_i$ are evaluated from a Jacobi–preconditioning on the input $\tilde{\Theta}_i$, w_i are the results of Algorithms A1/A2 on the n–vector y.
The hierarchical–like preconditioners for the elements (a) to (f) require a very small number of PCG–iterations as presented in the following Table 1.

Table 1. Number of iterations for examples (a) to (f)

L	n^1	(a)	(b)	(c)	(d)	(e)	(f)[1]
1	289	17	13	19	14	13	19
2	1,089	20	15	22	17	15	21
3	4,225	23	18	24	19	17	24
4	16,641	25	20	26	20	19	26
5	66,049	26	21	27	23	21	27
6	283,169	27	23	28	24	22	28
7	1,050,625	28	25	29	–	23	–

6 Crack growth

The finite element simulation of crack growth is well understood from the point of modeling this phenomenon. Around an actual crack tip, we are able to calculate stress intensity factors approximately, which give information on the potential growth (or stop) of the existing crack. Additionally we obtain the direction of further crack propagation from approximating the J–integral. For precise results with these approximations based on finite element calculations of the deformation field (at a fixed actual crack situation), a proper mesh with refinement around the crack tip is necessary. From the well established error estimators/error indicators, we are able to control this mesh refinement in using adaptive finite element method.

This means, at the fixed actual crack situation we perform some steps of following adaptive loop given in Fig. 1.

After three or four sub-calculations in this loop, we end at an approximation of the actual situation that allows us to decide the crack behavior precise enough. After calculating the direction of propagation $\mathbf{J}$, we are able to return this adaptive loop at a slightly changed mesh with longer crack (for details see [8]).

This new adaptive calculation on a slightly changed mesh was the original challenge for the adaptive solver strategy: All modern iterative solvers such as preconditioned conjugate gradients (PCGM) use hierarchical techniques for efficient preconditioners. Examples are the Multi-grid method, the hierarchical–basis preconditioner [16] or (especially for 3D) the BPX–preconditioner [3] as in Sect. 3. The implementation of these multi level techniques requires (among others) a hierarchical order of the unknowns. From the adaptive mesh refinement such a hierarchical node ordering is given by the way, if we store the full edge tree. From this reason we cannot allow the introduction of some extra edges ("double" the edges) along the new crack line. A reasonable way out of this problem is discussed in the next section.

[1]note that n is the number of nodes in the finest mesh $\mathcal{T}_L$,
which is equal to the dimension N of the linear system in cases (a) to (d).
In (e) $N \approx \frac{3}{4}n$ but in (f) $N = 3n$.

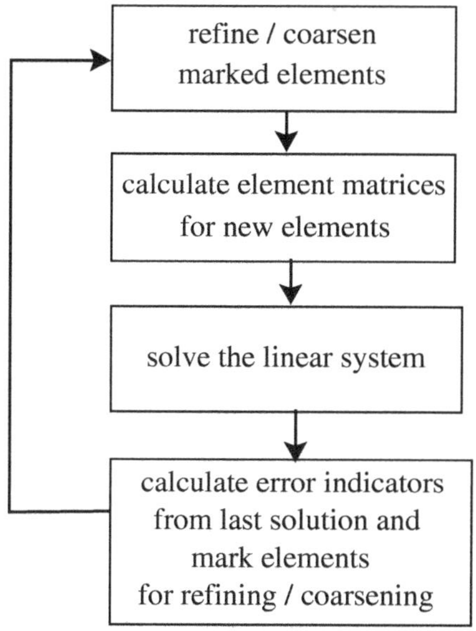

Fig. 1. The adaptive solution loop

7 Data structure for maintaining hierarchies

The idea of the extension of the crack line is given in [8]. We subdivide the existing edges that cut the crack line at the cutting point. Then the usual mesh refinement creates (during "red" or "green" subdivision of some elements) new edges along the crack line together with a proper slightly refined mesh and the correct hierarchies in the new edge tree.

So, there are no "double" edges along the crack line. For defining the double number of unknowns at the so called "crack–nodes", we introduce a copy of each crack–node and call the old one "–"-node on one side of the crack and the other "+"-node at the other side. Now, the hierarchical preconditioner

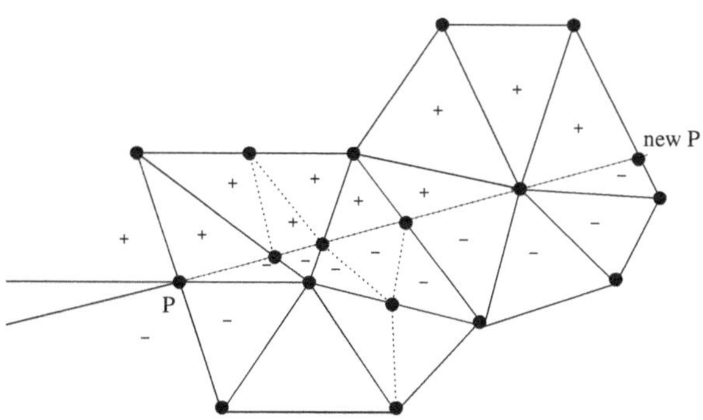

Fig. 2. Mesh handling after crack extension from the old crack tip P to "*new* P"

can access from its edge tree information only the usual and the "–"-values but never the new additional "+"-values. An efficient preconditioner has to combine both informations as it was expected in [8] from a simple averaging technique of the results of two preconditioners (first with usual and "–"-values, then with usual and "+"-values). This simple approach never can lead to a spectrally equivalent preconditioner which is clearly seen in relative high numbers of PCG–iterations. Here, another approach with a basis change and a domain decomposition like method will be given in the next chapter.

8 Two kind of basis functions for crack growth finite elements

The fact that we have a coherent continuum before the growth of the crack that changes into a slit–domain after growing requires a special finite element treatment. One possibility is the construction of a **new mesh** after crack growing, where the new free crack shores are usual free boundaries (with zero traction boundary condition). This is far away from being efficient because we have a mesh from Sect. 6 and the error indicators will drive the future refinement and coarsening to the required mesh for good approximating this new situation. So, we start to work with the **existing mesh** with double degrees of freedom at the crack nodes.

From the technique in [8], the crack–line in the undeformed domain is represented by some edges. Each edge refers to its end–nodes. For the calculation of the crack opening these nodes carry twice as much degrees of freedom and are called "crack–nodes".

Let the total number of nodes $N = n + d$ of the actual mesh be split into n usual nodes and d crack–nodes. The degrees of freedom of the crack–nodes are called "–"-values on one shore of the crack and (different) "+"-values on the other shore. A finite element which contains at least one crack–node is called "–"-element, if it refers to "–"-values (lays at "–"-side of the crack) and conversely, a "+"-element refers to "+"-values and lays at the other side of the crack–line (as indicated in Fig. 2).

From the usual element by element calculation of the stiffness matrix, using these 2d double unknowns along the d crack–nodes, we understand the resulting stiffness matrix as usual finite element matrix that belongs to the following basis of (vector) ansatz functions:

$$\Phi = \left(\varphi_1 I, \dots, \varphi_n I,\ \varphi^-_{n+1} I, \dots, \varphi^-_{n+d} I,\ \varphi^+_{n+1} I, \dots, \varphi^+_{n+d} I\right).$$

Here, we write $\varphi_k I = (\varphi_k \mathbf{e}_1 \vdots \varphi_k \mathbf{e}_2)$ to specify the two usual vector ansatz functions. Moreover, φ_k $(k = 1 \dots n)$ denotes a usual hat–function on the usual node k. In contrast to that, φ^-_{n+j} and φ^+_{n+j} are half hat functions with its support around the crack–node $(n+j)$ at the "–"-elements ("+"-elements resp.) only. (If a usual node k lays on the remaining free boundary of the domain, the function φ_k is such a "half" hat function as well). The total number of ansatz functions is $2 \cdot (n+d+d)$, which is the dimension of the resulting stiffness matrix K.

For the efficient preconditioning of this matrix K, we introduce another basis of possible ansatz functions, that span the same $2(n+2d)$–dimensional finite element space:

$$\tilde{\Phi} = (\varphi_1 I, \ldots, \varphi_n I, \varphi_{n+1} I, \ldots, \varphi_{n+d} I, \tilde{\varphi}_{n+1} I, \ldots, \tilde{\varphi}_{n+d} I).$$

Here, φ_{n+j} is the usual full (continuous) hat function at the node $(n+j)$ and $\tilde{\varphi}_{n+j}$ is the product of φ_{n+j} with the Heaviside function of the crack–line. This means:

$$\begin{aligned} \varphi_{n+j} &:= \varphi^-_{n+j} + \varphi^+_{n+j} \quad \text{(a.e.)} \\ \tilde{\varphi}_{n+j} &:= \varphi^-_{n+j} - \varphi^+_{n+j} \quad \text{(a.e.)}\ . \end{aligned} \tag{24}$$

Hence, $\tilde{\varphi}_{n+j}$ has a jump from -1 to $+1$ over the crack–line at the crack–node $(n+j)$.

Theoretically, we can use $\tilde{K}$, the stiffness matrix of $\tilde{\Phi}$ instead of K for the same finite element computation of the new crack opening. Note the following differences between these two basis definitions:

Advantages of Φ:

- usual element routines
- usual post–processing (direct calculation of the displacements of both crack shores)
- usual error estimators / error indicators (at the crack the same data as at free boundaries)

Disadvantage of Φ:

- requires special preconditioner for K

For the basis $\tilde{\Phi}$ the reverse properties are true. The use of $\tilde{\Phi}$ would require some special treatment in element routines, post–processing and a new error control.

For $\tilde{K}$ an efficient preconditioner can be found, so by use of the basis transformation (24) we construct an efficient preconditioner for K as well. Obviously

$$\tilde{\Phi} = \Phi D$$

with the block–diagonal matrix

$$D = \text{ blockdiag } (D_1, D_2),$$

where

$$D_1 = I, \quad (2n \times 2n),$$

and

$$D_2 = \begin{pmatrix} I & I \\ I & -I \end{pmatrix}, \quad (2 \cdot (2d) \times 2 \cdot (2d)).$$

This leads to

$$\tilde{K} = DKD. \tag{25}$$

So, if $\tilde{C}$ is a good preconditioner for $\tilde{K}$, then $C = D^{-1}\tilde{C}D^{-1}$ is as good for K.

9 An efficient DD–preconditioner for $\tilde{K}$

From the special structure of the matrix $\tilde{K}$ a domain decomposition approach leads to a very efficient preconditioner. Let us recall the structure of K from the definition of the basis Φ:

$$K = \begin{pmatrix} A & B^- & B^+ \\ (B^-)^T & T^- & \mathbb{O} \\ (B^+)^T & \mathbb{O} & T^+ \end{pmatrix}$$

with the blocks:
A $(2n \times 2n)$ of the energy inner products of all usual basis functions $\varphi_i \mathbf{e}_k$ with itself,
T^- $(2d \times 2d)$ of the energy inner products of all $\varphi^-_{n+i}\mathbf{e}_k$ with itself,
T^+ $(2d \times 2d)$ of the energy inner products of all $\varphi^+_{n+i}\mathbf{e}_k$ with itself and
B^-, B^+ contain all energy inner products of $\varphi_i \mathbf{e}_k (i \leq n)$ with $\varphi^-_{n+j}\mathbf{e}_l$ (resp. $\varphi^+_{n+j}\mathbf{e}_l$).

No "+"-node is coupled to a "–"-node, this leads to both zero blocks (the crack-tip is understood as "usual node").

If we perform the transformation (25), the new matrix $\tilde{K}$ possesses a much more simple structure:

$$\tilde{K} = \begin{pmatrix} A & B^- + B^+ & B^- - B^+ \\ \text{sym.} & T^- + T^+ & T^- - T^+ \\ \text{sym.} & T^- - T^+ & T^- + T^+ \end{pmatrix} \tag{26}$$

which is abbreviated as

$$\tilde{K} = \begin{pmatrix} A_0 & B \\ B^T & T \end{pmatrix}$$

with A_0 the leading $2(n+d)$–block and

$$B = \begin{pmatrix} B^- - B^+ \\ T^- - T^+ \end{pmatrix}, \quad T = T^- + T^+.$$

Then the matrix A_0 is defined from all energy inner products of all first $2(n+d)$ usual hat functions of the basis $\tilde{\Phi}$. Hence, A_0 is the usual stiffness matrix for the actual mesh without any crack opening.
All the $n+d$ nodes are represented in a hierarchical structure of the edge–tree, so any kind of multi–level preconditioners can easily be applied for preconditioning A_0.
Especially the most simple hierarchical–basis preconditioner as explained in Sect. 3 (for details see [16]) would be very cheap but effective.
A resulting good preconditioner for the whole matrix follows from the fine structure of T:
If all crack–nodes are ordered in a 1–dimensional chain, then T has block–tridiagonal form for linear elements or block–pentadiagonal form for quadratic elements. Hence, the storage of the sub block T (upper triangle) can be arranged with

8d values (linear elements, fixed bandwidth scheme),
7d values (linear elements, variable bandwidth scheme),
12d values (quadratic elements, fixed bandwidth scheme) or
9d values (quadratic elements, variable bandwidth scheme),

and the Cholesky decomposition of T is of optimal order of complexity due to the fixed bandwidth.
This is exploited best in a domain decomposition–like preconditioner, which results from a simple block factorization of $\tilde{K}$:

$$\tilde{K} = \begin{pmatrix} I & BT^{-1} \\ \mathbb{O} & I \end{pmatrix} \begin{pmatrix} S & \mathbb{O} \\ \mathbb{O} & T \end{pmatrix} \begin{pmatrix} I & \mathbb{O} \\ T^{-1}B^T & I \end{pmatrix}$$

with the Schur–complement matrix

$$S = A_0 - BT^{-1}B^T.$$

The inverse of $\tilde{K}$ can be approximated by the inverse preconditioner $\tilde{C}^{-1}$:

$$\tilde{C}^{-1} = \begin{pmatrix} I & \mathbb{O} \\ -T^{-1}B^T & I \end{pmatrix} \begin{pmatrix} C_0^{-1} & \mathbb{O} \\ \mathbb{O} & T^{-1} \end{pmatrix} \begin{pmatrix} I & -BT^{-1} \\ \mathbb{O} & I \end{pmatrix},$$

when C_0 represents a good preconditioner for S. Note that all other inverts are exact (especially T^{-1}), so the spectrum of $\tilde{C}^{-1}\tilde{K}$ coincides with the spectrum of $C_0^{-1}S$ and additional unities. With proper scaling, we have $\tilde{C}$ for $\tilde{K}$ as good as C_0 for S. The Schur complement is a rank–2d–perturbation of A_0, so we use the same preconditioner C_0 for S as it was explained for A_0. The resulting preconditioner for K follows from (24),(25) as the formula

$$C^{-1} = D \begin{pmatrix} I & 0 \\ -T^{-1}B^T & I \end{pmatrix} \begin{pmatrix} C_0^{-1} & \mathbb{O} \\ \mathbb{O} & I \end{pmatrix} \begin{pmatrix} I & -BT^{-1} \\ 0 & T^{-1} \end{pmatrix} D.$$

For the action $\underline{w} := C^{-1}\underline{r}$ within the PCGM iterations, we need once the action of the (hierarchical) preconditioner C_0 and twice a solver with T, the remaining effort is a multiplication with parts of the stiffness matrix only, which determines a preconditioner of optimal complexity.

10 A numerical example to the crack growth preconditioner

The power of the above technique can be demonstrated at the example in [8]. Here, we start with a domain of size $(0,4)\times(-1,1)$ with a crack $(0,1)\times\{0\}$ at the beginning. Then, after each 3 adaptive mesh refinements we let the crack grow with constant direction of $(1,0)^T$. Each new crack extension is bounded by 0.25, so it matches with initial finite element boundaries. This results in relative fine meshes near the actual crack tips, but coarsening along the crack path. In the solution method proposed in [8], this example produced relative high number of PCG–iterations in the succeeding steps, which arrived near 200. This demonstrates that the preconditioner in [8] cannot be (near) spectrally equivalent to the stiffness matrix. This situation is drastically improved by introducing the preconditioner of Sect. 9. Now, the total numbers of necessary iterations are bounded near about 30 over all calculations of a fixed crack (compare Fig. 3). After each new crack hop the necessary PCG–iterations are higher due to the fact that no good starting vector for this new changed mesh is available. This leads to the nine peaks in Fig.3. Here, the preconditioner of [8] exceeded more than 100 iterations in this example and over 300 in the other experiment in [8]. Now, the averaged iteration numbers are far less 100 for both examples.

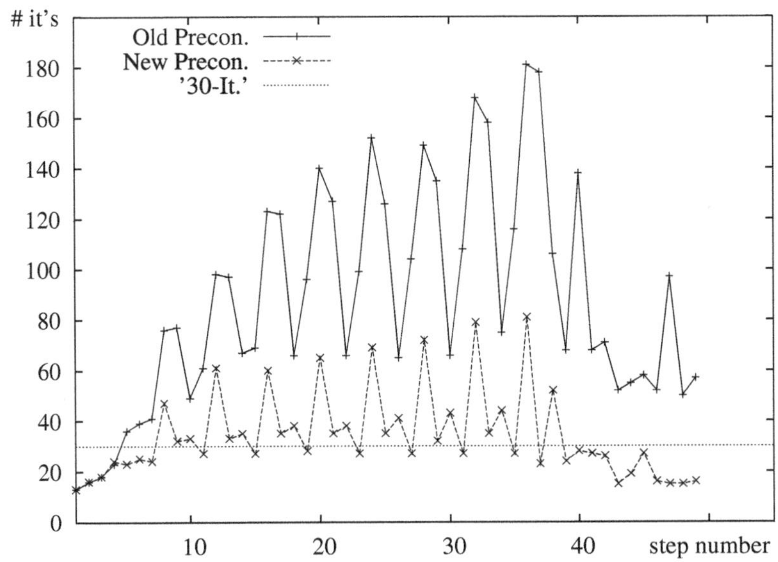

Fig. 3. Development of the number of necessary PCG–iterations for the solution method in [8] and the new method

References

1. E. Bänsch: *Local mesh refinement in 2 and 3 dimensions.* IMPACT of Computing in Science and Engineering 3, pp. 181–191, 1991.
2. J. H. Bramble, J. E. Pasciak, A. H. Schatz, "The Construction of Preconditioners for Elliptic Problems by Substructuring" I – IV, *Mathematics of Computation*, 47:103–134,1986, 49:1–16,1987, 51:415–430,1988, 53:1–24,1989.
3. I. H. Bramble, J. E. Pasciak, J. Xu, Parallel Multilevel Preconditioners, *Math. Comp.*, 55:1-22,1990.
4. G. Haase, U. Langer, A. Meyer,The Approximate Dirichlet Domain Decomposition Method, Part I: An Algebraic Approach, Part II: Application to 2nd-order Elliptic B.V.P.s, *Computing*, 47:137-151/153-167,1991.
5. G. Haase, U. Langer, A. Meyer, Domain Decomposition Preconditioners with Inexact Subdomain Solvers, *J. Num. Lin. Alg. with Appl.*, 1:27-42,1992.
6. G. Haase, U. Langer, A. Meyer, S.V. Nepommnyaschikh, Hierarchical Extension Operators and Local Multigrid Methods in Domain Decomposition Preconditioners, *East-West J. Numer. Math.*, 2:173-193,1994.
7. A. Meyer, A Parallel Preconditioned Conjugate Gradient Method Using Domain Decomposition and Inexact Solvers on Each Subdomain, *Computing*, 45:217-234,1990.
8. A. Meyer, F.Rabold, M.Scherzer: Efficient Finite Element Simulation of Crack Propagation Using Adaptive Iterative Solvers, *Commun.Numer.Meth.Engng.*, 22,93-108, 2006. Preprint-Reihe des Chemnitzer Sonderforschungsbereiches 393, Nr.: 04-01, TU Chemnitz, 2004.
9. S.V. Nepommnyaschikh, Ficticious components and subdomain alternating methods, *Sov.J.Numer.Anal.Math.Model.*, 5:53-68,1991.
10. Rolf Rannacher and Franz-Theo Suttmeier: *A feed-back approach to error control in finite element methods: Application to linear elasticity.* Reaktive Strömungen, Diffusion und Transport SFB 359, Interdiziplinäres Zentrum für Wissenschaftliches Rechnen der Universität Heidelberg, 1996.
11. P.Oswald, *Multilevel Finite Element Approximation: Theory and Applications*, Teubner Skripten zur Numerik, B.G.Teubner Stuttgart 1994.
12. M. Scherzer and A. Meyer: *Zur Berechnung von Spannungs- und Deformationsfeldern an Interface-Ecken im nichtlinearen Deformationsbereich auf Parallelrechnern.* Preprint-Reihe des Chemnitzer Sonderforschungsbereiches 393, Nr.: 96-03, TU Chemnitz, 1996.
13. M.Thess, Parallel Multilevel Preconditioners for Thin Smooth Shell Finite Element Analysis, *Num.Lin.Alg.with Appl.*, 5:5,401-440,1998.
14. M.Thess, Parallel Multilevel Preconditioners for Thin Smooth Shell Problems, Diss.A, TU Chemnitz,1998.
15. P. Wriggers: *Nichtlineare Finite-Element-Methoden.* Springer, Berlin, 2001.
16. H. Yserentant, Two Preconditioners Based on the Multilevel Splitting of Finite Element Spaces, *Numer. Math.*, 58:163-184,1990.

Nitsche Finite Element Method for Elliptic Problems with Complicated Data

Bernd Heinrich and Kornelia Pönitz

Technische Universität Chemnitz, Fakultät für Mathematik
09107 Chemnitz, Germany
bernd.heinrich@mathematik.tu-chemnitz.de
kornelia.poenitz@mathematik.tu-chemnitz.de

1 Introduction

For the efficient numerical treatment of boundary value problems (BVPs), domain decomposition methods are widely used in science and engineering. They allow to work in parallel: generating the mesh in subdomains, calculating the corresponding parts of the stiffness matrix and of the right-hand side, and solving the system of finite element equations.
Moreover, there is a particular interest in triangulations which do not match at the interface of the subdomains. Such non-matching meshes arise, for example, if the meshes in different subdomains are generated independently from each other, or if a local mesh with some structure is to be coupled with a global unstructured mesh, or if an adaptive remeshing in some subdomain is of primary interest. This is often caused by extremely different data (material properties or right-hand sides) of the BVP in different subdomains or by a complicated geometry of the domain, which have their response in a solution with anisotropic or singular behaviour. Furthermore, non-matching meshes are also applied if different discretization approaches are used in different subdomains.
There are several approaches to work with non-matching meshes, e.g., the Lagrange multiplier mortar technique, see [1–3] and the literature cited therein. Here, new unknowns (the Lagrange multipliers) occur and the stability of the problem has to be ensured by satisfying some inf-sup-condition or by stabilization techniques.

Another approach which is of particular interest in this paper is related to the Nitsche method [4] originally employed for treating essential boundary conditions. This approach has been worked out more generally in [5] and is transferred to interior continuity conditions by Stenberg [6] (Nitsche type mortaring), cf. also [7–14]. The Nitsche finite element method (or Nitsche mortaring) can be interpreted as a stabilized variant of the mortar method based on a saddle point problem.

Compared with the classical mortar method, the Nitsche type mortaring has several advantages. Thus, the saddle point problem, the inf-sup-condition as well as the calculation of additional variables (the Lagrange multipliers) are circumvented. The method employs only a single variational equation which is, compared with the usual equations (without any mortaring), slightly modified by an interface term. This allows to apply existing software tools by slight modifications. Moreover, the Nitsche finite element method yields symmetric and positive definite discretization matrices in correspondence to symmetry and ellipticity of the operator of the BVP. Although the approach involves a stabilizing parameter γ, it is not a penalty method since it is consistent with the solution of the BVP. The parameter γ can be estimated easily. In

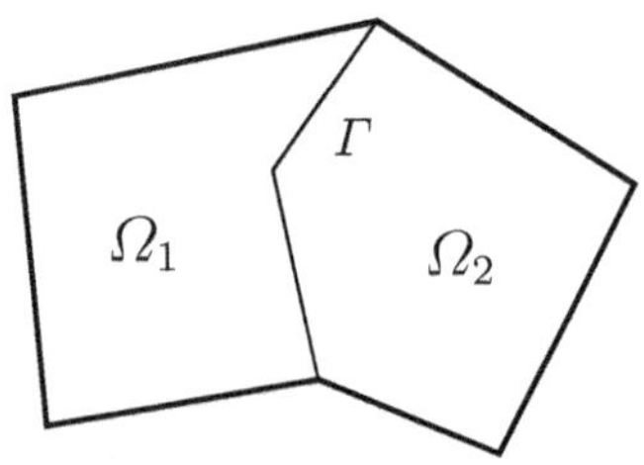

Fig. 1. Decomposition of Ω

most papers, mortar methods are founded for BVPs with solutions being sufficiently regular and without boundary layers. Moreover, quasi-uniform triangulations $\mathcal{T}_h$ with "shape regular" elements T are employed. "Shape regularity" for triangles $T \in \mathcal{T}_h$ means here that the relation $\frac{h_T}{\varrho_T} \leq C < \infty$ is satisfied (h_T: diameter of T, ϱ_T: radius of incircle in T), where C is independent of h ($h := \max_{T \in \mathcal{T}_h} h_T$) and of some pertubation parameter ε (if h_T, ϱ_T depend on ε). We call such triangles also "isotropic" in contrast to "anisotropic" triangles where $\frac{h_T}{\varrho_T} \to \infty$ as $h \to 0$ or $\varepsilon \to 0$.

Basic aspects of the Nitsche type mortaring and error estimates for regular solutions $u \in H^k(\Omega)$ ($k \geq 2$) on quasi-uniform meshes are published in [6,7,9]. Compared with these papers, we extend the application of the Nitsche mortaring to BVPs with non-regular solutions and with boundary layers caused by complicated data. In particular, we apply meshes being locally refined near corners and not quasi-uniform as well as anisotropic meshes. So we shall consider linear reaction-diffusion problems with small diffusion parameter ε^2 ($0 < \varepsilon < 1$). It is well-known that for small values of ε singularly perturbed problems with boundary layers occur and that isotropic finite elements are not convenient for the efficient treatment of such problems.

First we derive the Nitsche mortaring approach. Consider the model problem

$$\begin{aligned} Lu \; := \; -\varepsilon^2 \Delta u + cu &= f \quad \text{in } \Omega \subset \mathbb{R}^2 \\ u &= 0 \quad \text{on } \partial\Omega, \end{aligned} \tag{1}$$

where Ω is assumed to be a polygonal domain with Lipschitz-boundary $\partial\Omega$ consisting of straight segments. The following two cases are taken into account:

a) $\varepsilon = 1$ and $c = 0$ on Ω (the Poisson equation, corners on $\partial\Omega$),

b) $0 < \varepsilon < 1$ and $0 < c_0 \leq c$ on Ω (singularly perturbed problem),

with some constant c_0, and $c \in L_\infty(\Omega)$. Furthermore, assume $f \in L_2(\Omega)$ at least.

For simplicity the domain Ω is decomposed into two non-overlapping, polygonal subdomains Ω_1 and Ω_2 such that $\overline{\Omega} = \overline{\Omega}_1 \cup \overline{\Omega}_2$ and $\Omega_1 \cap \Omega_2 = \emptyset$ hold, cf. Fig. 1. Introduce $\Gamma := \overline{\Omega}_1 \cap \overline{\Omega}_2$ to be the interface of Ω_1 and Ω_2. In this context, we utilize the restrictions $v^i := v|_{\Omega_i}$ of some function v on Ω_i as well as the vectorized form $v = \left(v^1, v^2\right)$, i.e., we have $v^i(x) = v(x)$ for $x \in \Omega_i$ $(i = 1, 2)$. So we shall use here the same symbol v for denoting the function v on Ω as well as the vector $v = \left(v^1, v^2\right)$, which should not lead to confusion.

Using the domain decomposition, BVP (1) is equivalent to the following problem. Find $u = \left(u^1, u^2\right)$ such that

$$\begin{aligned} -\varepsilon^2 \Delta u^i + c u^i &= f^i \quad \text{in } \Omega_i, \qquad u^i = 0 \quad \text{on } \partial\Omega_i \cap \partial\Omega, \qquad \text{for } i = 1, 2, \\ \frac{\partial u^1}{\partial n_1} + \frac{\partial u^2}{\partial n_2} &= 0 \quad \text{on } \Gamma, \qquad u^1 = u^2 \text{ on } \Gamma, \end{aligned} \tag{2}$$

are satisfied, where n_i $(i = 1, 2)$ denotes the outward normal to $\partial\Omega_i \cap \Gamma$. Define $H_0^1(\Omega) := \left\{v \in H^1(\Omega) : v|_{\partial\Omega} = 0\right\}$ and use the variational equation of (1). Find $u \in H_0^1(\Omega)$ such that

$$a(u, v) \; = \; f(v) \quad \text{for any } v \in H_0^1(\Omega), \tag{3}$$

$$\text{with} \quad a(u, v) := \varepsilon^2 \int_\Omega (\nabla u, \nabla v) \, \mathrm{d}x + \int_\Omega c u v \, \mathrm{d}x, \quad f(v) := \int_\Omega f v \, \mathrm{d}x.$$

It is well-known that there is a unique solution $u \in H^{\frac{3}{2}+\delta}(\Omega)$ of (3), with some $\delta > 0$. Moreover, we have $\Delta u \in L_2(\Omega)$ (see [15]) and, owing to the trace theorem, $\frac{\partial u}{\partial n} \in L_2(\Gamma)$. If Ω is convex, the solution is regular, i.e., $u \in H^2(\Omega)$. Nevertheless, because of small values of ε, the solution u exhibits boundary layers, in general.

The BVP (2) can be formulated also in a weak form (see [16]). Clearly, there is weak unique solution of (2) satisfying $\left(u^1, u^2\right) \in V$, with $Lu^i \in L_2(\Omega_i)$ for $i = 1, 2$. Here, the space V is defined by $V := V^1 \times V^2$, with $V^i := \{v^i : v^i \in H^1(\Omega_i),\, v^i|_{\partial\Omega \cap \partial\Omega_i} = 0\}$ for $\partial\Omega \cap \partial\Omega_i \neq \emptyset$, $V^i := H^1(\Omega_i)$ for $\partial\Omega \cap \partial\Omega_i = \emptyset$ $(i = 1, 2)$. The continuity of the solution u and of its normal derivative $\frac{\partial u^i}{\partial n}$ on Γ $(n = n_1$ or $n = n_2)$ is to be required in the sense of $H_*^{\frac{1}{2}}(\Gamma)$ and $H_*^{-\frac{1}{2}}(\Gamma)$ (the dual space of $H_*^{\frac{1}{2}}(\Gamma)$), respectively. For Γ like in Fig. 1 $(\partial\Omega \cap \Gamma \neq \emptyset)$ we define $H_*^{\frac{1}{2}}(\Gamma)$ as the trace space $H_{00}^{\frac{1}{2}}(\Gamma)$ of $H_0^1(\Omega)$ provided with the quotient norm, see e.g. [2,15]. In the case $\partial\Omega \cap \Gamma = \emptyset$ we employ $H_*^{\frac{1}{2}}(\Gamma) := H^{\frac{1}{2}}(\Gamma)$. In the following, $\langle .\, , . \rangle_\Gamma$ denotes the $H_*^{-\frac{1}{2}}$-$H_*^{\frac{1}{2}}$ duality pairing. With this notation, (2) implies formally

$$\sum_{i=1}^{2}\left(\int_{\Omega_i}\varepsilon^2\left(\nabla u^i,\nabla v^i\right)\mathrm{d}x+\int_{\Omega_i}cuv\,\mathrm{d}x-\left\langle\varepsilon^2\frac{\partial u^i}{\partial n_i},v^i\right\rangle_\Gamma\right)=\sum_{i=1}^{2}\int_{\Omega_i}f^iv^i\,\mathrm{d}x \tag{4}$$

$\forall v\in V$, or equivalently, owing to $\frac{\partial u^1}{\partial n_1}=\alpha_1\frac{\partial u^1}{\partial n_1}-\alpha_2\frac{\partial u^2}{\partial n_2}=-\frac{\partial u^2}{\partial n_2}$ for any $\alpha_i\geq 0$ $(i=1,2)$ such that $\alpha_1+\alpha_2=1$,

$$\begin{aligned}&\sum_{i=1}^{2}\left(\int_{\Omega_i}\varepsilon^2\left(\nabla u^i,\nabla v^i\right)\mathrm{d}x+\int_{\Omega_i}cuv\,\mathrm{d}x\right)-\left\langle\alpha_1\varepsilon^2\frac{\partial u^1}{\partial n_1}-\alpha_2\varepsilon^2\frac{\partial u^2}{\partial n_2},v^1-v^2\right\rangle_\Gamma\\&-\left\langle\alpha_1\varepsilon^2\frac{\partial v^1}{\partial n_1}-\alpha_2\varepsilon^2\frac{\partial v^2}{\partial n_2},u^1-u^2\right\rangle_\Gamma+\int_\Gamma\sigma\left(u^1-u^2\right)\left(v^1-v^2\right)\mathrm{d}s\\&=\sum_{i=1}^{2}\int_{\Omega_i}f^iv^i\,\mathrm{d}x.\end{aligned} \tag{5}$$

Note that the two additional terms (both equal to zero) containing u^1-u^2 and introduced artificially have the following purpose. The first one ensures the symmetry (in u, v) of the left-hand side, the second one penalizes (after the discretization) the jump of the trace of the approximate solution and guarantees the stability for appropriately chosen weighting function $\sigma>0$. The Nitsche mortar finite element method is the Galerkin discretization of equation (5) in the sense of (8) given subsequently, using a finite element subspace V_h of V allowing non-matching meshes and discontinuity of the finite element approximation along Γ. The function σ is taken as $\gamma\varepsilon^2h^{-1}(x)$, where $\gamma>0$ is a sufficiently large constant, $h(x)$ is a mesh parameter function on Γ.

2 Non-matching mesh finite element discretization

Let $\mathcal{T}_h^i$ be a triangulation of $\overline{\Omega}_i$ $(i=1,2)$ consisting of triangles T $(T=\overline{T})$. The triangulations $\mathcal{T}_h^1$ and $\mathcal{T}_h^2$ are independent of each other, in general, i.e. the nodes of $T\in\mathcal{T}_h^i$ $(i=1,2)$ do not match along $\Gamma:=\partial\Omega_1\cap\partial\Omega_2$. Let h denote the mesh parameter of the triangulation $\mathcal{T}_h:=\mathcal{T}_h^1\cup\mathcal{T}_h^2$, with $0<h\leq h_0$ and sufficiently small h_0. Take e.g. $h:=\max_{T\in\mathcal{T}_h}h_T$. Admit that h_T and ϱ_T may depend on the parameter $\varepsilon\in(0,1)$ (cf. (14)). Furthermore, employ F $(F=\overline{F})$ for denoting any side of a triangle, h_F its length. Sometimes we use T_F in order to indicate that F is a side of $T=T_F$. Throughout the paper let the following assumption on the geometrical conformity of $\mathcal{T}_h^i$ $(i=1,2)$ be satisfied. Assume that for $i=1,2$, it holds $\overline{\Omega}_i=\bigcup_{T\in\mathcal{T}_h^i}T$, and two arbitrary triangles $T,T'\in\mathcal{T}_h^i$ $(T\neq T')$ are either disjoint or have a common vertex, or a common side. Since anisotropic triangles (see [17]) will be applied, they are not shape regular and, therefore, the mesh is not quasi-uniform (with respect to ε).

Consider further some triangulation $\mathcal{E}_h$ of the interface Γ by intervals E $(E=\overline{E})$, i.e., $\Gamma=\bigcup_{E\in\mathcal{E}_h}E$, where h_E denotes the diameter of E. Here, two segments $E, E' \in \mathcal{E}_h$ are either disjoint or have a common endpoint. A natural choice for the triangulation $\mathcal{E}_h$ of Γ is $\mathcal{E}_h := \mathcal{E}_h^1$ or $\mathcal{E}_h := \mathcal{E}_h^2$ (cf. Fig. 2), where $\mathcal{E}_h^1$ and $\mathcal{E}_h^2$ denote the triangulations of Γ defined by the traces of $\mathcal{T}_h^1$ and $\mathcal{T}_h^2$ on Γ, respectively, viz. $\mathcal{E}_h^i := \{E : E = \partial T \cap \Gamma, \text{ if } E \text{ is a segment}, T \in \mathcal{T}_h^i\}$ for $i = 1, 2$, i.e., here $E = F = T_F \cap \Gamma$ for some $T_F \in \mathcal{T}_h^i$ holds. Subsequently

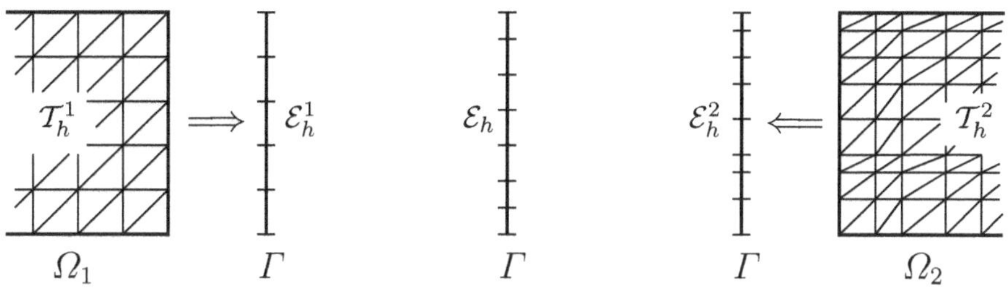

Fig. 2. Choice of $\mathcal{E}_h$

we use real parameters α_1, α_2 with

$$0 \leq \alpha_i \leq 1 \ (i = 1, 2), \quad \alpha_1 + \alpha_2 = 1, \tag{6}$$

and require the asymptotic behaviour of the triangulations $\mathcal{T}_h^1, \mathcal{T}_h^2$ and of $\mathcal{E}_h$ to be consistent on Γ in the sense of the following assumption. According to different cases of chosing $\mathcal{E}_h$ and α_1, α_2 from (6) assume that there are constants C_1, C_2 independent of $h \in (0, h_0]$ and $\varepsilon \in (0, 1)$ such that, uniformly with respect to E and F, the following relations hold (use $\mathring{E}$, $\mathring{F}$ as 'interior' of E, F, resp.).
(i) case $\mathcal{E}_h$ arbitrary: for any $E \in \mathcal{E}_h$ and $F \in \mathcal{E}_h^i$ $(i = 1, 2)$ with $\mathring{E} \cap \mathring{F} \neq \emptyset$, we have $C_1 h_F \leq h_E \leq C_2 h_F$,
(ii) case $\mathcal{E}_h := \mathcal{E}_h^i$ and $\alpha_i = 1$ $(i = 1$ or $i = 2)$: for any $E \in \mathcal{E}_h^i$ and $F \in \mathcal{E}_h^{3-i}$ with $\mathring{E} \cap \mathring{F} \neq \emptyset$, we have $C_1 h_F \leq h_E$.
This ensures that the asymptotics of segments E and sides F which touch each other is locally the same, uniformly with respect to h and ε, in case (ii) with some weakening which admits different asymptotics of triangles $T_1 \in \mathcal{T}_h^1$ and $T_2 \in \mathcal{T}_h^2$, with $T_1 \cap T_2 \neq \emptyset$.
For getting stability of the method in the case of anisotropic meshes we require that the following assumption is satisfied, which restricts the orientation of anisotropy of the triangles T at Γ: If $h_F^\perp$ denotes the height of the triangle $T_F \in \mathcal{T}_h^i$ over the side $F \in \mathcal{E}_h^i$ with length h_F, then for $i \in \{1, 2\}$ with $0 < \alpha_i \leq 1$ assume that

$$\frac{h_F}{h_F^\perp} \leq C_3 \qquad \forall\, F \in \mathcal{E}_h^i,$$

is satisfied, where C_3 is independent of $h \in (0, h_0]$ and $\varepsilon \in (0, 1)$. This guarantees that anisotropic triangles $T = T_F \in \mathcal{T}_h^i$ touching Γ along the whole side

F and being "active" (for $i : \alpha_i \neq 0$) in the approximation have their "short side" F on Γ.
For $i = 1, 2$, introduce the finite element space V_h^i of functions on Ω_i by

$$V_h^i := \{v_h^i \in H^1(\Omega_i) : v_h^i\big|_T \in \mathbb{P}_k(T) \ \forall\, T \in \mathcal{T}_h^i,\ v_h^i\big|_{\partial\Omega_i \cap \partial\Omega} = 0\},$$

where $\mathbb{P}_k(T)$ denotes the set of all polynomials on T with degree $\leq k$. The finite element space V_h of functions v_h with components v_h^i on Ω_i is given by $V_h := V_h^1 \times V_h^2$. In general, $v_h \in V_h$ is not continuous across Γ. For the approximation of (5) on V_h let us fix a positive constant γ (to be specified subsequently) and real parameters α_1, α_2 from (6), and introduce the forms $\mathcal{B}_h(.,.)$ on $V_h \times V_h$ and $\mathcal{F}_h(.)$ on V_h as follows:

$$\begin{aligned}
&\mathcal{B}_h(u_h, v_h) := \\
&\sum_{i=1}^{2} \left(\varepsilon^2 (\nabla u_h^i, \nabla v_h^i)_{\Omega_i} + (c u_h^i, v_h^i)_{\Omega_i} \right) - \left\langle \alpha_1 \varepsilon^2 \frac{\partial u_h^1}{\partial n_1} - \alpha_2 \varepsilon^2 \frac{\partial u_h^2}{\partial n_2}, v_h^1 - v_h^2 \right\rangle_\Gamma \\
&- \left\langle \alpha_1 \varepsilon^2 \frac{\partial v_h^1}{\partial n_1} - \alpha_2 \varepsilon^2 \frac{\partial v_h^2}{\partial n_2}, u_h^1 - u_h^2 \right\rangle_\Gamma + \varepsilon^2 \gamma \sum_{E \in \mathcal{E}_h} h_E^{-1} \left(u_h^1 - u_h^2, v_h^1 - v_h^2 \right)_E, \\
&\mathcal{F}_h(v_h) := \sum_{i=1}^{2} (f, v_h^i)_{\Omega_i}. \qquad (7)
\end{aligned}$$

Here, $(.,.)_\Lambda$ denotes the scalar product in $L_2(\Lambda)$ for $\Lambda \in \{\Omega_i, E\}$, and $\langle .,. \rangle_\Gamma$ is taken from (5). The weights in the fourth term of $\mathcal{B}_h$ are introduced in correspondence to $\sigma = \gamma\varepsilon^2 h^{-1}(x)$ at (5) and ensure the stability of the method, if γ is a sufficiently large positive constant (cf. Theorem 2 below).

According to (5), but with the discrete forms $\mathcal{B}_h$ and $\mathcal{F}_h$ from (7), the Nitsche mortar finite element approximation u_h of the solution u of equation (3), with respect to the space V_h, is defined by $u_h = (u_h^1, u_h^2) \in V_h^1 \times V_h^2$ being the solution of

$$\mathcal{B}_h(u_h, v_h) = \mathcal{F}_h(v_h) \quad \forall\, v_h \in V_h. \qquad (8)$$

In the following, we quote some important properties of the discretization (8). First we have the consistency: If u is the weak solution of (1), then $u = (u^1, u^2)$ satisfies $\mathcal{B}_h(u, v_h) = \mathcal{F}_h(v_h) \ \forall\, v_h \in V_h$. Then, owing to the consistency and to (8) we obtain the $\mathcal{B}_h$-orthogonality of the error $u - u_h$ on V_h, i.e. $\mathcal{B}_h(u - u_h, v_h) = 0 \ \forall\, v_h \in V_h$. For getting stability and convergence of the method, we need the following theorem.

Theorem 1. *For $v_h \in V_h$ the inequality*

$$\sum_{E \in \mathcal{E}_h} h_E \left\| \alpha_1 \frac{\partial v_h^1}{\partial n_1} - \alpha_2 \frac{\partial v_h^2}{\partial n_2} \right\|_{0,E}^2 \leq C_I \sum_{i=1}^{2} \left\| \nabla v_h^i \right\|_{0,\Omega_i}^2,$$

holds, where C_I is independent of h ($h \leq h_0$) and of ε ($\varepsilon < 1$).

The proof is given in [14]. For the ellipticity of the discrete form $\mathcal{B}_h(.,.)$ we introduce the following discrete energy-like norm $\|\,.\,\|_{1,h}$,

$$\|v_h\|_{1,h}^2 = \sum_{i=1}^{2}\left(\varepsilon^2\,\|\nabla v_h^i\|_{0,\Omega_i}^2 + \|\sqrt{c}v_h^i\|_{0,\Omega_i}^2\right) + \varepsilon^2 \sum_{E\in\mathcal{E}_h} h_E^{-1}\,\|v_h^1 - v_h^2\|_{0,E}^2\,. \tag{9}$$

For $\varepsilon = 1$ and $c \equiv 0$ this norm is also applied e.g. in [2,6,9,10,12,13]. Using Young's inequality and Theorem 1 we can prove the following theorem.

Theorem 2. *If the constant γ in (7) is chosen (independently of h and ε) such that $\gamma > C_I$ is valid, C_I from Theorem 1, then*

$$\mathcal{B}_h(v_h,v_h) \geq \mu_1\,\|v_h\|_{1,h}^2 \qquad \forall\, v_h \in V_h$$

holds, with a positive constant μ_1 independent of h ($h \leq h_0$) and ε ($\varepsilon < 1$).

3 Numerical treatment of corner singularities

We now study the finite element approximation with non-matching meshes for the case that the domain Ω has re-entrant corners and that the endpoints of the interface Γ are vertices of such corners. It is well-known that such corners generate singularities enlarging the dicretization error and diminishing the rate of convergence of the finite element approximation. If Ω has re-entrant corners with angles $\varphi_{0j} : \pi < \varphi_{0j} < 2\pi$ ($j = 1,\ldots,I$), then the solution u can be represented by

$$u = \sum_{j=1}^{I} \eta_j\, a_j\, r_j^{\lambda_j}\,\sin(\lambda_j\varphi_j) + w,$$

with a regular remainder $w \in H^2(\Omega)$. Here, (r_j,φ_j) denote the local polar coordinates of a point $P \in \Omega$ with respect to the vertex $P_j \in \partial\Omega$, where $0 < r_j \leq r_{0j}$ and $0 < \varphi_j < \varphi_{0j}$ hold. Moreover, we have $\lambda_j = \frac{\pi}{\varphi_{0j}}$ ($\frac{1}{2} < \lambda_j < 1$), a_j is some constant, and η_j is a local cut-off function at P_j. Obviously, $u \in H^s(\Omega)$ for some $s > 3/2$ is satisfied. For approaches to improve the approximation properties and to treat corner singularities, see e.g. [15,18–21]. Since the influence region of corner singularities is a local one, it suffices to consider one corner. Here we employ piecewise linear elements ($k = 1$ in V_h^i) and triangular meshes with appropriate local refinement at one corner.

Let (x_0,y_0) be the coordinates of the vertex P_0 of the corner, (r,φ) the local polar coordinates with center at P_0, i.e. $x - x_0 = r\cos(\varphi + \varphi_r)$, $y - y_0 = r\sin(\varphi + \varphi_r)$, cf. Fig. 3. Define some circular sector $\overline{G}$ around P_0, with the radius $r_0 > 0$ and the angle φ_0 (here: $\pi < \varphi_0 < 2\pi$): $\overline{G} := \{(x,y) \in \overline{\Omega} : 0 \leq r \leq r_0,\, 0 \leq \varphi \leq \varphi_0\}$, $G := \overline{G} \setminus \partial G$, ∂G boundary

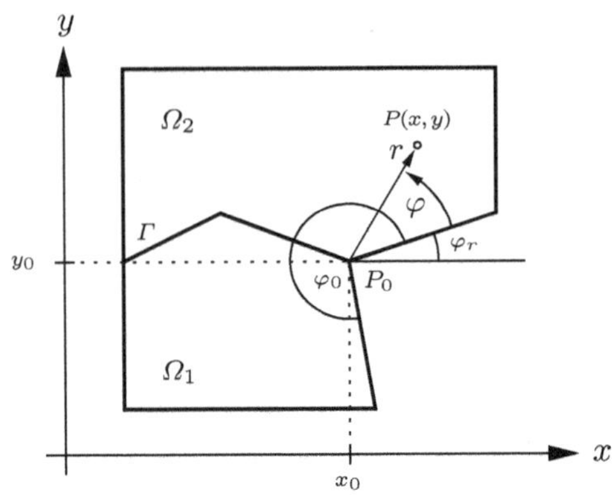

Fig. 3.

of G. For defining a mesh with grading, we employ the real grading parameter μ, $0 < \mu \le 1$, the grading function R_i $(i = 0, 1, \ldots, n)$ with some real constant $b > 0$, and the step size h_i for the mesh associated with layers $[R_{i-1}, R_i] \times [0, \varphi_0]$ around P_0:

$$R_i := b\,(ih)^{\frac{1}{\mu}} \quad (i = 0, 1, \ldots, n), \qquad h_i := R_i - R_{i-1} \quad (i = 1, 2, \ldots, n).$$

Here $n := n(h)$ denotes an integer of the order h^{-1}, $n := \left[\beta h^{-1}\right]$ for some real $\beta > 0$ ($[\,.\,]$: integer part). We shall choose the numbers β, $b > 0$ such that $\frac{2}{3} r_0 < R_n < r_0$ holds, i.e., the mesh grading is located within $\overline{G}$. For h, h_i, R_i, and μ $(0 < h \le h_0,\ 0 < \mu < 1)$ the relation $b^\mu h R_i^{1-\mu} \le h_i \le \frac{b^\mu}{\mu} h R_i^{1-\mu}$ $(i = 1, 2, \ldots, n)$ holds.

Using the step size h_i $(i = 1, 2, \ldots, n)$, define in the neighbourhood of the vertex P_0 of the corner a mesh with grading such that h_T depends on the distance R_T of T from P_0 ($R_T := \operatorname{dist}(T, P_0) := \inf_{P \in T} |P_0 - P|$) in the same way like h_i on R_i. Outside of the corner neighbourhood a quasi-uniform mesh is employed. The triangulation is now characterized by the mesh size h and the grading parameter μ, denoted by $\mathcal{T}_{h\mu}$, with $0 < h \le h_0$ and fixed μ: $0 < \mu \le 1$. We summarize the properties of $\mathcal{T}_{h\mu}$ and assume the following one: The triangulation $\mathcal{T}_{h\mu}$ is shape regular and provided with a grading around the vertex P_0 of the corner such that $h_T := \operatorname{diam} T$ depends on the distance R_T and on μ in the following way:

$$\begin{aligned} \rho_1 h^{\frac{1}{\mu}} &\le h_T \le \rho_1^{-1} h^{\frac{1}{\mu}} && \text{for } T \in \mathcal{T}_{h\mu} : R_T = 0, \\ \rho_2 h R_T^{1-\mu} &\le h_T \le \rho_2^{-1} h R_T^{1-\mu} && \text{for } T \in \mathcal{T}_{h\mu} : 0 < R_T < R_g, \\ \rho_3 h &\le h_T \le \rho_3^{-1} h && \text{for } T \in \mathcal{T}_{h\mu} : R_g \le R_T, \end{aligned} \tag{10}$$

with some constants ρ_i, $0 < \rho_i \le 1$ $(i = 1, 2, 3)$ and some real R_g, $0 < \underline{R}_g < R_g < \overline{R}_g$, where $\underline{R}_g, \overline{R}_g$ are fixed and independent of h. Here, R_g is the radius of the sector with mesh grading, we put e.g. $R_g := R_n$. Outside of this sector the mesh is quasi-uniform. The value $\mu = 1$ yields a quasi-uniform mesh in the whole region Ω, i.e., $\frac{\max_{T\in\mathcal{T}_{h\mu}} h_T}{\min_{T\in\mathcal{T}_{h\mu}} \rho_T} \leq C$ holds. In [18,20,21] related types of mesh grading are described. In [22] a mesh generator is given which automatically generates a mesh of type (10).

A final error estimate in the norm $\|\cdot\|_{1,h}$ for $\varepsilon = 1$ and $c = 0$ is given in the next theorem, where a proof can be found in [13].

Theorem 3. *Let u and u_h be the solutions of the BVP (1) ($\varepsilon = 1$, $c = 0$) with one re-entrant corner (λ: singularity exponent) and of the finite element equation (8), respectively. For $\mathcal{T}_{h\mu}$ let the assumptions on the mesh be satisfied. Then the error $u - u_h$ in the norm $\|\,.\,\|_{1,h}$ (9) is bounded by*

$$\|u - u_h\|_{1,h} \leq c\kappa(h,\mu)\,\|f\|_{0,\Omega}\,, \tag{11}$$

$$\text{with}\;\; \kappa(h,\mu) = \begin{cases} h^{\frac{\lambda}{\mu}} & \text{for } \lambda < \mu \leq 1 \\ h\,|\ln h|^{\frac{1}{2}} & \text{for } \mu = \lambda \\ h & \text{for } 0 < \mu < \lambda < 1. \end{cases}$$

Remark: Under the assumption of Theorem 3 and for the error in the L_2-norm, the estimate

$$\|u - u_h\|_{0,\Omega} \leq C\kappa^2(h,\mu)\,\|f\|_{0,\Omega} \tag{12}$$

is satisfied, with $\kappa(h,\mu)$ from (11). In particular, $\|u - u_h\|_{0,\Omega} = \mathcal{O}(h^2)$ holds for meshes with appropriate grading.

4 Numerical treatment of boundary layers

In correspondence to the anisotropic behaviour of the solution u in the boundary layers, we shall apply anisotropic triangular meshes for improving accuracy and rate of convergence of the finite element method like treated e.g. in [17]. Introduce vectors $\underline{h}_{T,i}$ with length $h_{T,i} := |\underline{h}_{T,i}|$ $(i = 1,2)$ as follows:

$\underline{h}_{T,1}$: vector of the longest side of T,

$\underline{h}_{T,2}$: vector of the height of T over $\underline{h}_{T,1}$.

Apply the multiindex $\beta = (\beta_1, \beta_2) \in \mathbb{R}^2$, with $|\beta| = \beta_1 + \beta_2$, $\beta_i \geq 0$ $(i = 1,2)$, and write shortly $D^\beta := \frac{\partial^{\beta_1}}{\partial x_1^{\beta_1}} \frac{\partial^{\beta_2}}{\partial x_2^{\beta_2}}$.

For the estimation of the interpolation error on anisotropic triangles we need the so-called 'maximal angle condition' and the 'coordinate system condition' (according to [17]) given subsequently, cf. also Fig. 4.

'Maximal angle condition': The interior angles θ of any triangle $T \in \mathcal{T}_h$ satisfy $0 < \theta \leq \pi - \theta_0$ where the constant $\theta_0 > 0$ is independent of T, $h \in (0, h_0]$ and $\varepsilon \in (0,1)$.

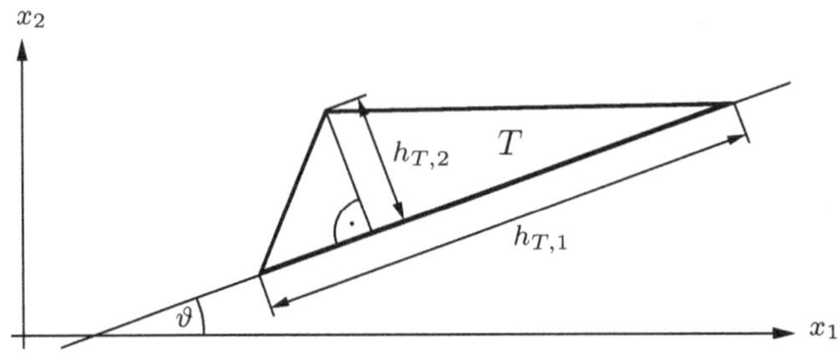

Fig. 4. Anisotropic triangle T

'Coordinate system condition': The position of the triangle T in the x_1-x_2-coordinate system is such that the angle ϑ between $\underline{h}_{T,1}$ and the x_1-axis is bounded by

$$|\sin \vartheta| \leq C_4 \frac{h_{T,2}}{h_{T,1}},$$

where C_4 is independent of T, $h \in (0, h_0]$ and $\varepsilon \in (0, 1)$.

In order to present the treatment of boundary layers and the application of anisotropic meshes, we consider for simplicity the BVP (1) on a rectangle, w.l.o.g. for $\Omega = (0,1)^2$. For describing the behaviour of the solution u and the mesh adapted to this solution, we split the domain as given by Fig. 5 into an interior part Ω^{i}, boundary layers $\Omega^{\mathrm{b},j}$ ($j = 1, 2, 3, 4$) of width a and corner neighbourhoods $\Omega^{\mathrm{c},j}$ ($j = 1, 2, 3, 4$). Employ also $\Omega^{\mathrm{b}} = \bigcup_{j=1}^{4} \Omega^{\mathrm{b},j}$ and $\Omega^{\mathrm{c}} = \bigcup_{j=1}^{4} \Omega^{\mathrm{c},j}$. It should be noted that depending on the data and according to the solution properties some of the boundary layer parts $\Omega^{\mathrm{b},j}$ may be empty (cf. numerical example); then Ω^{i} is extended to the corresponding part of the boundary. In each $\Omega^{\mathrm{b},j}$ choose a local coordinate system (x_1, x_2), where x_1 goes with the tangent of the boundary $\partial\Omega$ and x_2 with its interior normal such that $x_2 = \mathrm{dist}(x, \partial\Omega)$ (distance of $x \in \Omega$ to $\partial\Omega$) holds. The derivatives $D^\beta u$, $\beta = (\beta_1, \beta_2)$, in the boundary layers are taken with respect to these coordinates. According to [14], cf. also [23, Lemma 2], the L_2-norms of the second order derivatives $D^\beta u$ ($|\beta| = 2$) considered in the subdomains Ω^{i}, Ω^{b} and Ω^{c} satisfy at least the estimates

$$\left\|D^\beta u\right\|_{0,\Omega^{\mathrm{i}}}^2 \leq C, \; \left\|D^\beta u\right\|_{0,\Omega^{\mathrm{b}}}^2 \leq C\, a\, \varepsilon^{-2\beta_2}, \; \left\|D^\beta u\right\|_{0,\Omega^{\mathrm{c}}}^2 \leq C\, a^2 \varepsilon^{-2|\beta|}, \tag{13}$$

with $a := \frac{a_0}{\underline{c}_0} \varepsilon \, |\ln \varepsilon|$ and $a_0 \geq 2$, for $c(x) = c_0 = const > \underline{c}_0 > 0$.

Introduce triangulations $\mathcal{T}_h(\Omega^{\mathrm{i}})$, $\mathcal{T}_h(\Omega^{\mathrm{c}})$ and $\mathcal{T}_h(\Omega^{\mathrm{b}})$ of the subdomains Ω^{i}, Ω^{b} and Ω^{c}, respectively. For each of these subdomains employ $\mathcal{O}(h^{-1}) \times \mathcal{O}(h^{-1})$ triangles T with mesh sizes $h_{T,1}$ and $h_{T,2}$ according to the asymptotics (14) of the subtriangulations given by

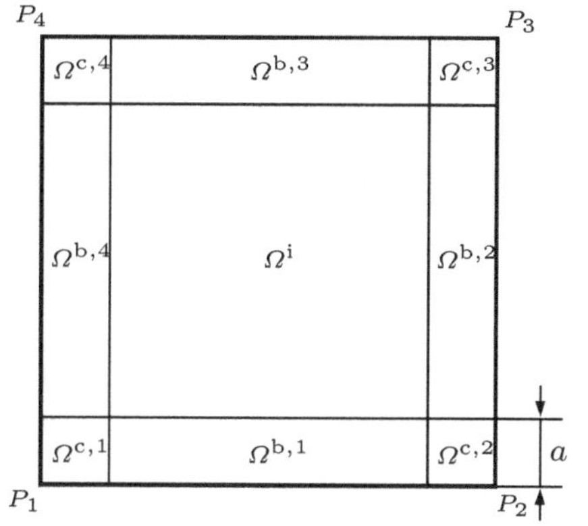

Fig. 5. Subdomains of Ω: Ω^{i} and the parts $\Omega^{\mathrm{b},j}$ and $\Omega^{\mathrm{c},j}$ $(j = 1, 2, 3, 4)$

$$\begin{array}{ll} h_{T,1} \sim h_{T,2} \sim h & \text{for } T \in \mathcal{T}_h(\Omega^{\mathrm{i}}), \\ h_{T,1} \sim h, \quad h_{T,2} \sim ah & \text{for } T \in \mathcal{T}_h(\Omega^{\mathrm{b}}), \\ h_{T,1} \sim h_{T,2} \sim ah & \text{for } T \in \mathcal{T}_h(\Omega^{\mathrm{c}}). \end{array} \tag{14}$$

Here, for brevity the symbol $\sim$ is used for equivalent mesh asymptotics (see e.g. [17]). In particular, assumption (14) means that we apply isotropic triangles in Ω^{i} and Ω^{c}, but anisotropic triangles in Ω^{b}.

An important application is the following one. We decompose Ω into Ω_1, Ω_2 such that the interface Γ is formed by the interior boundary part of the boundary layer. Then, cover Ω_1 and Ω_2 by axiparallel rectangles. They are formed by putting $\mathcal{O}(h^{-1})$ points on the axiparallel edges of the subdomains defining axiparallel mesh lines. Finally, we obtain rectangular triangles by dividing the rectangles in the usual way, see e.g. Fig. 12. The triangles in Ω^{b} have a 'long' side $\underline{h}_{T,1}$ being parallel to $\partial\Omega$, a 'short' side $\underline{h}_{T,2}$ perpendicular to $\partial\Omega$, with $h_{T,1} \sim h$ and $h_{T,2} \sim \varepsilon|\ln\varepsilon|h$. Then we are able to state the following theorem, where a proof is given in [14].

Theorem 4. *Assume that u is the solution of BVP (1) over the domain $\Omega = (0,1)^2$, with $0 < \varepsilon < 1$,$0 < c_0 \le c(x) = const$ and the smoothness assumptions (13). Furthermore, suppose that all mesh assumptions quoted previously are satisfied and that u_h denotes the Nitsche mortar finite element approximation according to (8), with $\gamma > C_I$. Then, the error $u - u_h$ can be bounded in the norm $\|\,.\,\|_{1,h}$ (9) as follows:*

$$\|u - u_h\|_{1,h}^2 \le C\left(\varepsilon\,|\ln\varepsilon|^3\,h^2 + h^4\right),$$

where C is independent of $h \in (0, h_0]$ and $\varepsilon \in (0,1)$.

5 Computational experiments with corner singularities

We shall give some computations using the Nitsche type mortaring in presence of some corner singularity, with application of local mesh refinement near the corner. Consider the BVP $-\Delta u = f$ in Ω, $u = 0$ on $\partial\Omega$, with Ω is the L-shaped domain of Fig. 6. The right-hand side f is chosen such that the exact solution u is of the form

$$u(x,y) = (a^2 - x^2)(b^2 - y^2)r^{\frac{2}{3}} \sin(\frac{2}{3}\varphi), \tag{15}$$

where $r^2 = x^2 + y^2$, $0 \le \varphi \le \varphi_0$, $\varphi_0 = \frac{3}{2}\pi$. Clearly, $u|_{\partial\Omega} = 0$, $\lambda = \frac{\pi}{\varphi_0} = \frac{2}{3}$ and, therefore, $u \in H^{\frac{5}{3}-\delta}(\Omega)$ $(\delta > 0)$ is satisfied. We apply the Nitsche finite element method to this BVP and use two subdomains Ω_i $(i = 1, 2)$ as well as initial meshes shown as in Fig. 7 and 8. The approximate solution u_h is visualized in Fig. 10.

The initial meshes covering Ω_i $(i = 1, 2)$ are refined globally by dividing each triangle into four equal triangles such that the mesh parameters form a sequence $\{h_1, h_2, \ldots\}$ given by $\{h_1, \frac{h_1}{2}, \ldots\}$. The ratio of the number of mesh segments (without grading) on the mortar interface Γ is given by $2:3$ (see Fig. 7) and $2:5$ (see Fig. 8). In the computational experiments, different values of α_1 $(\alpha_2 := 1 - \alpha_1)$ are chosen, e.g. $\alpha_1 = 0, 0.5, 1$. For $\alpha_i = 1$ $(i = 1$ or $i = 2)$, the trace $\mathcal{E}_h^i$ of the triangulation $\mathcal{T}_h^i$ of Ω_i on the interface Γ was chosen to form the partition $\mathcal{E}_h$ (for Ω_1, Ω_2, cf. Fig. 6), and for $\alpha_i \neq 0$ $(i = 1$ and $i = 2)$, the partition $\mathcal{E}_h$ was defined by the intersection of $\mathcal{E}_h^1$ and $\mathcal{E}_h^2$. For the examples the choice $\gamma = 3$ was sufficient to ensure stability. Moreover, we also applied local refinement by grading the mesh around the vertex P_0 of the corner, according to Sect. 3. The parameter μ was chosen by $\mu := 0.7\lambda$.

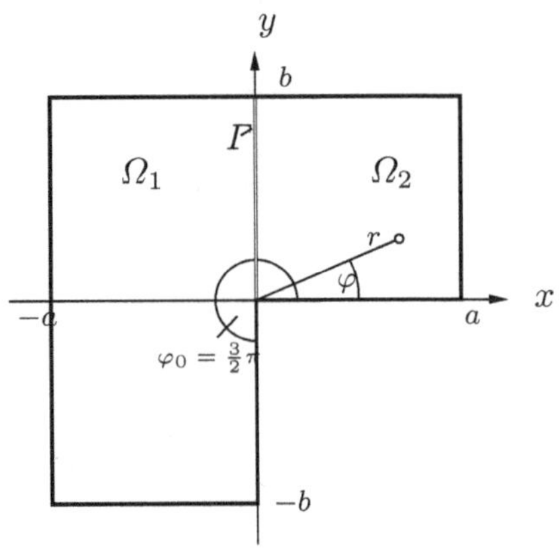

Fig. 6. The L-shaped domain Ω

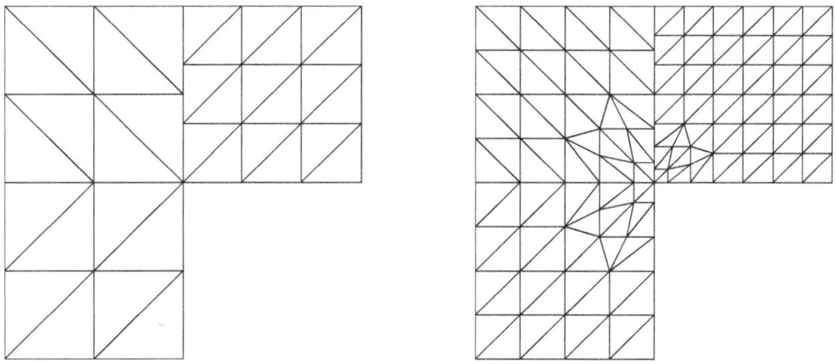

Fig. 7. Triangulations with mesh ratio 2 : 3, h_1–mesh (left) and h_2–mesh with refinement (right)

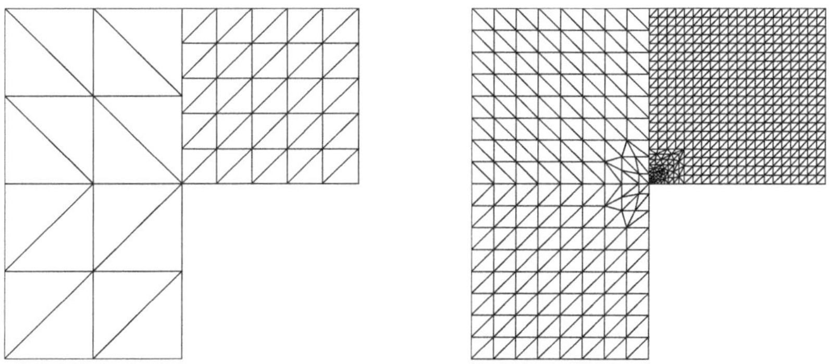

Fig. 8. Triangulations with mesh ratio 2 : 5, h_1–mesh (left) and h_3–mesh with refinement (right)

Let u_h denote the finite element approximation according to (8) of the exact solution u from (15). Then the error estimate in the discrete norm $||\,.\,||_{1,h}$ is given by (11). We assume that h is sufficiently small such that

$$\|u - u_h\|_{1,h} \approx Ch^{\alpha} \tag{16}$$

holds with some constant C which is approximately the same for two consecutive levels h_i, h_{i+1} of the mesh parameter h, like h, $h/2$. Then $\alpha = \alpha_{obs}$ (observed value) is derived from (16) by $\alpha_{obs} := \log_2 q_h$, where $q_h := \|u - u_h\|(\|u - u_{h/2}\|)^{-1}$. The same is carried out for the L_2–norm, where $\|u - u_h\|_{0,\Omega} \approx Ch^{\beta}$ is supposed, cf. (12). The observed and expected values of α and β are given in Table 1. The computational experiments show that the observed rates of convergence are approximately equal to the values expected by the theory: $\alpha = \frac{2}{3}$, $\beta = 2\alpha$ for quasi-uniform meshes, and $\alpha = 1$, $\beta = 2$ for meshes with appropriate mesh grading. Furthermore, it can be seen that local mesh grading is suited to overcome the diminishing of the rate of convergence

Table 1. Observed convergence rates α_{obs} and β_{obs} for the level pair (h_5, h_6), for $\mu = 1$ and for $\mu = 0.7\lambda$ ($\lambda = \frac{3}{2}$) in the norms $||\,.\,||_{1,h}$ and $||\,.\,||_{0,\Omega}$, resp.

	mesh ratio 2 : 3	mesh ratio 2 : 5	
norm $\|\|\,.\,\|\|_{1,h}$	α (observed)		α (expected)
$\alpha_{obs} : \mu = 1$	0.73	0.73	0.67
$\alpha_{obs} : \mu = 0.7\lambda$	1.00	0.99	1
norm $\|\|\,.\,\|\|_{0,\Omega}$	β (observed)		β (expected)
$\beta_{obs} : \mu = 1$	1.39	1.38	1.33
$\beta_{obs} : \mu = 0.7\lambda$	1.97	1.87	2

on non-matching meshes and the loss of accuracy (cf. error representation in Fig. 10) caused by corner singularities.

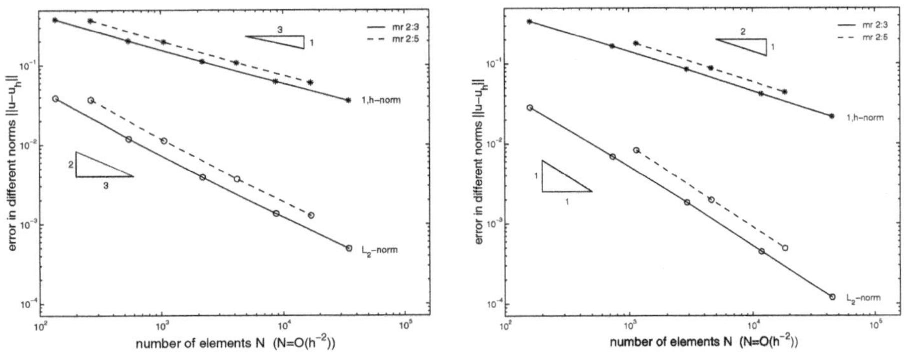

Fig. 9. The error in the norms $\|\cdot\|_{1,h}$ and $\|\cdot\|_{L_2}$ on quasi-uniform meshes (left) and on meshes with grading (right), for mesh ratios 2:3 and 2:5

6 Computational experiments with boundary layers

In the following, we consider the BVP $-\varepsilon^2 \Delta u + u = 0$ in Ω, $u = -e^{-\frac{x}{\varepsilon}} - e^{-\frac{y}{\varepsilon}}$ on $\partial\Omega$, where Ω is defined by $\Omega = (0,1)^2 \subset \mathbb{R}^2$, and the solution u is given by $u = -e^{-\frac{x}{\varepsilon}} - e^{-\frac{y}{\varepsilon}}$. For small values of $\varepsilon \in (0,1)$, boundary layers near $x = 0$ and $y = 0$ occur. The L_2-norms of the second order derivatives of u satisfy the inequalities at (13). Here, according to the position of the boundary layers the definition of Ω^{i}, Ω^{b} and Ω^{c} is to be modified, cf. Figs. 5 and 12. In correspondence to the boundary layers, we subdivide Ω into subdomains $\Omega_1 = (a,1)\times(a,1)$ and $\Omega_2 = \Omega \backslash \overline{\Omega}_1$, define Γ by $\Gamma = \overline{\Omega}_1 \cap \overline{\Omega}_2$. The triangulations of the subdomains $\overline{\Omega}_1$ and $\overline{\Omega}_2$ are partially independent from each other. Choose

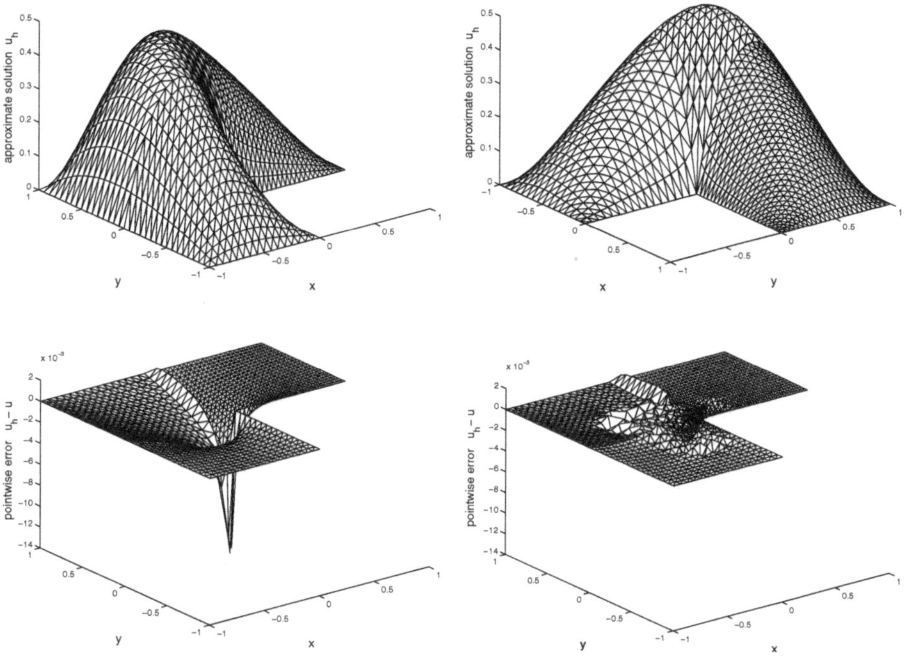

Fig. 10. The approximate solution u_h in two different perspectives (top), the local pointwise error on the quasi-uniform mesh for $\mu = 1$ (bottom left) and the local pointwise error on the mesh with grading for $\mu = 0.7\lambda$ (bottom right)

an initial mesh like in Fig. 12, where the nodes of triangles $T \in \mathcal{T}_h(\Omega_1)$ and $T \in \mathcal{T}_h(\Omega_2)$ do not coincide on Γ. The initial mesh is refined by subdividing a triangle T into four equal triangles such that new vertices coincide with the old ones or with the midpoints of the old triangle sides. Therefore, the mesh sequence parameters $\{h_1, h_2, h_3, \ldots\}$ are given by $\{h_1, \frac{h_1}{2}, \ldots\}$. Let u_h be the Nitsche mortar finite element approximation of u defined by (8). Since u is known, the error $u - u_h$ in the $\|\,.\,\|_{1,h}$-norm can be calculated. Then, the convergence rates with respect to h will be estimated as follows. We fix ε and assume that the constant C in the relation $\|u - u_h\|_{1,h} \approx Ch^\alpha$ is nearly the same for a pair of mesh sizes $h_i = h$ and $h_{i+1} = \frac{h}{2}$. Under this assumption we derive observed values α_{obs} of α like in Sect. 5. In Table 2 the error norms $\|u - u_h\|_{1,h}$ and the convergence rates α_{obs} are given, with the settings $\mathcal{E}_h = \mathcal{E}_h^1$, $\alpha_1 = 1$ and $\gamma = 2.5$. The results of Table 2 show that for appropriate choice of the mesh layer parameter a, here e.g. for $a = \varepsilon|\ln \varepsilon|$ and $a = 2\varepsilon\,|\ln \varepsilon|$, optimal convergence rates $\mathcal{O}(h)$ like in Theorem 4 stated can be observed for a wide range of mesh parameters h. In Fig. 13, for $\varepsilon = 10^{-2}$ and on a mesh of level h_3, the local error $u_h - u$ for two different values of

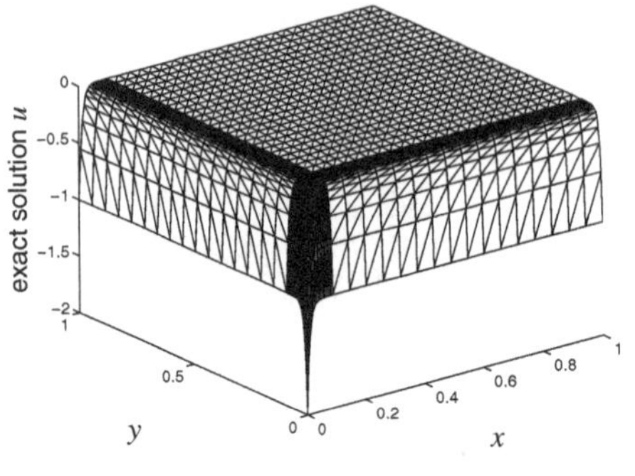

Fig. 11. Solution u on the h_3-mesh for $\varepsilon = 10^{-2}$ and $a = 2\varepsilon\,|\ln \varepsilon|$

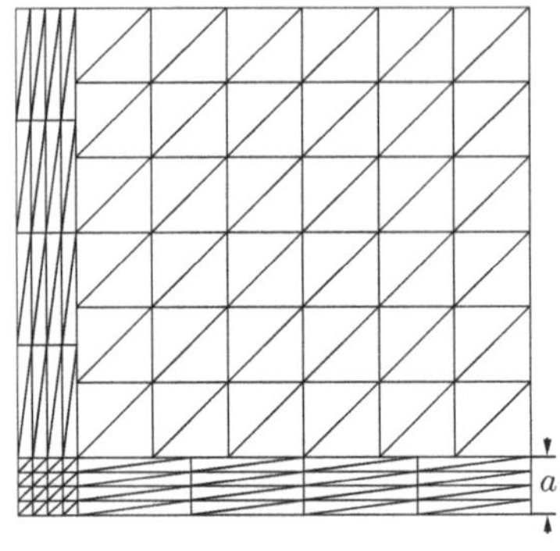

Fig. 12. h_1-mesh with layer thickness a, for $a = \frac{1}{2}\varepsilon\,|\ln \varepsilon|$ and $\varepsilon = 10^{-1}$

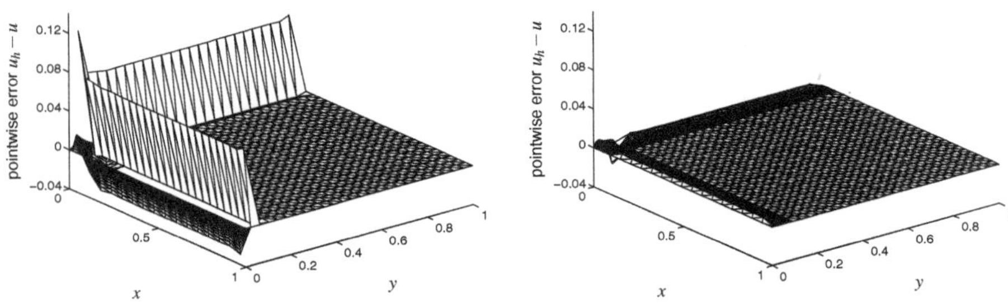

Fig. 13. Pointwise error $u_h - u$ for $\varepsilon = 10^{-2}$ on meshes with $a = \frac{1}{2}\varepsilon\,|\ln \varepsilon|$ (left) and $a = 2\varepsilon\,|\ln \varepsilon|$ (right)

the parameter a is represented. The influence of the parameter a is visible, in particular, the local error is significantly smaller for the value $a = 2\varepsilon\,|\ln \varepsilon|$ compared with that of $a = \frac{1}{2}\varepsilon\,|\ln \varepsilon|$. For constant $a = 0.5$ the observed rate is far from $\mathcal{O}(h)$ (cf. also Fig. 14, left-hand side) and, for $a = \frac{1}{2}\varepsilon\,|\ln \varepsilon|$, the rates are not appropriate for small h.

In particular, the computational experiments show that non-matching isotropic and anisotropic meshes can be applied to singularly perturbed problems without loss of the optimal convergence rate. The appropriate choice of the width a of the strip with anisotropic triangles is important for diminishing the error and getting optimal convergence rates.

Summary. The paper is concerned with the Nitsche finite element method as a mortar method in the framework of domain decomposition. The approach is applied to elliptic problems in 2D with corner singularities and boundary layers. Non-matching meshes of triangles with grading near corners and being anisotropic in the boundary layers are employed. The numerical treatment of such problems and computational experiments are described.

Table 2. Observed errors in the $\|\,.\,\|_{1,h}$-norm on the level h_i ($i = 5,6,7$) and the convergence rate α_{obs} assigned to the levelpair (h_i, h_{i+1}) ($i = 5,6$); case $\mathcal{E}_h = \mathcal{E}_h^1$, $\alpha_1 = 1$ and $\gamma = 2.5$

Parameter a	$\varepsilon = 10^{-1}$		$\varepsilon = 10^{-3}$		$\varepsilon = 10^{-5}$	
	$\|u - u_h\|_{1,h}$	α_{obs}	$\|u - u_h\|_{1,h}$	α_{obs}	$\|u - u_h\|_{1,h}$	α_{obs}
0.5	7.132e-03	1.00	5.260e-02	0.74	6.738e-02	0.50
	3.566e-03	1.00	3.154e-02	0.88	4.757e-02	0.50
	1.783e-03		1.718e-02		3.361e-02	
$\frac{1}{2}\varepsilon\,\|\ln\varepsilon\|$	3.656e-03	1.08	1.660e-03	0.15	8.472e-05	0.95
	1.733e-03	1.04	1.500e-03	0.41	4.394e-05	0.82
	8.397e-04		1.127e-03		2.483e-05	
$\varepsilon\,\|\ln\varepsilon\|$	3.396e-03	1.01	9.863e-04	1.00	1.653e-04	1.00
	1.692e-03	1.00	4.948e-04	0.99	8.271e-05	1.00
	8.446e-04		2.488e-04		4.136e-05	
$2\varepsilon\,\|\ln\varepsilon\|$	6.569e-03	1.00	1.969e-03	1.00	3.275e-04	1.00
	3.284e-03	1.00	9.851e-04	1.00	1.639e-04	1.00
	1.642e-03		4.926e-04		8.200e-05	

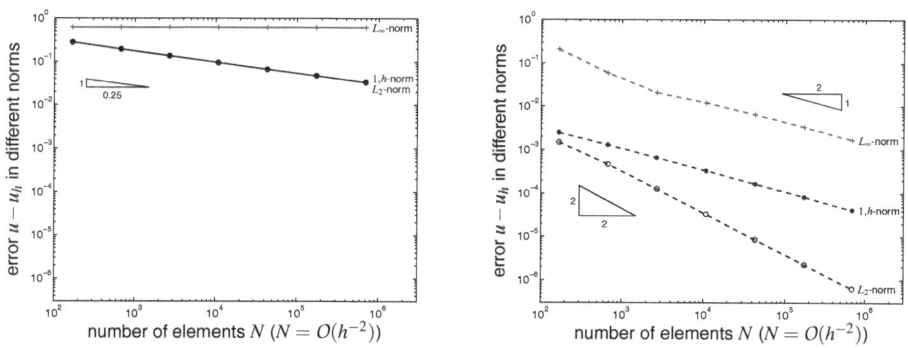

Fig. 14. Observed error $u - u_h$ in the L_2-, L_∞- and $\|\,.\,\|_{1,h}$-norm for $\varepsilon = 10^{-5}$, $a = 0.5$ (left) and $a = \varepsilon\,|\ln\varepsilon|$ (right)

References

1. F. Ben Belgacem. The mortar finite element method with Lagrange multipliers. *Numer. Math.*, 84:173–197, 1999.
2. D. Braess, W. Dahmen, and Ch. Wieners. A Multigrid Algorithm for the Mortar Finite Element Method. *SIAM J. Numer. Anal.*, 37(1):48–69, 1999.
3. B. I. Wohlmuth. *Discretization methods and iterative solvers based on domain decomposition*, volume 17 of *Lecture Notes in Computational Science and Engineering*. Springer, Berlin, Heidelberg, 2001.
4. J. Nitsche. Über ein Variationsprinzip zur Lösung von Dirichlet-Problemen bei Verwendung von Teilräumen, die keinen Randbedingungen unterworfen sind.

Abhandlung aus dem Mathematischen Seminar der Universität Hamburg, 36:9–15, 1970/1971.
5. V. Thomeé. *Galerkin Finite Element Methods for Parabolic Problems*, volume 25 of *Springer Series in Computational Mathematics*. Springer Verlag, Berlin, 1997.
6. R. Stenberg. Mortaring by a method of J. A. Nitsche. In S. Idelsohn, E. Onate, and E. Dvorkin, editors, *Computational Mechanics, New Trends and Applications*. ©CIMNE, Barcelona, 1998.
7. D. N. Arnold. An interior penalty finite element method with discontinuous elements. *SIAM J. Numer. Anal.*, 19(4):742–760, August 1982.
8. D. N. Arnold, F. Brezzi, B. Cockburn, and D. Marini. Discontinuous galerkin methods for elliptic problems. In B. Cockburn, G. E. Karniadakis, and C.-W. Shu, editors, *Discontinuous Galerkin Methods*, volume 11 of *Lecture Notes in Computational Science and Engineering*, pages 89–101. Springer, 2000.
9. R. Becker and P. Hansbo. A Finite Element Method for Domain Decomposition with Non-matching Grids. Technical Report INRIA 3613, 1999.
10. R. Becker, P. Hansbo, and R. Stenberg. A finite element method for domain decomposition with non-matching grids. *M²AN Math. Model. Numer. Anal.*, 37(2):209–225, 2003.
11. A. Fritz, S. Hüeber, and B. I. Wohlmuth. A comparison of mortar and Nitsche techniques for linear elasticity. IANS Preprint 2003/008, Universität Stuttgart, 2003.
12. B. Heinrich and S. Nicaise. Nitsche mortar finite element method for transmission problems with singularities. *IMA J. Numer. Anal.*, 23(2):331–358, 2003.
13. B. Heinrich and K. Pietsch. Nitsche type mortaring for some elliptic problem with corner singularities. *Computing*, 68:217–238, 2002.
14. B. Heinrich and K. Pönitz. Nitsche type mortaring for singularly perturbed reaction-diffusion problems. *Computing*, 75(4):257–279, 2005.
15. P. Grisvard. *Elliptic Problems in Nonsmooth Domains*. Pitman, Boston, 1985.
16. Franco Brezzi and Michel Fortin. *Mixed and hybrid finite element methods*, volume 15 of *Springer series in computational mathematics*. Springer, New York, Berlin, Heidelberg, 1991.
17. Th. Apel. *Anisotropic Finite Elements: Local Estimates and Applications*. Advances in Numerical Mathematics. Teubner, Stuttgart, 1999.
18. I. Babuška, R. B. Kellogg, and J. Pitkäranta. Direct and inverse error estimates for finite elements with mesh refinement. *Numer. Math.*, 33:447–471, 1979.
19. H. Blum and M. Dobrowolski. On finite element methods for elliptic equations on domains with corners. *Computing*, 28:53–63, 1982.
20. L. A. Oganesyan and L. A. Rukhovets. *Variational-Difference Methods for Solving Elliptic Equations*. Izdatel'stvo Akad. Nauk Arm. SSR, Jerevan, 1979. (In Russian).
21. G. Raugel. Résolution numérique par une méthode d'éléments finis du problème Dirichlet pour le Laplacian dans un polygone. *C. R. Acad. Sci. Paris Sér. A*, 286:A791–A794, 1978.
22. M. Jung. On adaptive grids in multilevel methods. In S. Hengst, editor, *GAMM-Seminar on Multigrid-Methods*, Rep. No. 5, pages 67–80. Inst. Angew. Anal. Stoch., Berlin, 1993.
23. Th. Apel and G. Lube. Anisotropic mesh refinement for a singularly perturbed reaction diffusion model problem. *Appl. Numer. Math.*, 26:415–433, 1998.

Hierarchical Adaptive FEM at Finite Elastoplastic Deformations

Reiner Kreißig, Anke Bucher, and Uwe-Jens Görke

Technische Universität Chemnitz, Institut für Mechanik
09107 Chemnitz, Germany
reiner.kreissig@mb.tu-chemnitz.de

1 Introduction

The simulation of non-linear problems of continuum mechanics was a crucial point within the framework of the subproject "Efficient parallel algorithms for the simulation of the deformation behaviour of components of inelastic materials". Nonlinearity appears with the occurence of finite deformations as well as with special material behaviour as e.g. elastoplasticity.

To solve such kind of problems by means of a Finite Element simulation for reliable predictions of the response of components and structures to mechanical loading, the insertion of suitable material models into FE codes is a basic requirement. Additionally, their efficient numerical implementation plays also an important role.

Likewise, a realistic numerical simulation essentially depends on appropriate FE meshes. According to the motto "as fine as necessary, as coarse as possible", several techniques of mesh adaptation have been developed in the last decades. These days the feature of adaptivity is a high-performance attribute for FE codes.

We want to present the results of our research within the framework of the subproject D1 concerning the efficient treatment of non-linear problems of continuum mechanics:

- The theoretical development and numerical realization of appropriate material laws for elastoplasticity and,
- the implementation of hierarchical mesh adaptation in case of non-linear material behaviour.

All contrived theoretical descriptions and numerical algorithms had been implemented and tested with the in-house code SPC-PM2AdNl which was developed at the Chemnitz University of Technology within the context of our collaborative research centre 393.

2 Material model for finite elastoplastic deformations

This section deals with the formulation and efficient numerical implementation of a material model decsribing finite elastoplasticity.

2.1 Theoretical foundation and thermodynamical consistency

The presented approach for the establishment of elastoplastic material models considering finite deformations is based on the multiplicative split of the deformation gradient $\boldsymbol{F} = \boldsymbol{F}^e \boldsymbol{F}^p$ into an elastic part $\boldsymbol{F}^e$ and a plastic part $\boldsymbol{F}^p$. The plastic part of the deformation gradient realizes the mapping of the deformed body from the reference to the plastic intermediate configuration, the elastic part the remaining mapping from the plastic intermediate to the current configuration (see Fig. 1). It has to be mentioned that the plastic intermediate configuration is an incompatible one.

In the literature different approaches to model plastic anisotropy are presented:

- Hill developed a quadratic yield condition for the special case of plastic orthotropy [16].
- General plastic anisotropy can be treated in a phenomenological manner introducing special (tensorial) internal variables describing the hardening behaviour [2, 6, 15, 26, 30].
- Another possibility to describe general plastic anisotropy consists in the application of microstructural approaches [3, 13, 25, 27, 32].
- Several authors propose phenomenological models based on so-called substructure concepts [12, 18, 24, 28]. This procedure is based on the consideration of microstructural characteristics of the material (substructure) using macroscopic constitutive approaches (definition of special internal variables). The continuum and the underlying substructure are supposed to have different characteristics of orientation.

The substructure approach has been developed particularely by Mandel [24] and Dafalias [12]. The last mentioned distinguished between the kinematics of the continuum and the kinematics of the substructure in a phenomenological manner introducing a special deformation gradient $\boldsymbol{F}_S = \boldsymbol{F}^e \boldsymbol{\beta}$ related to the substructure, where the tensor $\boldsymbol{\beta}$ is supposed to be an orthogonal one. Just as Mandel he defined special objective time derivatives connected with the substructure using its spin $\boldsymbol{\omega}_D$.

The spin of the continuum is defined as

$$\boldsymbol{w} = \frac{1}{2}\left\{ \overset{\triangle}{\boldsymbol{F}}{}^e \boldsymbol{F}^{e-1} - \boldsymbol{F}^{e-T} \overset{\triangle}{\boldsymbol{F}}{}^{eT} \right\} + \tag{1}$$

$$\frac{1}{2}\left\{ \boldsymbol{F}^e \overset{\triangle}{\boldsymbol{F}}{}^p \boldsymbol{F}^{p-1} \boldsymbol{F}^{e-1} - \boldsymbol{F}^{e-T} \boldsymbol{F}^{p-T} \overset{\triangle}{\boldsymbol{F}}{}^{pT} \boldsymbol{F}^{eT} \right\}$$

$$= \boldsymbol{w}_D^e + \boldsymbol{w}_D^p. \tag{2}$$

$$\text{with}\quad \overset{\triangle}{\boldsymbol{F}}{}^p = \dot{\boldsymbol{F}}^p - \boldsymbol{\omega}_D \boldsymbol{F}^p, \quad \overset{\triangle}{\boldsymbol{F}}{}^e = \dot{\boldsymbol{F}}^e + \boldsymbol{F}^e \boldsymbol{\omega}_D \quad \text{and} \quad \boldsymbol{\omega}_D = \dot{\boldsymbol{\beta}}\boldsymbol{\beta}^{-1}. \tag{3}$$

The crucial point of Dafalias' approach consists in the assumption of an evolutional equation for the plastic spin $\boldsymbol{w}_D^p$, which represents the difference between the spin of the continuum and the substructure spin:

$$\boldsymbol{w}_D^p = \boldsymbol{w} - \bar{\boldsymbol{\omega}}_D \quad \text{with} \quad \bar{\boldsymbol{\omega}}_D = \dot{\bar{\boldsymbol{\beta}}}\bar{\boldsymbol{\beta}}^{-1} \quad \text{and} \quad \bar{\boldsymbol{\beta}} = \boldsymbol{R}^e \boldsymbol{\beta}, \tag{4}$$

where $\boldsymbol{R}^e$ represents the rotation tensor of the elastic part of the deformation gradient. Dafalias did not give a thermodynamical explanation for the suggested evolutional equation for $\boldsymbol{w}_D^p$.

In the following we define a substructure configuration apart from the configurations describing the deformation of a continuum. A mapping tensor $\boldsymbol{H}$ is supposed to map tensorial variables from the reference into this substructure configuration. Accordingly, the tensor combination $\boldsymbol{F}\boldsymbol{H}^{-1}$ serves the mapping from the substructure configuration into the current configuration. The substructure configuration is assumed to be an incompatible intermediate configuration characterizing a special constitutively motivated decomposition of the deformation gradient. It differs from the plastic intermediate configuration only by a rotation characterized by the tensor $\boldsymbol{\beta}$ (see Fig. 1). In the following, variables related to the substructure configuration are denoted by capital letters with a hat.

In the framework of the presented phenomenological material model, the evolution of stresses and strains is considered with respect to the continuum. The internal variables describing the plastic anisotropy are related to the substructure. Therefore, the substructure configuration is connected with a newly established objective time derivative. This Lie-type derivative is defined in case of a contravariant tensor as follows:

$$\begin{aligned} \overset{\triangledown}{\boldsymbol{a}} &= \boldsymbol{F}\boldsymbol{H}^{-1}\Big(\boldsymbol{H}\,\underbrace{\boldsymbol{F}^{-1}\boldsymbol{a}\boldsymbol{F}^{-T}}_{\boldsymbol{A}}\,\boldsymbol{H}^T\Big)^{\boldsymbol{\cdot}}\,\boldsymbol{H}^{-T}\boldsymbol{F}^T \\ &= \boldsymbol{F}\boldsymbol{H}^{-1}\Big(\dot{\boldsymbol{H}}\boldsymbol{A}\boldsymbol{H}^T + \boldsymbol{H}\dot{\boldsymbol{A}}\boldsymbol{H}^T + \boldsymbol{H}\boldsymbol{A}\dot{\boldsymbol{H}}^T\Big)\boldsymbol{H}^{-T}\boldsymbol{F}^T \\ &= \boldsymbol{F}\boldsymbol{H}^{-1}\dot{\hat{\boldsymbol{A}}}\boldsymbol{H}^{-T}\boldsymbol{F}^T. \end{aligned} \tag{5}$$

It should be mentioned that the definition of this objective time derivative is closely connected with the supposition of the existence of a material derivative in the substructure configuration.

If the mapping tensor is supposed to be defined as

$$\boldsymbol{H} = \boldsymbol{\beta}^T \boldsymbol{F}^p, \tag{6}$$

we get

$$\boldsymbol{F}\boldsymbol{H}^{-1} = \boldsymbol{F}^e \boldsymbol{\beta} = \boldsymbol{F}_S \tag{7}$$

and following the plastic spin at the reference configuration can be written

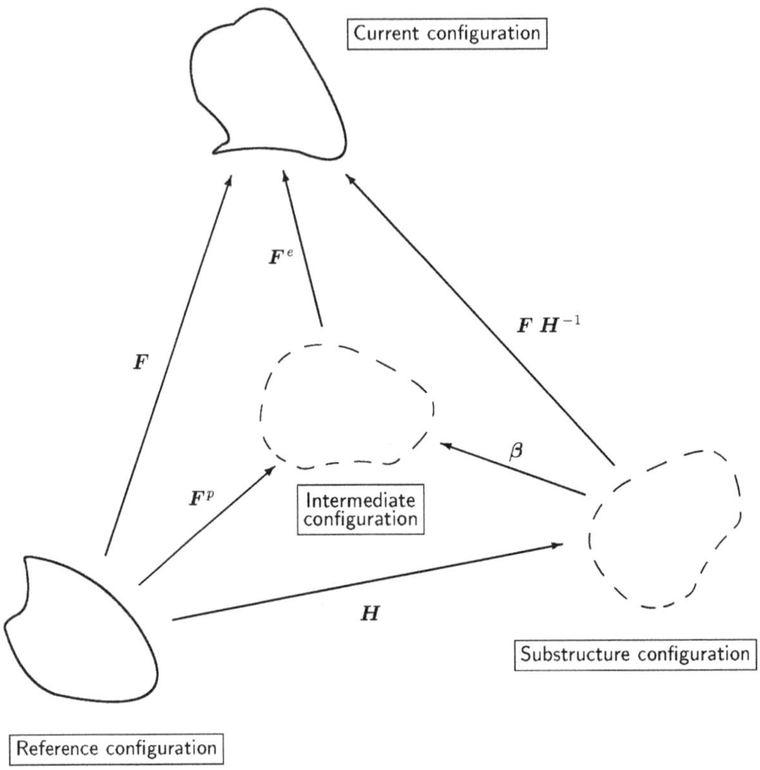

Fig. 1. Mappings between the configurations

$$\boldsymbol{W}^p_D = \frac{1}{2}\left\{\boldsymbol{C}\boldsymbol{H}^{-1}\dot{\boldsymbol{H}} - \dot{\boldsymbol{H}}^T\boldsymbol{H}^{-T}\boldsymbol{C}\right\} \tag{8}$$

using the right Cauchy-Green tensor $\boldsymbol{C} = \boldsymbol{F}^T\boldsymbol{F}$. As the material model has been implemented into the in-house code SPC-PM2AdNl based on a total Lagrangean description, the equations of the material law are defined in the reference configuration. Considering the concept of conjugate variables, firstly established by Ziegler and Mac Vean [35], the Clausius-Duhem inequality for isothermal processes in the reference configuration can be written as

$$-\varrho_0\dot{\psi} + \frac{1}{2}\boldsymbol{T}\cdot\cdot\dot{\boldsymbol{C}} \geq 0 \tag{9}$$

with the second Piola-Kirchhoff stress tensor $\boldsymbol{T}$ and the material mass density ϱ_0. The free energy density ψ is chosen to be additively splitted into an elastic part and a plastic part. We propose its following special representation [7]:

$$\psi = \tilde{\psi}_e\left(\tilde{\boldsymbol{E}}^e\right) + \psi_p\left(\hat{\boldsymbol{A}}_1, \hat{\boldsymbol{A}}_2\right) = \bar{\psi}_e\left(\boldsymbol{C}\boldsymbol{B}^p\right) + \psi_p\left(\hat{\boldsymbol{A}}_1, \hat{\boldsymbol{A}}_2\right) \tag{10}$$

$$\text{with} \quad \boldsymbol{B}^p = \boldsymbol{C}^{p-1} = \boldsymbol{F}^{p-1}\boldsymbol{F}^{p-T} \tag{11}$$

introducing a covariant symmetric second-order tensorial internal variable $\hat{\boldsymbol{A}}_1$ and a covariant skew-symmetric one $\hat{\boldsymbol{A}}_2$, both defined in the substructure configuration. Based on relation (9) we get the following inequality considering the time derivative of equation (10):

$$-\varrho_o \left\{ \frac{\partial \bar{\psi}_e}{\partial (\boldsymbol{C}\boldsymbol{B}^p)} \cdot\cdot (\boldsymbol{B}^p\boldsymbol{C})^{\cdot} + \frac{\partial \psi_p}{\partial \hat{\boldsymbol{A}}_1} \cdot\cdot \dot{\hat{\boldsymbol{A}}}_1 - \frac{\partial \psi_p}{\partial \hat{\boldsymbol{A}}_2} \cdot\cdot \dot{\hat{\boldsymbol{A}}}_2 \right\} + \frac{1}{2}\boldsymbol{T} \cdot\cdot \dot{\boldsymbol{C}} \geq 0 \,. \tag{12}$$

In the case of pure elastic material behaviour no dissipation does appear, and a hyperelastic material law can be defined easily from (12):

$$\boldsymbol{T} = 2\,\varrho_o \frac{\partial \bar{\psi}_e}{\partial (\boldsymbol{C}\boldsymbol{B}^p)} \boldsymbol{B}^p = 2\,\frac{\partial \psi_e}{\partial \boldsymbol{C}} \,. \tag{13}$$

Describing the material behaviour of metals we propose a modified compressible Neo-Hookean model for the elastic part of the free energy density [7] with material parameters which can be estimated based on the Young's modulus and the Poisson's ratio.

Considering the assumptions

$$\hat{\boldsymbol{\alpha}} = \varrho_0 \frac{\partial \psi_p}{\partial \hat{\boldsymbol{A}}_1}, \qquad \hat{\boldsymbol{T}}^p = -\varrho_0 \frac{\partial \psi_p}{\partial \hat{\boldsymbol{A}}_2} \tag{14}$$

for the backstress tensor $\hat{\boldsymbol{\alpha}}$ work-conjugated to $\hat{\boldsymbol{A}}_1$ and for a stress-like tensor $\hat{\boldsymbol{T}}^p$ work-conjugated to $\hat{\boldsymbol{A}}_2$ in the remaining part of the Clausius-Duhem inequality (12) after the elimination of the hyperelastic relation, we get the plastic dissipation inequality formulated in the reference configuration

$$\mathcal{D}^p = -\boldsymbol{\alpha} \cdot\cdot \overset{\triangle}{\boldsymbol{A}}_1 - \boldsymbol{T}^p \cdot\cdot \overset{\triangle}{\boldsymbol{A}}_2 - \frac{1}{2}\boldsymbol{C}\boldsymbol{T}\boldsymbol{C}^p \cdot\cdot \dot{\boldsymbol{B}}^p \geq 0 \tag{15}$$

with

$$\begin{aligned} \boldsymbol{\alpha} &= \boldsymbol{H}^{-1}\hat{\boldsymbol{\alpha}}\boldsymbol{H}^{-T}, & \overset{\triangle}{\boldsymbol{A}}_1 &= \boldsymbol{H}^T\dot{\hat{\boldsymbol{A}}}_1\boldsymbol{H}, \\ \boldsymbol{T}^p &= \boldsymbol{H}^{-1}\hat{\boldsymbol{T}}^p\boldsymbol{H}^{-T}, & \overset{\triangle}{\boldsymbol{A}}_2 &= \boldsymbol{H}^T\dot{\hat{\boldsymbol{A}}}_2\boldsymbol{H}. \end{aligned} \tag{16}$$

Considering the postulate that the plastic dissipation has to achieve a maximum under the constraint of satisfying an appropriate yield condition

$$F(\boldsymbol{T}, \boldsymbol{\alpha}, \boldsymbol{T}^p) \leq 0 \tag{17}$$

(see e.g. [31]), a corresponding constrained optimization problem based on the method of Lagrangean multipliers can be defined as follows:

$$\mathcal{M} = \mathcal{D}^p(\boldsymbol{T}, \boldsymbol{\alpha}, \boldsymbol{T}^p) - \lambda\, G(\boldsymbol{T}, \boldsymbol{\alpha}, \boldsymbol{T}^p, y) \quad \rightarrow \quad \text{stat} \tag{18}$$

where λ represents the plastic multiplier and y a slip variable with

$$G = F(\boldsymbol{T}, \boldsymbol{\alpha}, \boldsymbol{T}^p) + y^2. \tag{19}$$

The analysis of the necessary conditions (also known as Kuhn-Tucker conditions) for the objective function $\mathcal{M}$ to be a stationary point results in the associated flow rule

$$\dot{\boldsymbol{B}}^p = -2\lambda \boldsymbol{B} \frac{\partial F}{\partial \boldsymbol{T}} \boldsymbol{B}^p \qquad \text{with} \quad \boldsymbol{B} = \boldsymbol{C}^{-1} \tag{20}$$

and the evolutional equations for the internal variables

$$\overset{\triangle}{\boldsymbol{A}}_1 = -\lambda \frac{\partial F}{\partial \boldsymbol{\alpha}}, \qquad \overset{\triangle}{\boldsymbol{A}}_2 = \lambda \frac{\partial F}{\partial \boldsymbol{T}^p}. \tag{21}$$

The tensors $\boldsymbol{\alpha}$ and $\boldsymbol{A}_1$ respectively $\boldsymbol{T}^p$ and $\boldsymbol{A}_2$ are variables assumed to be connected by the following free energy density relation defined in the substructure configuration with its metric tensor $\hat{\boldsymbol{G}}$

$$\psi_p = \frac{1}{2}\bar{c}_1 \hat{\boldsymbol{G}} \hat{\boldsymbol{A}}_1 \cdot\cdot \hat{\boldsymbol{G}} \hat{\boldsymbol{A}}_1 - \frac{1}{2}\bar{c}_2 \hat{\boldsymbol{G}} \hat{\boldsymbol{A}}_2 \cdot\cdot \hat{\boldsymbol{G}} \hat{\boldsymbol{A}}_2. \tag{22}$$

This approach with (14) and a subsequent mapping into the reference configuration leads to

$$\boldsymbol{\alpha} = c_1 \boldsymbol{X} \boldsymbol{A}_1 \boldsymbol{X}, \qquad \boldsymbol{T}^p = -c_2 \boldsymbol{X} \boldsymbol{A}_2 \boldsymbol{X} \tag{23}$$

$$\overset{\triangledown}{\boldsymbol{\alpha}} = c_1 \boldsymbol{X} \overset{\triangle}{\boldsymbol{A}}_1 \boldsymbol{X}, \qquad \overset{\triangledown}{\boldsymbol{T}}{}^p = -c_2 \boldsymbol{X} \overset{\triangle}{\boldsymbol{A}}_2 \boldsymbol{X} \tag{24}$$

with

$$\boldsymbol{X} = \boldsymbol{H}^{-1} \hat{\boldsymbol{G}} \boldsymbol{H}^{-T}. \tag{25}$$

Due to the dependency of the yield condition on the equivalent stress T_F which on its part represents a function of the plastic arc length E_v^p it is necessary to consider an evolutional equation for E_v^p. Following the usual representation the rate of the plastic arc length is given by

$$\dot{E}_v^p = \sqrt{\frac{2}{3} \boldsymbol{B}^p \dot{\boldsymbol{E}}^p \cdot\cdot \boldsymbol{B}^p \dot{\boldsymbol{E}}^p} \quad \text{with} \quad \boldsymbol{E}^p = \frac{1}{2}(\boldsymbol{C}^p - \boldsymbol{G}). \tag{26}$$

The tensor $\boldsymbol{G}$ represents the metric of the reference configuration. Finally we get the following system of differential and algebraic equations (DAE) as a material model describing anisotropic finite elastoplastic deformations considering an underlying substructure:

$$\dot{\boldsymbol{T}} + \lambda \underset{4}{\boldsymbol{D}} \cdot\cdot \frac{\partial F}{\partial \boldsymbol{T}} - \frac{1}{2} \underset{4}{\boldsymbol{D}} \cdot\cdot \dot{\boldsymbol{C}} + \lambda \left(\boldsymbol{T} \frac{\partial F}{\partial \boldsymbol{T}} \boldsymbol{B} + \boldsymbol{B} \frac{\partial F}{\partial \boldsymbol{T}} \boldsymbol{T} \right) = \boldsymbol{0} \tag{27}$$

$$\dot{\boldsymbol{\alpha}} + \boldsymbol{Q}_1 (\boldsymbol{T}, \boldsymbol{\alpha}, \boldsymbol{T}^p, \lambda) = \boldsymbol{0} \tag{28}$$

$$\dot{\boldsymbol{T}}^p + \boldsymbol{Q}_2 (\boldsymbol{T}, \boldsymbol{\alpha}, \boldsymbol{T}^p, \lambda) = \boldsymbol{0} \tag{29}$$

$$\dot{E}^p_v + Q_3 (\boldsymbol{T}, \boldsymbol{\alpha}, \boldsymbol{T}^p, \lambda) = 0 \tag{30}$$

$$F (\boldsymbol{T}, \boldsymbol{\alpha}, \boldsymbol{T}^p) = 0 \tag{31}$$

with

$$\underset{4}{\boldsymbol{D}} = 4 \frac{\partial^2 \psi_e}{\partial \boldsymbol{C} \partial \boldsymbol{C}} \tag{32}$$

$$\boldsymbol{Q}_1 = c_1 \boldsymbol{X} \left(\lambda \frac{\partial F}{\partial \boldsymbol{\alpha}} \right) \boldsymbol{X} + \boldsymbol{H}^{-1} \dot{\boldsymbol{H}} \boldsymbol{\alpha} + \boldsymbol{\alpha} \dot{\boldsymbol{H}}^T \boldsymbol{H}^{-T} \tag{33}$$

$$\boldsymbol{Q}_2 = c_2 \boldsymbol{X} \left(\lambda \frac{\partial F}{\partial \boldsymbol{T}^p} \right) \boldsymbol{X} + \boldsymbol{H}^{-1} \dot{\boldsymbol{H}} \boldsymbol{T}^p + \boldsymbol{T}^p \dot{\boldsymbol{H}}^T \boldsymbol{H}^{-T} \tag{34}$$

$$Q_3 = -\lambda \sqrt{\frac{2}{3} \boldsymbol{B} \frac{\partial F}{\partial \boldsymbol{T}} \cdot\cdot \boldsymbol{B} \frac{\partial F}{\partial \boldsymbol{T}}}. \tag{35}$$

The DAE (27)-(31) represents the initial value problem which has to be solved within the equilibrium iterations at each Gauss point.

2.2 Numerical solution of the initial value problem

Generally, initial boundary value problems of finite elastoplasticity can not be solved analytically. Therefore numerical solution methods have to be applied.

Within the context of numerical solution strategies there exist different kinds of proceeding. In our case we have chosen a method implying the simultaneous integration of the complete system of differential and algebraic equations. This procedure has some advantages we want to elaborate after having given its numerical implementation in the following.

Using the implizit single step discretization scheme

$$y_{n+1} = y_n + (\alpha \, f_{n+1} + (1 - \alpha) \, f_n) \, \Delta t \tag{36}$$

for the solution of the differential equation

$$\frac{\mathrm{d}y}{\mathrm{d}t} = \dot{y} = f(t, y) \tag{37}$$

with the time increment $\Delta t = t_{n+1} - t_n$ and the weighting factor $\alpha \in [0, 1]$ we get the following time discretization of equations (27)-(31):

$$\boldsymbol{T}_{n+1} - \boldsymbol{T}_n - \left[\alpha \frac{1}{2} \underset{4}{\boldsymbol{D}}_{n+1} \cdot\cdot\, \dot{\boldsymbol{C}}_{n+1} + (1-\alpha) \frac{1}{2} \underset{4}{\boldsymbol{D}}_{n} \cdot\cdot\, \dot{\boldsymbol{C}}_{n}\right] \Delta t$$

$$+ \left[\alpha \lambda_{n+1} \underset{4}{\boldsymbol{D}}_{n+1} \cdot\cdot \left.\frac{\partial F}{\partial \boldsymbol{T}}\right|_{n+1} + (1-\alpha)\lambda_n \underset{4}{\boldsymbol{D}}_{n} \cdot\cdot \left.\frac{\partial F}{\partial \boldsymbol{T}}\right|_{n}\right] \Delta t$$

$$+ \alpha\lambda_{n+1} \left[\boldsymbol{T}_{n+1} \left.\frac{\partial F}{\partial \boldsymbol{T}}\right|_{n+1} \boldsymbol{B}_{n+1} + \boldsymbol{B}_{n+1} \left.\frac{\partial F}{\partial \boldsymbol{T}}\right|_{n+1} \boldsymbol{T}_{n+1}\right] \Delta t$$

$$+ (1-\alpha)\lambda_n \left[\boldsymbol{T}_{n} \left.\frac{\partial F}{\partial \boldsymbol{T}}\right|_{n} \boldsymbol{B}_{n} + \boldsymbol{B}_{n} \left.\frac{\partial F}{\partial \boldsymbol{T}}\right|_{n} \boldsymbol{T}_{n}\right] \Delta t = \boldsymbol{0} \quad (38)$$

$$\boldsymbol{\alpha}_{n+1} - \boldsymbol{\alpha}_n + \left[\alpha\, \boldsymbol{Q}_{1_{n+1}} + (1-\alpha)\, \boldsymbol{Q}_{1_n}\right] \Delta t = \boldsymbol{0} \quad (39)$$

$$\boldsymbol{T^p}_{n+1} - \boldsymbol{T^p}_n + \left[\alpha\, \boldsymbol{Q}_{2_{n+1}} + (1-\alpha)\, \boldsymbol{Q}_{2_n}\right] \Delta t = \boldsymbol{0} \quad (40)$$

$$E^p_{v_{n+1}} - E^p_{v_n} + \left[\alpha\, Q_{3_{n+1}} + (1-\alpha)\, Q_{3_n}\right] \Delta t = 0 \quad (41)$$

$$F(\boldsymbol{T}_{n+1}, \boldsymbol{\alpha}_{n+1}, \boldsymbol{T^p}_{n+1}) = 0 \quad (42)$$

To eliminate the material time derivative $\dot{\boldsymbol{C}}_{n+1}$, which can not be calculated from other available values, relation (36) is applied once again:

$$\dot{\boldsymbol{C}}_{n+1} = \frac{1}{\alpha \Delta t} \left[\Delta \boldsymbol{C}_{n+1} - (1-\alpha)\, \dot{\boldsymbol{C}}_n \Delta t\right] \quad (43)$$

with

$$\Delta \boldsymbol{C}_{n+1} = \boldsymbol{C}_{n+1} - \boldsymbol{C}_n \qquad (\alpha = 1 \quad \text{for the load step} \quad n = 0) \quad (44)$$

and equation (38) becomes:

$$\boldsymbol{T}_{n+1} - \boldsymbol{T}_n + \left[\alpha \lambda_{n+1} \underset{4}{\boldsymbol{D}}_{n+1} \cdot\cdot \left.\frac{\partial F}{\partial \boldsymbol{T}}\right|_{n+1} + (1-\alpha)\lambda_n \underset{4}{\boldsymbol{D}}_{n} \cdot\cdot \left.\frac{\partial F}{\partial \boldsymbol{T}}\right|_{n}\right] \Delta t$$

$$- \frac{1}{2} \underset{4}{\boldsymbol{D}}_{n+1} \cdot\cdot\, \Delta \boldsymbol{C}_{n+1} - \frac{1}{2}(1-\alpha) \left(\underset{4}{\boldsymbol{D}}_{n} - \underset{4}{\boldsymbol{D}}_{n+1}\right) \cdot\cdot\, \dot{\boldsymbol{C}}_n \Delta t$$

$$+ \alpha\lambda_{n+1} \left[\boldsymbol{T}_{n+1} \left.\frac{\partial F}{\partial \boldsymbol{T}}\right|_{n+1} \boldsymbol{B}_{n+1} + \boldsymbol{B}_{n+1} \left.\frac{\partial F}{\partial \boldsymbol{T}}\right|_{n+1} \boldsymbol{T}_{n+1}\right] \Delta t$$

$$+ (1-\alpha)\lambda_n \left[\boldsymbol{T}_{n} \left.\frac{\partial F}{\partial \boldsymbol{T}}\right|_{n} \boldsymbol{B}_{n} + \boldsymbol{B}_{n} \left.\frac{\partial F}{\partial \boldsymbol{T}}\right|_{n} \boldsymbol{T}_{n}\right] \Delta t = \boldsymbol{0}. \quad (45)$$

Relations (45), (39)–(42) represent a non-linear system of algebraic equations with respect to $\boldsymbol{T}_{n+1}$, $\boldsymbol{\alpha}_{n+1}$, $\boldsymbol{T}^p_{n+1}$, $E^p_{v\,n+1}$ and λ_{n+1}. In the following a vector of variables

$$\boldsymbol{z} = (\boldsymbol{T}, \boldsymbol{\alpha}, \boldsymbol{T^p}, E^p_v, \lambda)^T \quad (46)$$

is introduced. The left-hand side of the system of equations (45),(39)-(42) can be written using an operator $\mathcal{G}$

$$\mathcal{G} = \mathcal{G}\left(\boldsymbol{z}_n, \boldsymbol{z}_{n+1}, \Delta \boldsymbol{C}_{n+1}, \dot{\boldsymbol{C}}_n\right) . \tag{47}$$

As we can consider the vectors $\boldsymbol{z}_n$ and $\dot{\boldsymbol{C}}_n$ as known and therefore fixed quantities, the vector $\boldsymbol{z}_{n+1}$

$$\boldsymbol{z}_{n+1} = \left(\boldsymbol{T}_{n+1}, \boldsymbol{\alpha}_{n+1}, \boldsymbol{T^p}_{n+1}, E^p_{v_{n+1}}, \lambda_{n+1}\right)^T \tag{48}$$

represents the solution of the non-linear system of algebraic equations

$$\mathcal{G}_{n+1} = \mathcal{G}\left(\boldsymbol{z}_{n+1}\right) = \boldsymbol{0} \tag{49}$$

with respect to the load step $[t_n, t_{n+1}]$. For the calculation of $\boldsymbol{z}_{n+1}$ from (49) the Newton's method is applied. This kind of proceeding leads to a linear system of algebraic equations

$$\left(\nabla_{\boldsymbol{z}^i_{n+1}} \mathcal{G}\right) \Delta\, \boldsymbol{z}^{i+1}_{n+1} = -\, \mathcal{G}^i_{n+1}, \tag{50}$$

for the iterative determination of the increments of the solution vector $\boldsymbol{z}$

$$\Delta \boldsymbol{z}^{i+1}_{n+1} = \boldsymbol{z}^{i+1}_{n+1} - \boldsymbol{z}^{i}_{n+1}. \tag{51}$$

This kind of proceeding to solve the initial value problem has an important advantage: The consistent material tangent $\mathrm{d}\boldsymbol{T}/\mathrm{d}\boldsymbol{E}$ necessary for the generation of the element stiffnes matrices can be calculated very easily. From the implicit differentiation of equations (45),(39)-(42), cp. (47) and (49), with respect to the strain tensor $\boldsymbol{E}$ follows

$$2\frac{\mathrm{d}\,\mathcal{G}}{\mathrm{d}\,\boldsymbol{C}_{n+1}} = \frac{\mathrm{d}\,\mathcal{G}}{\mathrm{d}\,\boldsymbol{E}_{n+1}} = \frac{\partial\,\mathcal{G}}{\partial\,\boldsymbol{E}_{n+1}} + \left(\nabla_{\boldsymbol{z}_{n+1}} \mathcal{G}\right) \frac{\mathrm{d}\,\boldsymbol{z}_{n+1}}{\mathrm{d}\,\boldsymbol{E}_{n+1}} = \boldsymbol{0}\,. \tag{52}$$

Because of the Jacobi matrix in equation (52) is always known the consistent material tangent can be calculated immediately without any further efforts:

$$\left(\nabla_{\boldsymbol{z}_{n+1}} \mathcal{G}\right) \begin{pmatrix} \frac{\mathrm{d}\boldsymbol{T}}{\mathrm{d}\boldsymbol{E}}\big|_{n+1} \\ \frac{\mathrm{d}\boldsymbol{\alpha}}{\mathrm{d}\boldsymbol{E}}\big|_{n+1} \\ \frac{\mathrm{d}\boldsymbol{T}^p}{\mathrm{d}\boldsymbol{E}}\big|_{n+1} \\ \frac{\mathrm{d}E^p_v}{\mathrm{d}\boldsymbol{E}}\big|_{n+1} \\ \frac{\mathrm{d}\lambda}{\mathrm{d}\boldsymbol{E}}\big|_{n+1} \end{pmatrix} = \begin{pmatrix} \underset{4}{\boldsymbol{D}}_{n+1} + \underset{4}{\boldsymbol{M}}_{n+1} \\ \underset{4}{\boldsymbol{0}} \\ \underset{4}{\boldsymbol{0}} \\ \boldsymbol{0} \\ \boldsymbol{0} \end{pmatrix} \tag{53}$$

with

$$\underset{4}{\boldsymbol{M}}_{n+1} = 2\alpha\lambda_{n+1} \left(\boldsymbol{T}_{n+1} \left. \frac{\partial F}{\partial \boldsymbol{T}} \right|_{n+1} \boldsymbol{B}_{n+1} \underset{4}{\boldsymbol{I}} \boldsymbol{B}_{n+1} \right.$$

$$\left. + \boldsymbol{B}_{n+1} \underset{4}{\boldsymbol{I}} \boldsymbol{B}_{n+1} \left. \frac{\partial F}{\partial \boldsymbol{T}} \right|_{n+1} \boldsymbol{T}_{n+1} \right) \Delta t. \tag{54}$$

Finally we want to mention that once the numerical implementation of the initial value problem is accomplished it can be solved not only at the Gauss points but also at any other point of each element. This fact is used in the following especially within the context of mesh adaptation because field variables have to be determined for mesh refinements at the element nodes (see more in detail in Sect. 3).

3 Hierarchical adaptive strategy

As mentioned in the introduction, the quality of the numerical simulation of non-linear mechanical problems essentially depends on appropriately designed FE meshes as well. Within this context, spatial adaptivity becomes a more and more important feature of modern FE codes allowing local refinement in regions with large stress gradients. Automated mesh control has significant influence on the accuracy and the efficiency of the solution of the initial-boundary value problem as well as on the generation of well adapted meshes at critical areas of components and structures.

The already mentioned in-house FE code SPC-PM2AdNl implies the feature of adaptivity for 2D triangular and rectangular elements of the serendipity class. The implemented mesh adaptation strategy facilitates an improved simulation of the mechanical behaviour in regions with large stress gradients (e.g. near cracks, notches and in contact areas) and/or at the boundary between elastic and plastic zones [9].

Independent of the special material model the strategy for adaptivity of the space discretization generally consists of the following steps:

- error estimation,
- mesh refinement and/or coarsening,
- transfer of the field variables to newly generated nodes and integration points.

As shown in Fig. 2 the adaptive FE strategy for geometrically and physically non-linear problems is characterized by some particular features:

- Non-linear initial-boundary value problems are usually solved subdividing them into load steps. Within the context of adaptive strategies, each load increment can be regarded as a separate subproblem which has to be solved in view of a sufficiently accurate mesh before starting the next load step.

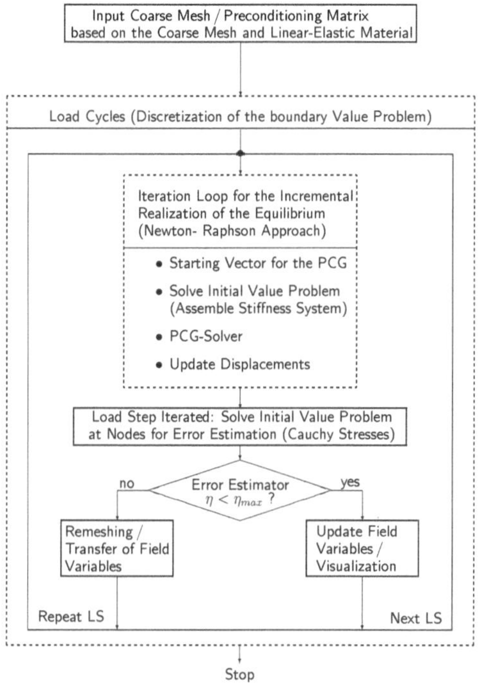

Fig. 2. General scheme of the adaptive FE algorithm: Non-linear approach

- Therewith the current load step $[t_n, t_{n+1}]$ has to be restarted with an adapted mesh if the error control indicates an unsatisfying solution of the displacement field.
- Due to the dependence of the non-linear solution on the load history, the transfer of all field variables at t_n is required for the newly generated elements. This mapping procedure generally leads to a violated equilibrium state and a possibly not overall satisfied yield condition.
- To overcome these difficulties an iterative correction procedure of the initial-boundary value problem at t_n applying a zero external load increment is proceeded additionally.

Commonly the field variables are known only at the Gauss points resulting from the solution of the initial value problem. Different strategies for their mapping to the Gauss points of newly created son elements have been developed. Here, the authors present a special mapping algorithm as a crucial point of their adaptive approach: In contrast to the majority of FE applications known from the literature the solution of the initial value problem is performed in SPC-PM2AdNl not only at the Gauss points but supplementary at the nodes of the elements. As for FE approaches usually only the displacement field is continuous over the element boundaries, this procedure leads to different values of the field variables at nodes pertaining to several elements.

Nevertheless, the presented adaptive algorithm is even primarily based on this non-smoothed nodal data.

3.1 Error control

The field of error estimation is the most important step for the control of the mesh adaptation. Several residual a posteriori error estimators are realized in the program SPC-PM2AdNl within the framework of finite elastoplasticity. Thereby the residua of the equilibrium including the edge jumps of inner forces between neighbouring elements as well as the residua of the yield condition are evaluated.

Error estimator with respect to the equilibrium

The used error estimator η_T for an element T

$$\eta_T^2 \approx \frac{h_T^2}{\lambda_D} \int_T | \operatorname{div} \boldsymbol{\sigma}(\boldsymbol{u}_h) + \boldsymbol{f}_h |^2 \, \mathrm{d}\Omega + \sum_{E \in \partial T} \frac{h_T}{\lambda_D} \int_E |[\boldsymbol{\sigma}(\boldsymbol{u}_h) \boldsymbol{n}_E]|^2 \, \mathrm{d}s_E \tag{55}$$

is based on Babuška et al. [4, 5] and further developments of Johnson and Hansbo [17], Kunert [19–23], Cramer et al. [11], Carstensen and Alberty [10], Stein et al. [33]. Here $\boldsymbol{u}_h$ and $\boldsymbol{f}_h$ denote the finite element representation of the displacement vector and the volume forces respectively, $\boldsymbol{\sigma}$ the Cauchy stress tensor, λ_D an interpolation parameter depending on the material, h_T the characteristic element length and $\boldsymbol{n}_E$ the normal vector of the element edge.

The first part of relation (55) represents the element related residuum and the second part describes the jumps of tractions over element edges E which are usually approximated by the midpoint integration rule.

In linear elasticity λ_D is usually approximated by the Young's modulus [1, 14]. In the non-linear case this interpolation parameter can not be given exactly. It is assumed that its order of magnitude is the same as for linear problems. For that reason, we approximate λ_D within the framework of the presented model by the Young's modulus too. Using this approach, it can happen that the error in plastic regions is underestimated. To overcome this, a supplementary heuristic error indicator with respect to the fulfillment of the yield condition was implemented (see equation (59)).

Error indicator with respect to the yield condition

In elastoplasticity it is necessary to describe the boundary between elastic and plastic zones as exact as possible. A special error indicator with respect to the yield condition presented in [33] reacts very sensitively in these regions. Therefore, this error indicator was modified for finite deformations and implemented into the FE-code SPC-PM2AdNl.

Derivating the DAE (27)-(31) from the optimization problem (18) (compare Sect. 2.1), we get additionally the so called Kuhn-Tucker conditions:

$$\lambda \geq 0 \tag{56}$$

$$F \leq 0 \tag{57}$$

$$\lambda F = 0 \tag{58}$$

Because of $\lambda = 0$ the condition (58) is always fulfilled in the elastic case, at unloading and neutral loading. During loading in the plastic case the algorithms for the solution of the initial value problem result in an "exact" fulfillment (with a given numerical accuracy) of the yield condition at the Gauss points of the applied integration scheme, and in the presented case additionally at the nodes of the elements. How accurate the plastic zone is mapped with the current FE mesh can be estimated using the error indicator

$$\eta_{KT}^2 = \|\lambda F - \lambda_h F_h\|_{L_2(T)}^2 = \|\lambda_h F_h\|_{L_2(T)}^2 \tag{59}$$

with respect to the condition (58).

The integrals (59) are taken for each element over integration points differing from the Gauss points applied for the solution of the initial value problem. Within this context the values λ_h and F_h at these points are approximated based on the shape functions.

The application of this error indicator leads to refined meshes especially at the boundary of the plastic zone, while in its core coarsening may appear up to the coarse mesh.

3.2 Mesh refinement

As mentioned before it is very advantageous for the mesh refinement procedure that the solution of the initial value problem is performed not only at the Gauss points but also at the nodes of the elements. This kind of proceeding facilitates the transfer of the nodal values from the father to the newly created son elements.

Generally, the refinement procedure passes off like follows:

1. Creation of the son elements (definition of edges and nodes).
2. Direct transfer of the nodal values from the father element to the son elements for all the points where son nodes coincide exactly with father nodes.
3. Calculating for newly created son nodes the corresponding nodal values using the shape functions of the father element. The transfer of the nodal displacements

$$\boldsymbol{u}^{Sonj}(\xi,\eta) = \sum_{k=1}^{Nel} h_k(\xi,\eta)\, \boldsymbol{u}_k^{Fath} \tag{60}$$

is consistent with the basic assumptions of the FE approach, and yields to continuous functions over the element edges. For all other field variables y_i with $\boldsymbol{y} = (y_1, y_2, ..., y_n)^T$ the transfer rule

$$\boldsymbol{y}^{Sonj}(\xi,\eta) = \sum_{k=1}^{Nel} h_k(\xi,\eta)\, \boldsymbol{y}_k^{Fath} \tag{61}$$

is a useful interpolation method as well. In contrast to the transfer of the nodal displacements the transfer of $\boldsymbol{y}$ results in noncontinuous functions at the element boundaries.

4. The new Gauss point values of the son elements are determined using the shape functions of the newly created son elements.

3.3 Mesh coarsening

The feature of mesh adaptation includes not only mesh refinements but also the possibility of mesh coarsening in regions of the structure in which no high stress gradients apppear. The FE code SPC-PM2AdNl permits to coarse only elements which are originated by a former element division. That means a back tracing from several leafs to branches with respect to the element tree (see Fig. 3).

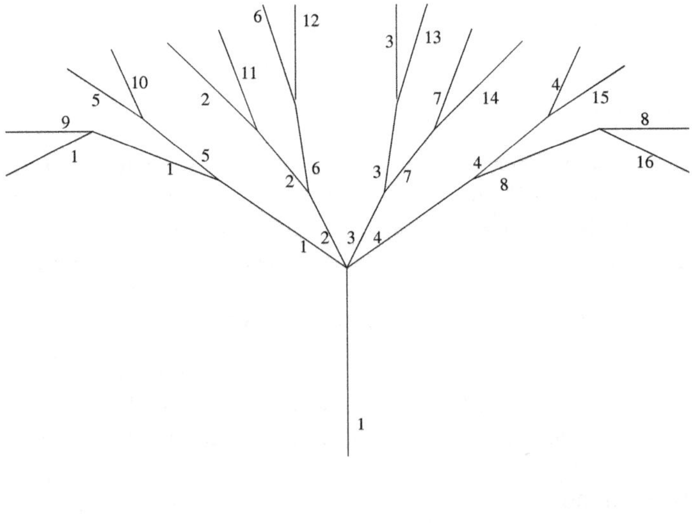

Fig. 3. Element tree in the case of triangular elements: Branches and leafs

Here, the transfer of the element based data from the sons to their common father has to be especially observed. It can be understood as the adjoint rule to the refinement procedure mentioned above. In detail, the values of nodes of the future father element which coincide with a node of only one son element are transferred directly. In contrast, nodal values of several son elements belonging to the same node of the father element have to be merged. In general there are different strategies for the transfer of the nodal results between son elements and the newly created father element. We investigated the following two approaches:

- The transfer of the son nodal values to the father element is performed using the arithmetical mean. This is the simplest way of transferring.
- Using the least squares method an improved algorithm to determine the father nodal values can be derived. Therewith it is possible to consider all available nodal values of the son elements. This procedure is supposed to be more accurate.

Finally, the Gauss point values of the newly created father element have to be calculated as well. The determination of these new values is executed using the shape functions of the corresponding element.

4 Numerical examples

In this section we would like to present some numerical examples calculated with the mentioned in-house code SPC-PM2AdNl. They show the efficiency of the developed material law and the implemented feature of adaptivity as well. But before presenting concrete results, we want to specify some other important characteristics of the Finite Element code under consideration. It is based on the following assumptions and numerical algorithms:

- Total Lagrangean description
- Damped Newton-Raphson method with a consistent linearization and a simple load step control for the numerical solution of the boundary value problem
- Conjugated gradient method with hierarchical preconditioning for the iterative solution of the FE-stiffness system
- Elastoplastic material behaviour considering finite deformations based on the multiplicative decomposition of the deformation gradient and the additive split of the free energy density, thermodynamically consistent material law as a system of differential and algebraic equations
- Discretization of the initial value problem with implicit single step standard methods and solution of the non-linear system of equations with damped Newton methods
- Efficient determination of the consistent material matrix

In the following we investigate the example of the tension of a plate with a hole under plane strain. The boundary conditions and two different coarse meshes with fully integrated eight node quadrilateral elements utilized for the presented calculations are shown in Fig. 4.

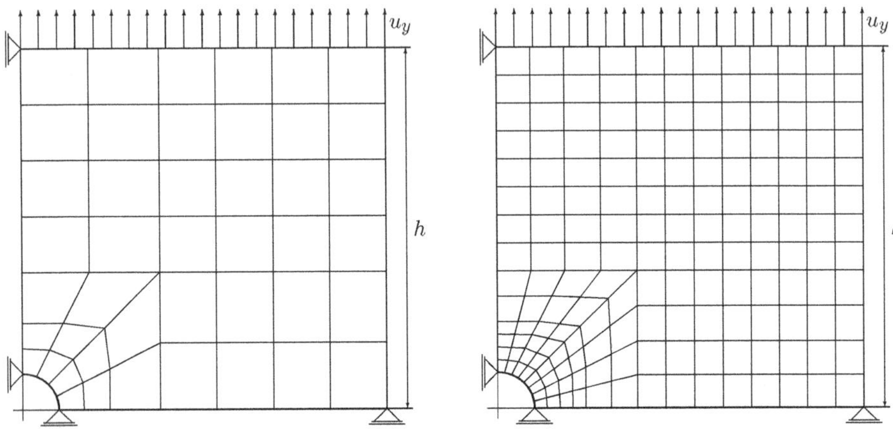

Fig. 4. Plate with a hole (one quarter of the plate). Boundary conditions. Coarse meshes with 44 and 176 elements. Edge length h = 100 mm, radius of the hole 10 mm

The elastic part of the material behaviour is described using a compressible Neo-Hookean hyperelastic approach with the following elastic part of the free Helmholtz energy density function

$$\psi_e = c_{10}(I - \ln III - 3) + D_2 (\ln III)^2 \tag{62}$$

resulting in

$$\underset{4}{\boldsymbol{D}} = 8D_2\boldsymbol{C}^{-1} \otimes \boldsymbol{C}^{-1} - 4\left[2D_2 \ln III - c_{10}\right] \boldsymbol{C}^{-1} \underset{4}{\boldsymbol{I}} \boldsymbol{C}^{-1} \tag{63}$$

Here I and III denote invariants of the elastic strain tensor $\tilde{C}^e = {F^e}^T F^e$. The coordinates of the fourth order tensor $\underset{4}{\boldsymbol{I}}$ are defined as

$$I_{IJKL} = \delta_{IK}\delta_{JL} \tag{64}$$

For details see [7, 8]. The material parameters c_{10} and D_2 are approximated with

$$c_{10} \approx \frac{E}{4(1+\nu)}, \qquad D_2 \approx \frac{c_{10}}{2}\frac{\nu}{1-2\nu} \tag{65}$$

using a Young's modulus $E = 2.1 \cdot 10^5$ MPa and a Poisson's ratio $\nu = 0.3$.

4.1 Numerical results with respect to the substructure approach

The substructure approach offers the possibility to describe plastic anisotropy using relatively simple formulations. Based on a special yield condition [7, 8],

$$F = \left(\hat{\boldsymbol{T}} - \hat{\boldsymbol{\alpha}}\right) \cdot\cdot \underset{4}{\boldsymbol{K}} \cdot\cdot \left(\hat{\boldsymbol{T}} - \hat{\boldsymbol{\alpha}}\right) + c_s \boldsymbol{M}_{S3} \boldsymbol{C} \cdot\cdot \boldsymbol{M}_{S3} \boldsymbol{C} - \frac{2}{3} T_F^2 = 0 \quad (66)$$

$$\text{with} \quad \boldsymbol{M}_{S3} = \left[(\boldsymbol{T} - \boldsymbol{T}^p) \boldsymbol{C} \boldsymbol{\alpha} + \boldsymbol{\alpha} \boldsymbol{C} (\boldsymbol{T} + \boldsymbol{T}^p) \right] \quad (67)$$

and on the following evolutional equation for the yield stress T_F [34]

$$T_F = T_{F0} + a \left[(E_v^p + \beta)^n + \beta^n \right] \quad (68)$$

isotropic and kinematic as well as distorsional hardening can be described.

Material parameters appearing in (24), (66) and (68) are $c_1 = 40$ MPa, $T_{F0} = 200$ MPa, $a = 100$ MPa, $\beta = 1.0 \cdot 10^{-8}$, $n = 0.3$, $c_2 = 1.0 \cdot 10^4$ MPa, $c_s = 5.0 \cdot 10^{-3}$ MPa^{-2}. The internal variables $\boldsymbol{\alpha}$ and $\boldsymbol{T}^p$ start with zero values.

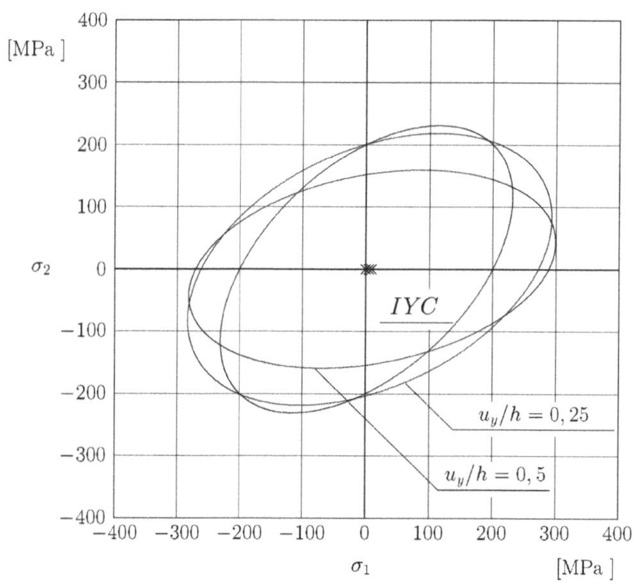

Fig. 5. Plate with a hole: Evolution of the yield condition using a substructure approach, initial yield condition (IYC) of von Mises type and subsequent yield conditions at different load level

In Fig. 5 the evolution of the yield condition observed at one point near the hole is presented. It can be seen easily that we have anisotropic material behaviour because of the changes of the axis ratio of the yield condition, its rotation and small translation.

4.2 Numerical results with respect to adaptivity

This subsection shows some examples of adapted meshes. As the proceeding of mesh adaptation is nearly independent from the applied elastoplastic material law, for reasons of simplicity we want to neglect in the following the extensive substructure approach and just focus on a von Mises type formulation for finite deformations. The yield condition considered is

$$F = \left(\dot{\boldsymbol{T}} - \dot{\boldsymbol{\alpha}}\right)\boldsymbol{C} \cdot\cdot \left(\dot{\boldsymbol{T}} - \dot{\boldsymbol{\alpha}}\right)\boldsymbol{C} - \frac{2}{3} T_F^2 = 0 \tag{69}$$

Material parameters for the evolution of the yield stress T_F (see (68)) are chosen as in Sect. 4.1 apart from $a = 1000$ MPa and the remaining parameter of equation (24) $c_1 = 500$ MPa.

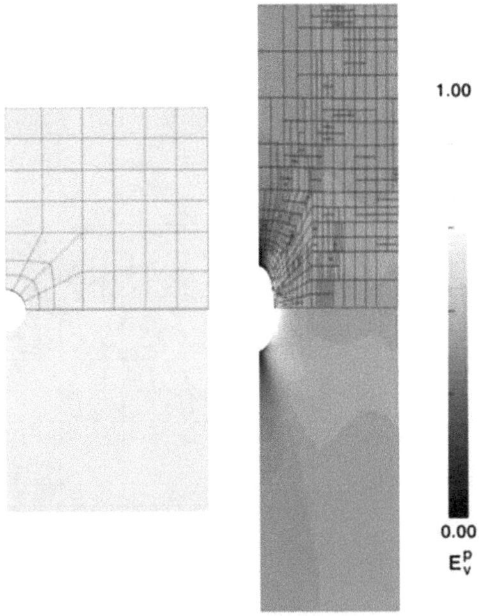

Fig. 6. Tension of a plate. Undeformed and deformed geometries with meshes. Distribution of the plastic arc length as contour bands on the deformed geometry

The initial geometry of the plate (right half) with the coarse mesh is shown in Fig. 6 on the left-hand side. In order to emphasize the performance of the FE code SPC-PM2AdNl in the case of an adaptive simulation of finite deformations, on the right-hand side the distribution of the plastic arc length on the deformed plate including the adaptive mesh after an elongation of 50% is presented. For the hierarchical adaptive strategy a combination of the error estimators η_E and η_{KT} was used.

The following numerical results are based on the finer coarse mesh with 176 elements.

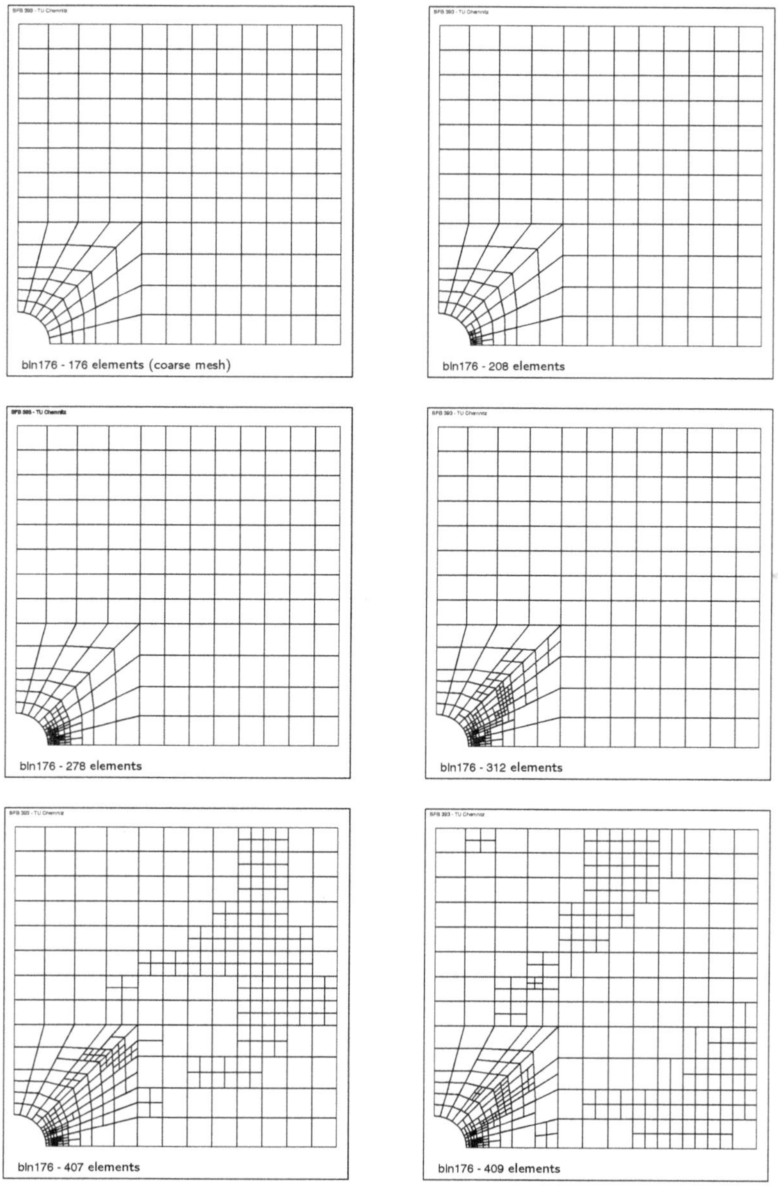

Fig. 7. Plate with a hole. Evolution of the mesh based on error indicator η_{KT}

A first investigation concerns the mesh evolution if exclusively the yield condition error indicator η_{KT} was applied. In Fig. 7 the mesh evolution is presented to demonstrate particularly that this kind of error estimation results in a mesh refinement especially at the boundary of the plastic zone. Already during the first load steps, where only small displacements but locally moderate deformations occur, the plastic zone propagates nearly over the whole sheet. Large deformations appearing due to a further increase of the external load do not result in significant changes of the adapted mesh.

Another analysis investigates the different mesh adaptation and its efficiency in dependence from the applied error estimator.

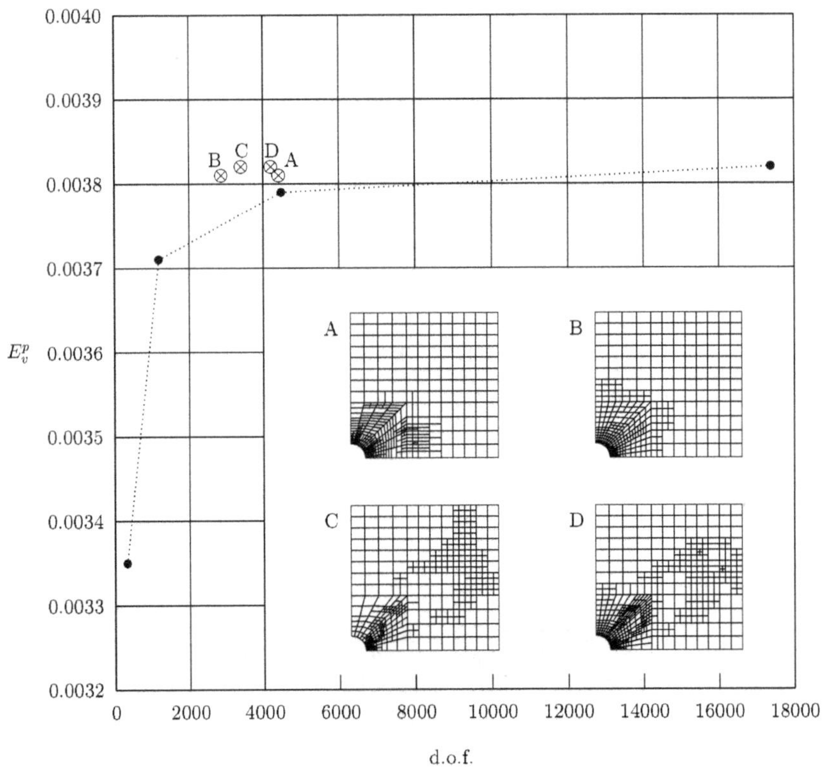

Fig. 8. Plate with a hole. Convergence of the solution in dependence on the mesh with respect to the example of the the maximum plastic arc length. (*Dashed line*) Global mesh refinement; (**A**) Adaptive remeshing based on edge oriented error estimator η_E; (**B**) Adaptive remeshing based on element oriented error estimator η_E; (**C**) Adaptive remeshing based on error estimator η_{KT}; (**D**) Adaptive remeshing based on element oriented error estimators η_E and η_{KT}

It can be seen that error estimation with respect to the equilibrium (η_E) yields to an element refinement at locations with high stress gradients (compare cases A and B in Fig. 8). In comparison, the yield condition error indicator η_{KT} leads to new elements especially at the boundary of the plastic zone. Generally it can be observed that for all kinds of error estimation the presented hierarchical adaptive strategy provides the asymptotic values of the field variables already at much lower element numbers compared with a global remeshing approach.

5 Outlook

Within the framework of the simulation of non-linear problems in continuum mechanics we focused our activities especially on the development of suitable material laws for finite elastoplasticity as well as on the implementation of adaptive hierarchical remeshing strategies.

Future developments will concern the development of a 3D code version as well as the extension to mixed finite FE approaches. The objective is to enlarge the application field of our developments to more technical problems and also to the field of biomechanics.

References

1. T. Apel, R. Mücke and J.R. Whiteman. An adaptive finite element technique with a-priori mesh grading. Report 9, BICOM Institute of Computational Mathematics, 1993.
2. N. Aravas. Finite strain anisotropic plasticity: Constitutive equations and computational issues. Advances in finite deformation problems in materials processing and structures. *ASME* 125 : 25–32, 1991.
3. R. Asaro. Crystal plasticity. *J Appl Mech* 50 : 921–934, 1983.
4. I. Babuška, W.C. Rheinboldt. A-posteriori error estimates for the finite element method. *Int J Numerical Meth Eng* 12 : 1597–1615, 1978.
5. I. Babuška, A. Miller. A-posteriori error estimates and adaptive techniques for the finite element method. *Technical report*, Institute for Physical Science and Technology, University of Maryland, 1981.
6. D. Besdo. Zur anisotropen Verfestigung anfangs isotroper starrplastischer Medien. *ZAMM* 51 : T97–T98, 1971.
7. A. Bucher. Deformationsgesetze für große elastisch-plastische Verzerrungen unter Berücksichtigung einer Substruktur. *PhD thesis*, TU Chemnitz, 2001.
8. A. Bucher, U.-J. Görke and R. Kreißig. A material model for finite elasto-plastic deformations considering a substructure. *Int. J. Plasticity* 20 : 619–642, 2004.
9. A. Bucher, A. Meyer, U.-J. Görke, R. Kreißig. A contribution to error estimation and mapping algorithms for a hierarchical adaptive FE-strategy in finite elastoplasticity. *Comput. Mech.* 36 : 182–195, 2005.
10. C. Carstensen, J. Alberty. Averaging techniques for reliable a posteriori FE-error control in elastoplasticity with hardening. *Comput. Method. Appl. Mech. Engrg.* 192 : 1435–1450, 2003.

11. H. Cramer, M. Rudolph, G. Steinl, W. Wunderlich. A hierarchical adaptive finite element strategy for elastic-plastic problems. *Comput. Struct.* 73 : 61–72, 1999.
12. Y.F. Dafalias. A missing link in the macroscopic constitutive formulation of large plastic deformations. In: *Plasticity today,* (Sawczuk, A., Bianchi, G. eds.), from the International Symposium on Recent Trends and Results in Plasticity, Elsevier, Udine, p.135, 1983.
13. Y.F. Dafalias. On the microscopic origin of the plastic spin. *Acta Mech.* 82 : 31–48, 1990.
14. J.P. Gago, D. Kelly, O.C. Zienkiewicz and I. Babuška. A-posteriori error analysis and adaptive processes in the finite element method. Part I: Error analysis. Part II: Adaptive processes. *Int. J. Num. Meth. Engng.*, 19/83 : 1593–1656, 1983.
15. P. Haupt. On the concept of an intermediate configuration and its application to a representation of viscoelastic-plastic material behavior. *Int. J. Plasticity,* 1 : 303–316.
16. R. Hill. A theory of yielding and plastic flow of anisotropic metals. *Proc. Roy. Soc. of London A 193*, 281–297, 1948.
17. C. Johnson, P. Hansbo. Adaptive finite element methods in computational mechanics. *Comput. Methods Appl. Mech. Engrg.* 101 : 143–181, 1992.
18. J. Kratochvil. Finite strain theory of cristalline elastic-plastic materials. *Acta Mech.* 16 : 127–142, 1973.
19. G. Kunert. Error estimation for anisotropic tetrahedral and triangular finite element meshes. *Preprint SFB393/97-16*, TU Chemnitz, 1997.
20. G. Kunert. An a posteriori residual error estimator for the finite element method on anisotropic tetrahedral meshes. *Numer. Math.*, 86(3) : 471–490, 2000.
21. G. Kunert. A local problem error estimator for anisotropic tetrahedral finite element meshes. *SIAM J. Numer. Anal.* 39 : 668–689, 2001.
22. G. Kunert. A posteriori l_2 error estimation on anisotropic tetrahedral finite element meshes. *IMA J. Numer. Anal.* 21 : 503–523, 2001.
23. G. Kunert and R. Verfürth. Edge residuals dominate a posteriori error estimates for linear finite element methods on anisotropic triangular and tetrahedral meshes. *Numer. Math.*, 86(2):283–303, 2000.
24. J. Mandel. Plasticité classique et viscoplasticité. *CISM Courses and Lectures No.97*, Springer-Verlag, Berlin, 1971.
25. C. Miehe, J. Schotte, J. Schröder. Computational micro-macro transitions and overall moduli in the analysis of polycristals at large strains. *Comp. Mat. Sci.* 16 : 372–382, 1999.
26. J. Ning, E.C. Aifantis. On anisotropic finite deformation plasticity, Part I. A two-back stress model. *Acta Mech.* 106 : 55–72, 1994.
27. J. Ning, E.C. Aifantis. Anisotropic yield and plastic flow of polycristalline solids. *Int. J. Plasticity* 12 : 1221–1240, 1996.
28. E.T. Onat. Representation of inelastic behaviour in the presence of anisotropy and of finite deformations. In: *Recent advances in creep and fracture of engineering materials and structures*, Pineridge Press, Swansea, 1982.
29. S.S. Rao. *Engineering Optimization.* John Wiley and Sons, New York, 1996.
30. C. Sansour. On anisotropy at the actual configuration and the adequate formulation of a free energy function. *IUTAM Symposium on Anisotropy, Inhomogenity and Nonlinearity in Solid Mechanics*, 43–50, 1995.

31. J.C. Simo. A framework for finite strain elastoplasticity based on maximum plastic dissipation and the multiplicative decomposition: Part I. Continuum formulation. *Comput. Methods Appl. Mech. Engrg.* 66 : 199–219, 1988.
32. E. Steck. Zur Berücksichtigung von Vorgängen im Mikrobereich metallischer Werkstoffe bei der Entwicklung von Stoffmodellen. *ZAMM* 75 : 331–341, 1995.
33. E. Stein (Ed.). *Error-controlled Adaptive Finite Elements in Solid Mechanics.* Wiley, Chichester, 2003.
34. V. Ulbricht, H. Röhle. Berechnung von Rotationsschalen bei nichtlinearem Deformationsverhalten, PhD thesis, TU Dresden, 1975.
35. H. Ziegler, D. Mac Vean. On the notion of an elastic solid. In: B. Broberg, J. Hult and F. Niordson, eds., *Recent Progress in Appl. Mech.* The Folke Odquist Volume, Eds. B. Broberg, J. Hult, F. Niordson, Almquist & Wiksell, Stockholm, 561–572, 1965.

Wavelet Matrix Compression for Boundary Integral Equations

Helmut Harbrecht[1], Ulf Kähler[2], and Reinhold Schneider[1]

[1] Christian–Albrechts–Universität zu Kiel
Institut für Informatik und Praktische Mathematik
Olshausenstr. 40, 24098 Kiel, Germany
hh,rs@numerik.uni-kiel.de

[2] Technische Universität Chemnitz, Fakultät für Mathematik
09107 Chemnitz, Germany
ulf.kaehler@mathematik.tu-chemnitz.de

1 Introduction

Many mathematical models concerning for example field calculations, flow simulation, elasticity or visualization are based on operator equations with *global operators*, especially *boundary integral operators*. Discretizing such problems will then lead in general to possibly very large linear systems with *densely populated* matrices. Moreover, the involved operator may have an order different from zero which means that it acts on different length scales in a different way. This is well known to entail the linear systems to become more and more ill-conditioned when the level of resolution increases. Both features pose serious obstructions to the efficient numerical treatment of such problems to an extent that desirable realistic simulations are still beyond current computing capacities.

Modern methods for the fast solution of boundary integral equations reduce the complexity to a nearly optimal rate or even an optimal rate. Denoting the number of unknowns by N_J, this means the complexity $\mathcal{O}(N_J \log^\alpha N_J)$ and $\mathcal{O}(N_J)$, respectively. Prominent examples for such methods are the *fast multipole method* [27], the *panel clustering* [30] or *hierarchical matrices* [1,29,52]. As introduced by [2] and improved in [13, 17, 18, 47], wavelet bases offer a further tool for the fast solution of boundary integral equations. In fact, a Galerkin discretization with wavelet bases results in quasi-sparse matrices, i.e., the most matrix entries are negligible and can be treated as zero. Discarding these nonrelevant matrix entries is called matrix compression. It has been shown first in [47] that only $\mathcal{O}(N_J)$ significant matrix entries remain.

Concerning boundary integral equations, a strong effort has been spent on the construction of appropriate wavelet bases on surfaces [15, 19, 20, 31, 43, 47]. In order to achieve the optimal complexity of the *wavelet Galerkin scheme*, wavelet bases are required that provide sufficiently many vanishing moments. Our realization is based on *biorthogonal* spline wavelets derived from the multiresolution developed in [9]. These wavelets are advantageous since the regularity of the duals is known [53]. Moreover, the duals are compactly supported which preserves the linear complexity of the fast wavelet transform also for its inverse. This is an important task for the coupling of FEM and BEM, cf. [33, 34]. Additionally, in view of the discretization of operators of positive order, for instance, the hypersingular operator, globally continuous wavelets are available [3, 10, 19, 38].

The efficient computation of the relevant matrix coefficients turned out to be an important task for the successful application of the wavelet Galerkin scheme [31, 44, 47]. We present a fully discrete Galerkin scheme based on numerical quadrature. Supposing that the given manifold is piecewise analytic we can use an *hp*-quadrature scheme [37, 45, 49] in combination with exponentially convergent quadrature rules. This yields an algorithm with asymptotically linear complexity without compromising the accuracy of the Galerkin scheme.

The outline is as follows. First, we introduce the class of problems under consideration. Then, in Sect. 3 we provide suitable wavelet bases on manifolds. With such bases at hand we are able to introduce the fully discrete wavelet Galerkin scheme in Sect. 4. We survey on practical issues like setting up the compression pattern, assembling the system matrix and preconditioning. Especially, we present numerical results with respect to a nontrivial domain geometry in order to demonstrate our scheme. Finally, in Sect. 5 we present recent developments concerning adaptivity and wavelet Galerkin schemes for complex geometries.

We shall frequently write $a \lesssim b$ to express that a is bounded by a constant multiple of b, uniformly with respect to all parameters on which a and b may depend. Then $a \sim b$ means $a \lesssim b$ and $b \lesssim a$.

2 Problem formulation and preliminaries

We consider a boundary integral equation on the closed boundary surface Γ of an $(n+1)$-dimensional domain Ω

$$(\mathcal{A}u)(\mathbf{x}) = \int_{\Gamma} k(\mathbf{x},\mathbf{y})u(\mathbf{y})\mathrm{d}\sigma_{\mathbf{y}} = f(\mathbf{x}), \quad \mathbf{x} \in \Gamma. \tag{1}$$

Herein, the boundary integral operator $\mathcal{A} : H^q(\Gamma) \to H^{-q}(\Gamma)$ is assumed to be an operator of order $2q$. Its kernel function will be specified below.

We assume that the boundary $\Gamma \subset \mathbb{R}^{n+1}$ is represented by piecewise parametric mappings. Let $\square$ denote the unit n-cube, i.e., $\square = [0,1]^n$. We subdivide the given manifold into several *patches*

$$\Gamma = \bigcup_{i=1}^{M} \Gamma_i, \qquad \Gamma_i = \gamma_i(\square), \qquad i = 1, 2, \ldots, M,$$

such that each $\gamma_i : \square \to \Gamma_i$ defines a diffeomorphism of $\square$ onto Γ_i. The intersection $\Gamma_i \cap \Gamma_{i'}$, $i \neq i'$, of the patches Γ_i and $\Gamma_{i'}$ is supposed to be either $\emptyset$ or a lower dimensional face.

A mesh of level j on Γ is induced by dyadic subdivisions of depth j of the unit cube into 2^{nj} cubes $C_{j,\mathbf{k}} \subseteq \square$, where $\mathbf{k} = (k_1, \ldots, k_n)$ with $0 \leq k_m < 2^j$. This generates $2^{nj}M$ *elements* (or elementary domains) $\Gamma_{i,j,\mathbf{k}} := \gamma_i(C_{j,\mathbf{k}}) \subseteq \Gamma_i$, $i = 1, \ldots, M$. In order to get a regular mesh of Γ the parametric representation is subjected to the following matching condition. A bijective, affine mapping $\Xi : \square \to \square$ exists such that for all $\mathbf{x} = \gamma_i(\mathbf{s})$ on a common interface of Γ_i and $\Gamma_{i'}$ it holds that $\gamma_i(\mathbf{s}) = (\gamma_{i'} \circ \Xi)(\mathbf{s})$. In other words, the diffeomorphisms γ_i and $\gamma_{i'}$ coincide at interfaces except for orientation.

The first fundamental tensor of differential geometry is given by the matrix

$$\mathbf{K}_i(\mathbf{s}) := \left[\left(\frac{\partial \gamma_i(\mathbf{s})}{\partial s_j}, \frac{\partial \gamma_i(\mathbf{s})}{\partial s_{j'}}\right)_{\ell^2(\mathbb{R}^{n+1})}\right]_{j,j'=1,\ldots,n} \in \mathbb{R}^{n \times n}.$$

Since γ_i is supposed to be a diffeomorphism, the matrix $\mathbf{K}_i(\mathbf{s})$ is symmetric and positive definite. The canonical inner product in $L^2(\Gamma)$ is given by

$$(u, v)_{L^2(\Gamma)} = \int_\Gamma u(\mathbf{x})v(\mathbf{x})\mathrm{d}\sigma_{\mathbf{x}} = \sum_{i=1}^{M} \int_\square u\big(\gamma_i(\mathbf{s})\big)v\big(\gamma_i(\mathbf{s})\big)\sqrt{\det\big(\mathbf{K}_i(\mathbf{s})\big)}\mathrm{d}\mathbf{s}.$$

The corresponding Sobolev spaces are indicated by $H^s(\Gamma)$. Of course, depending on the global smoothness of the surface, the range of permitted $s \in \mathbb{R}$ is limited to $s \in (-s_\Gamma, s_\Gamma)$. In case of general Lipschitz domains we have at least $s_\Gamma = 1$ since for all $0 \leq s \leq 1$ the spaces $H^s(\Gamma)$ consist of traces of functions $\in H^{s+1/2}(\Omega)$, cf. [11].

The present surface representation is in contrast to the usual approximation of the surface by panels. It has the advantage that the rate of convergence is not limited by this approximation. Notice that technical surfaces generated by CAD tools are represented in this form.

We can now specify the kernel functions. To this end, we denote by $\boldsymbol{\alpha} = (\alpha_1, \ldots, \alpha_n)$ and $\boldsymbol{\beta} = (\beta_1, \ldots, \beta_n)$ multi-indices of dimension n and define $|\boldsymbol{\alpha}| := \alpha_1 + \ldots + \alpha_n$. Moreover, we denote by $k_{i,i'}(\mathbf{s}, \mathbf{t})$ the transported kernel functions, that is

$$k_{i,i'}(\mathbf{s}, \mathbf{t}) := k\big(\gamma_i(\mathbf{s}), \gamma_{i'}(\mathbf{t})\big)\sqrt{\det\big(\mathbf{K}_i(\mathbf{s})\big)}\sqrt{\det\big(\mathbf{K}_{i'}(\mathbf{t})\big)}, \quad 1 \leq i, i' \leq M. \quad (2)$$

Definition 1. *A kernel $k(\mathbf{x}, \mathbf{y})$ is called standard kernel of the order $2q$, if the partial derivatives of the transported kernel functions $k_{i,i'}(\mathbf{s}, \mathbf{t})$, $1 \leq i, i' \leq M$, are bounded by*

$$\left|\partial_{\mathbf{s}}^{\boldsymbol{\alpha}} \partial_{\mathbf{t}}^{\boldsymbol{\beta}} k_{i,i'}(\mathbf{s}, \mathbf{t})\right| \leq c_{\boldsymbol{\alpha},\boldsymbol{\beta}} \big\|\gamma_i(\mathbf{s}) - \gamma_{i'}(\mathbf{t})\big\|^{-(n+2q+|\boldsymbol{\alpha}|+|\boldsymbol{\beta}|)}$$

provided that $n + 2q + |\boldsymbol{\alpha}| + |\boldsymbol{\beta}| > 0$.

We emphasize that this definition requires patchwise smoothness but *not* global smoothness of the geometry. The surface itself needs to be only Lipschitz. Generally, under this assumption, the kernel of a boundary integral operator $\mathcal{A}$ of order $2q$ is standard of order $2q$. Hence, we may assume this property in the sequel.

3 Wavelets and multiresolution analysis

In general, a multiresolution analysis consists of a nested family of finite dimensional subspaces

$$V_{j_0+1} \subset V_{j_0+2} \subset \cdots \subset V_j \subset V_{j+1} \cdots \subset \cdots \subset L^2(\Gamma), \tag{3}$$

such that $\dim V_j \sim 2^{jn}$ and $\overline{\bigcup_{j>j_0} V_j} = L^2(\Gamma)$. Each space V_j is defined by a single-scale basis $\Phi_j = \{\phi_{j,\mathbf{k}} : \mathbf{k} \in \Delta_j\}$, i.e., $V_j = \operatorname{span} \Phi_j$, where Δ_j denotes a suitable index set with cardinality $|\Delta_j| \sim 2^{nj}$. It is convenient to identify bases with row vectors, such that, for $\mathbf{v}_j = [v_{j,k}]_{k\in\Delta_j} \in \ell^2(\Delta_j)$, the function $v_j = \Phi_j \mathbf{v}_j$ is defined as $v_j = \sum_{k\in\Delta_j} v_{j,k}\varphi_{j,k}$. A final requirement is that the bases Φ_j are uniformly stable, i.e., $\|\mathbf{v}_j\|_{\ell^2(\Delta_j)} \sim \|\Phi_j \mathbf{v}_j\|_{L^2(\Gamma)}$ for all $\mathbf{v}_j \in \ell^2(\Delta_j)$ uniformly in j. Furthermore, the single-scale bases satisfy the locality condition $\operatorname{diam} \operatorname{supp} \phi_{j,\mathbf{k}} \sim 2^{-j}$.

If one is going to use the spaces V_j as trial spaces for the Galerkin scheme then additional properties are required. The trial spaces shall have *approximation order* $d \in \mathbb{N}$ and *regularity* $\gamma > 0$, that is

$$\gamma = \sup\{s \in \mathbb{R} : V_j \subset H^s(\Gamma)\},$$
$$d = \sup\Big\{s \in \mathbb{R} : \inf_{v_j \in V_j} \|v - v_j\|_{L^2(\Gamma)} \lesssim 2^{-js}\|v\|_{H^s(\Gamma)}\Big\}.$$

Note that conformity of the Galerkin scheme induces $\gamma > q$.

Instead of using only a single-scale j the idea of wavelet concepts is to keep track to increment of information between two adjacent scales j and $j+1$. Since $V_j \subset V_{j+1}$ one decomposes $V_{j+1} = V_j \oplus W_j$ with some complementary space W_j, $W_j \cap V_j = \{0\}$, not necessarily orthogonal to V_j. Of practical interest are the bases of the complementary spaces W_j in V_{j+1}

$$\Psi_j = \{\psi_{j,\mathbf{k}} : \mathbf{k} \in \nabla_j := \Delta_{j+1} \setminus \Delta_j\}.$$

It is supposed that the collections $\Phi_j \cup \Psi_j$ are also uniformly stable bases of V_{j+1}. If $\Psi = \bigcup_{j\geq j_0} \Psi_j$, where $\Psi_{j_0} := \Phi_{j_0+1}$, is a Riesz-basis of $L_2(\Gamma)$, we will call it a wavelet basis. We assume the functions $\psi_{j,\mathbf{k}}$ to be local with respect to the corresponding scale j, i.e., $\operatorname{diam} \operatorname{supp} \psi_{j,\mathbf{k}} \sim 2^{-j}$, and we will normalize them such that $\|\psi_{j,\mathbf{k}}\|_{L_2(\Gamma)} \sim 1$.

At first glance it would be very convenient to deal with a single orthonormal system of wavelets. But it was shown in [13, 18, 47] that orthogonal wavelets are not completely appropriate for the efficient solution of boundary integral equations. For that reason we use biorthogonal wavelet bases. Then, we have also a biorthogonal, or dual, multiresolution analysis, i.e., dual single-scale bases $\widetilde{\Phi}_j = \{\widetilde{\phi}_{j,\mathbf{k}} : \mathbf{k} \in \Delta_j\}$ and wavelets $\widetilde{\Psi}_j = \{\widetilde{\psi}_{j,\mathbf{k}} : \mathbf{k} \in \Delta_j\}$ which are coupled to the primal ones via $(\Phi_j, \widetilde{\Phi}_j)_{L^2(\Gamma)} = \mathbf{I}$ and $(\Psi_j, \widetilde{\Psi}_j)_{L^2(\Gamma)} = \mathbf{I}$. The associated spaces $\widetilde{V}_j := \operatorname{span} \widetilde{\Phi}_j$ and $\widetilde{W}_j := \operatorname{span} \widetilde{\Psi}_j$ satisfy

$$V_j \perp \widetilde{W}_j, \qquad \widetilde{V}_j \perp W_j. \tag{4}$$

Also the dual spaces shall have some approximation order $\widetilde{d} \in \mathbb{N}$ and regularity $\widetilde{\gamma} > 0$.

Denoting likewise to the primal side $\widetilde{\Psi} = \bigcup_{j \geq j_0} \widetilde{\Psi}_j$, where $\widetilde{\Psi}_{j_0} := \widetilde{\Phi}_{j_0+1}$, then every $v \in L^2(\Gamma)$ has a representation $v = \Psi(v, \widetilde{\Psi})_{L^2(\Gamma)} = \widetilde{\Psi}(v, \Psi)_{L^2(\Gamma)}$. Moreover, there hold the well known norm equivalences

$$\begin{aligned} \|v\|_{H^t(\Gamma)}^2 &\sim \sum_{j \geq j_0} 2^{2jt} \|(v, \widetilde{\Psi}_j)_{L^2(\Gamma)}\|_{\ell^2(\nabla_j)}^2, \qquad t \in (-\widetilde{\gamma}, \gamma), \\ \|v\|_{H^t(\Gamma)}^2 &\sim \sum_{j \geq j_0} 2^{2jt} \|(v, \Psi_j)_{L^2(\Gamma)}\|_{\ell^2(\nabla_j)}^2, \qquad t \in (-\gamma, \widetilde{\gamma}). \end{aligned} \tag{5}$$

The relation (4) implies that the wavelets provide *vanishing moments* of order $\widetilde{d}$

$$\left|(v, \psi_{j,\mathbf{k}})_{L^2(\Gamma)}\right| \lesssim 2^{-j(\widetilde{d}+n/2)} |v|_{W^{\widetilde{d},\infty}(\operatorname{supp} \psi_{j,\mathbf{k}})}. \tag{6}$$

Here $|v|_{W^{\widetilde{d},\infty}(\Omega)} := \sup_{|\boldsymbol{\alpha}|=\widetilde{d},\, \mathbf{x} \in \Omega} |\partial^{\boldsymbol{\alpha}} v(\mathbf{x})|$ denotes the semi-norm in $W^{\widetilde{d},\infty}(\Omega)$. We refer to [12] for further details.

For the current type of boundary surfaces Γ the $\Phi_j, \widetilde{\Phi}_j$ are generated by constructing first dual pairs of single-scale bases on the interval $[0, 1]$, using B-splines for the primal bases and the dual components from [9] adapted to the interval [16]. Tensor products yield corresponding dual pairs on $\square$. Using the parametric liftings γ_i and gluing across patch boundaries leads to globally continuous single-scale bases $\Phi_j, \widetilde{\Phi}_j$ on Γ, [3,10,19,38]. For B-splines of order d and duals of order $\widetilde{d} \geq d$ such that $d+\widetilde{d}$ is even the $\Phi_j, \widetilde{\Phi}_j$ have approximation orders $d, \widetilde{d}$, respectively. It is known that the respective regularity indices $\gamma, \widetilde{\gamma}$ (inside each patch) satisfy $\gamma = d - 1/2$ while $\widetilde{\gamma} > 0$ is known to increase proportionally to $\widetilde{d}$. Appropriate wavelet bases are constructed by projecting a *stable completion* into the correct complement spaces (see [4,19,31] for details). We refer the reader to [31, 36] for the details concerning the construction of biorthogonal wavelets on surfaces. Some illustrations are given in Fig. 1.

4 The wavelet Galerkin scheme

This section is devoted to a fully discrete wavelet Galerkin scheme for boundary integral equations. In the first subsection we discretize the given boundary

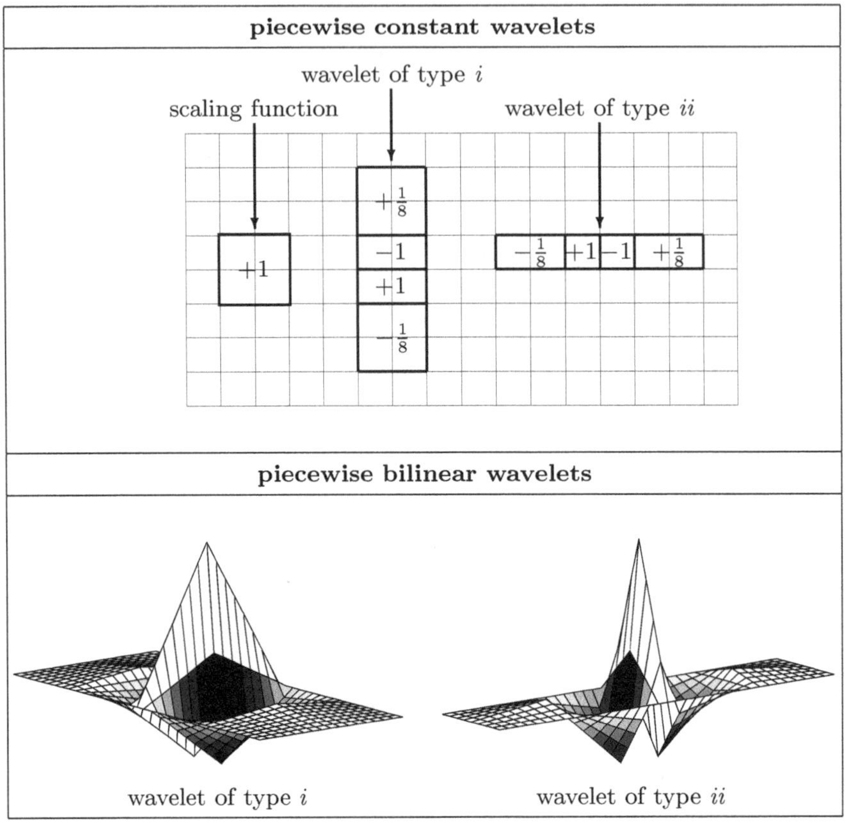

Fig. 1. (Interior) piecewise constant/bilinear wavelets with three/four vanishing moments

integral equation. In Subsect. 4.2 we introduce the a-priori matrix compression which reduces the relevant matrix coefficients to an asymptotically linear number. Then, in Subsect. 4.3 and Subsect. 4.4 we point out the computation of the compressed matrix. Next, in Subsect. 4.5 we introduce an a-posteriori compression which reduces again the number of matrix coefficients. Subsect. 4.6 is dedicated to the preconditioning of system matrices which arise from boundary integral operators of nonzero order. In the last subsection we present numerical results with respect to a nontrivial geometry.

In what follows, the collection Ψ_J with a capital J denotes the finite wavelet basis in the space V_J, i.e., $\Psi_J := \bigcup_{j=j_0}^{J-1} \Psi_j$. Further, $N_J := \dim V_J \sim 2^{Jn}$ indicates the number of unknowns.

4.1 Discretization

The variational formulation of the given boundary integral equation (1) reads:

$$\text{seek } u \in H^q(\Gamma): \quad (\mathcal{A}u, v)_{L^2(\Gamma)} = (f, v)_{L^2(\Gamma)} \quad \text{for all } v \in H^q(\Gamma). \tag{7}$$

It is well known, that the variational formulation (7) is equivalent to the boundary integral equation (1), see e.g. [28, 46] for details.

To gain the Galerkin method we replace the energy space $H^q(\Gamma)$ in the variational formulation (7) by the finite dimensional spaces V_J introduced in the previous section. Then, we arrive at the problem

$$\text{seek } u_J \in V_J: \quad (\mathcal{A}u_J, v_J)_{L^2(\Gamma)} = (f, v_J)_{L^2(\Gamma)} \quad \text{for all } v_J \in V_J.$$

Equivalently, employing the wavelet basis of V_J, the ansatz $u_J = \Psi_J \mathbf{u}_J$ yields the wavelet Galerkin scheme

$$\mathbf{A}_J \mathbf{u}_J = \mathbf{f}_J, \qquad \mathbf{A}_J = \big(\mathcal{A}\Psi_J, \Psi_J\big)_{L^2(\Gamma)}, \qquad \mathbf{f}_J = \big(f, \Psi_J\big)_{L^2(\Gamma)}. \tag{8}$$

The system matrix $\mathbf{A}_J$ is quasi-sparse and might be compressed to $\mathcal{O}(N_J)$ nonzero matrix entries if the wavelets provide enough vanishing moments.

Remark 1. Replacing above the wavelet basis Ψ_J by the single-scale basis Φ_J yields the traditional single-scale Galerkin scheme $\mathbf{A}_J^{\phi}\mathbf{u}_J^{\phi} = \mathbf{f}_J^{\phi}$, where $\mathbf{A}_J^{\phi} := \big(\mathcal{A}\Phi_J, \Phi_J\big)_{L^2(\Gamma)}$, $\mathbf{f}_J^{\phi} := \big(f, \Phi_J\big)_{L^2(\Gamma)}$ and $u_J = \Phi_J \mathbf{u}_J^{\phi}$. This scheme is related to the wavelet Galerkin scheme by

$$\mathbf{A}_J^{\psi} = \mathbf{T}_J \mathbf{A}_J^{\phi} \mathbf{T}_J^T, \qquad \mathbf{u}_J^{\psi} = \mathbf{T}_J^{-T} \mathbf{u}_J^{\phi}, \qquad \mathbf{f}_J^{\psi} = \mathbf{T}_J \mathbf{f}_J^{\phi},$$

where $\mathbf{T}_J$ denotes the wavelet transform. Since the system matrix $\mathbf{A}_J^{\phi}$ is densely populated, the costs of solving a given boundary integral equation traditionally in the single-scale basis is at least $\mathcal{O}(N_J^2)$.

4.2 A-priori compression

In a first compression step all matrix entries, for which the distances of the supports of the corresponding ansatz and test functions are bigger than a level dependent cut-off parameter $\mathcal{B}_{j,j'}$, are set to zero. In the second compression step also some of those matrix entries are neglected, for which the corresponding ansatz and test functions have overlapping supports.

First, we introduce the abbreviation

$$\Theta_{j,\mathbf{k}} := \operatorname{conv\,hull}(\operatorname{supp} \psi_{j,\mathbf{k}}), \qquad \Xi_{j,\mathbf{k}} := \operatorname{sing\,supp} \psi_{j,\mathbf{k}}.$$

Note that $\Theta_{j,\mathbf{k}}$ denotes the convex hull to the support of $\psi_{j,\mathbf{k}}$ while $\Xi_{j,\mathbf{k}}$ denotes the so-called *singular support* of $\psi_{j,\mathbf{k}}$, i.e., those points where $\psi_{j,\mathbf{k}}$ is not smooth.

We define the compressed system matrix $\mathbf{A}_J$ by

$$[\mathbf{A}_J]_{(j,\mathbf{k}),(j',\mathbf{k}')} := \begin{cases} 0, & \operatorname{dist}\big(\Theta_{j,\mathbf{k}}, \Theta_{j',\mathbf{k}'}\big) > \mathcal{B}_{j,j'},\ j, j' > j_0, \\ 0, & \operatorname{dist}\big(\Xi_{j,\mathbf{k}}, \Theta_{j',\mathbf{k}'}\big) > \mathcal{B}'_{j,j'},\ j' > j, \\ 0, & \operatorname{dist}\big(\Theta_{j,\mathbf{k}}, \Xi_{j',\mathbf{k}'}\big) > \mathcal{B}'_{j,j'},\ j > j', \\ \big(\mathcal{A}\psi_{j',\mathbf{k}'}, \psi_{j,\mathbf{k}}\big)_{L^2(\Gamma)}, & \text{otherwise.} \end{cases} \tag{9}$$

Herein, choosing

$$a > 1, \qquad \delta \in (d, \tilde{d} + 2q), \tag{10}$$

the cut-off parameters $\mathcal{B}_{j,j'}$ and $\mathcal{B}'_{j,j'}$ are set as follows

$$\begin{aligned} \mathcal{B}_{j,j'} &= a \, \max\left\{2^{-\min\{j,j'\}}, 2^{\frac{2J(\delta-q)-(j+j')(\delta+\tilde{d})}{2(\tilde{d}+q)}}\right\}, \\ \mathcal{B}'_{j,j'} &= a \max\left\{2^{-\max\{j,j'\}}, 2^{\frac{2J(\delta-q)-(j+j')\delta-\max\{j,j'\}\tilde{d}}{\tilde{d}+2q}}\right\}. \end{aligned} \tag{11}$$

The resulting structure of the compressed matrix is called *finger structure*, cf. Fig. 2. It is shown in [13, 47] that this compression strategy does not compromise the stability and accuracy of the underlying Galerkin scheme.

Theorem 1. *Let the system matrix* $\mathbf{A}_J$ *be compressed in accordance with* (9), (10) *and* (11). *Then, the wavelet Galerkin scheme is stable and the error estimate*

$$\|u - u_J\|_{H^{2q-d}(\Gamma)} \lesssim 2^{-2J(d-q)} \|u\|_{H^d(\Gamma)} \tag{12}$$

holds, where $u \in H^d(\Gamma)$ *denotes the exact solution of* (1) *and* $u_J = \Psi_J \mathbf{u}_J$ *is the numerically computed solution, i.e.,* $\mathbf{A}_J \mathbf{u}_J = \mathbf{f}_J$.

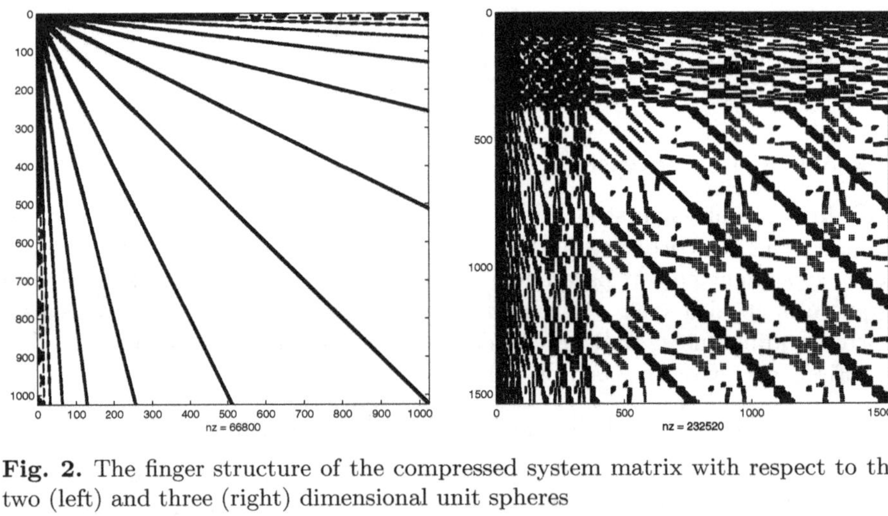

Fig. 2. The finger structure of the compressed system matrix with respect to the two (left) and three (right) dimensional unit spheres

The next theorem shows that the over-all complexity of assembling the compressed system matrix is $\mathcal{O}(N_J)$ even if each entry is weighted by a logarithmical penalty term [13, 31]. Particularly the choice $\alpha = 0$ proves that the a-priori compression yields $\mathcal{O}(N_J)$ relevant matrix entries in the compressed system matrix.

Theorem 2. *The complexity of computing the compressed system matrix* $\mathbf{A}_J$ *is* $\mathcal{O}(N_J)$ *if the calculation of its relevant entries* $(\mathcal{A}\psi_{j',\mathbf{k}'}, \psi_{j,\mathbf{k}})_{L^2(\Gamma)}$ *is performed in* $\mathcal{O}\big(\big[J - \frac{j+j'}{2}\big]^\alpha\big)$ *operations with some* $\alpha \geq 0$.

4.3 Setting up the compression pattern

Checking the distance criterion (9) for each matrix coefficient, in order to assemble the compressed matrix, would require $\mathcal{O}(N_J^2)$ function calls. To realize linear complexity, we exploit the underlying tree structure with respect to the supports of the wavelets, to predict negligible matrix coefficients. We will call a wavelet $\psi_{j+1,\text{son}}$ a son of $\psi_{j,\text{father}}$ if $\Theta_{j+1,\text{son}} \subset \Theta_{j,\text{father}}$. The following observation is an immediate consequence of the relations $\mathcal{B}_{j,j'} \geq \mathcal{B}_{j+1,j'} \geq \mathcal{B}_{j+1,j+1'}$, and $\mathcal{B}'_{j,j'} \geq \mathcal{B}'_{j+1,j'}$ if $j > j'$.

Lemma 1. *We consider* $\Theta_{j+1,\text{son}} \subseteq \Theta_{j,\text{father}}$ *and* $\Theta_{j'+1,\text{son}} \subseteq \Theta_{j',\text{father}}$.

1. *If*

$$\operatorname{dist}\big(\Theta_{j,\text{father}}, \Theta_{j',\text{father}'}\big) > \mathcal{B}_{j,j'}$$

then there holds

$$\operatorname{dist}\big(\Theta_{j+1,\text{son}}, \Theta_{j',\text{father}'}\big) > \mathcal{B}_{j+1,j'},$$
$$\operatorname{dist}\big(\Theta_{j+1,\text{son}}, \Theta_{j'+1,\text{son}'}\big) > \mathcal{B}_{j+1,j+1'}.$$

2. *For* $j > j'$ *suppose*

$$\operatorname{dist}\big(\Theta_{j,\text{father}}, \Xi_{j',\text{father}'}\big) > \mathcal{B}'_{j,j'}$$

then we can conclude that

$$\operatorname{dist}\big(\Theta_{j+1,\text{son}}, \Xi_{j',\text{father}'}\big) > \mathcal{B}'_{j+1,j'}$$

With the aid of this lemma we have to check the distance criteria only for coefficients which stem from subdivisions of calculated coefficients on a coarser level, cf. Fig. 3. Therefore, the resulting procedure of checking the distance criteria is still of linear complexity.

4.4 Computation of matrix coefficients

Of course, the significant matrix entries $(\mathcal{A}\psi_{j',\mathbf{k}'}, \psi_{j,\mathbf{k}})_{L^2(\Gamma)}$ retained by the compression strategy can generally neither be determined analytically nor be computed exactly. Therefore we have to approximate the matrix coefficients by quadrature rules. This causes an additional error which has to be controlled with regard to our overall objective of realizing asymptotically optimal accuracy while preserving efficiency. Thm. 2 describes the maximal allowed computational expenses for the computation of the individual matrix coefficients so as to realize still overall linear complexity.

The following lemma tells us that sufficient accuracy requires only a level dependent precision of quadrature for computing the retained matrix coefficients, see e.g. [13, 31, 47].

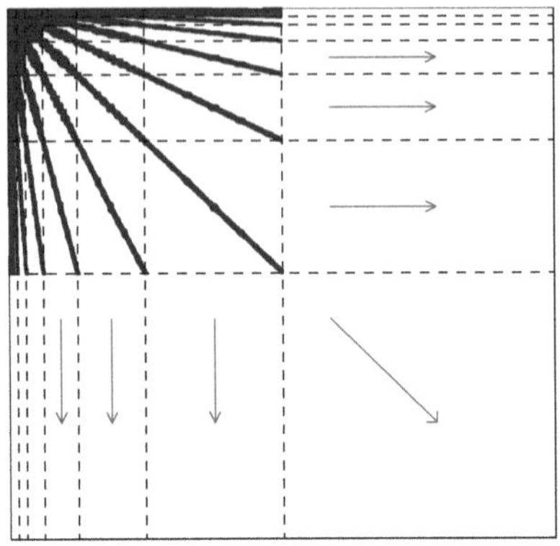

Fig. 3. The compression pattern are computed successively by starting from the coarse grids

Lemma 2. *Let the error of quadrature for computing the relevant matrix coefficients* $(\mathcal{A}\psi_{j',\mathbf{k}'}, \psi_{j,\mathbf{k}})_{L^2(\Gamma)}$ *be bounded by the level dependent accuracy*

$$\varepsilon_{j,j'} \sim \min\left\{2^{-|j-j'|n/2}, 2^{-2n(J-\frac{j+j'}{2})\frac{\delta-q}{\tilde{d}+q}}\right\} 2^{2Jq} 2^{-2\delta(J-\frac{j+j'}{2})} \tag{13}$$

with $\delta \in (d, \tilde{d}+r)$ *from* (10). *Then, the Galerkin scheme is stable and converges with the optimal order* (12).

From (13) we conclude that the entries on the coarse grids have to be computed with the full accuracy while the entries on the finer grids are allowed to have less accuracy. Unfortunately, the domains of integration are very large on coarser scales.

According to the fact that a wavelet is a linear combination of scaling functions the numerical integration can be reduced to interactions of polynomial

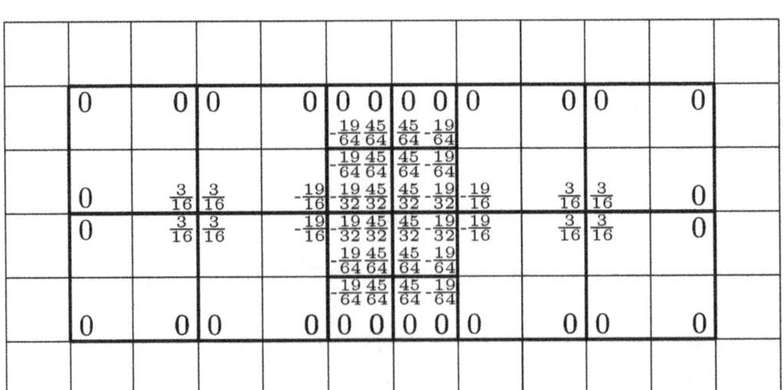

Fig. 4. The element-based representation of a piecewise bilinear wavelet with four vanishing moments

shape functions on certain elements. This suggests to employ an element-based representation of the wavelets like illustrated in Fig. 4 in the case of a piecewise bilinear wavelet. Consequently, we have only to deal with integrals of the form

$$I_{\ell,\ell'}(\Gamma_{i,j,\mathbf{k}}, \Gamma_{i',j',\mathbf{k}'}) := \int_{C_{j,\mathbf{k}}} \int_{C_{j',\mathbf{k}'}} k_{i,i'}(\mathbf{s},\mathbf{t})\, p_\ell(\mathbf{s})\, p_{\ell'}(\mathbf{t})\, \mathrm{d}\mathbf{t}\, \mathrm{d}\mathbf{s} \qquad (14)$$

with p_ℓ denoting the polynomial shape functions. This is quite similar to the traditional Galerkin discretization. The main difference is that in the wavelet approach the elements may appear on different levels due to the multilevel nature of wavelet bases.

Difficulties arise if the domains of integration are very close together relatively to their size. We have to apply numerical integration with some care in order to keep the number of evaluations of the kernel function at the quadrature nodes moderate and to fulfill the requirements of Thm. 2. The necessary accuracy can be achieved within the allowed expenses if we employ an exponentially convergent quadrature method.

In [31, 37, 47] a geometrically graded subdivision of meshes is proposed in combination with varying polynomial degrees of approximation in the integration rules, cf. Fig. 5. Exponential convergence is shown for boundary integral operators which are *analytically standard.*

Definition 2. *We call the kernel* $k(\mathbf{x},\mathbf{y})$ *an analytically standard kernel of the order* $2q$ *if the partial derivatives of the transported kernel functions* $k_{i,i'}(\mathbf{s},\mathbf{t})$, $1 \le i, i' \le M$, *satsify*

$$\left|\partial_{\mathbf{s}}^{\boldsymbol{\alpha}} \partial_{\mathbf{t}}^{\boldsymbol{\beta}} k_{i,i'}(\mathbf{s},\mathbf{t})\right| \le \frac{(|\boldsymbol{\alpha}| + |\boldsymbol{\beta}|)!}{\left(r\left\|\gamma_i(\mathbf{s}) - \gamma_{i'}(\mathbf{t})\right\|\right)^{n+2q+|\boldsymbol{\alpha}|+|\boldsymbol{\beta}|}}$$

for some $r > 0$ *provided that* $n + 2q + |\boldsymbol{\alpha}| + |\boldsymbol{\beta}| > 0$.

Generally, the kernels of boundary integral operators are analytically standard under the assumption that the underlying manifolds are patchwise analytic. It is shown in [31, 37] that an *hp*-quadrature scheme based on tensor product Gauß-Legendre quadrature rules leads to a number of quadrature points satisfying the assumptions of Thm. 2 with $\alpha = 2n$. Since the proofs are rather technical we refer the reader to [31, 37, 44, 47, 49] for further details.

Since the kernel function has a singularity on the diagonal we are still confronted with singular integrals if the domains of integration live on the same level and have some points in common. This happens if the underlying elements are identical or share a common edge or vertex. When we do not deal with weakly singular integral operators, the operators can be regularized, e.g. by integration by parts [41]. So we end up with weakly singular integrals. Such weakly singular integrals can be treated by the so-called *Duffy-trick* [22, 31, 45] in order to transform the singular integrands into analytical ones.

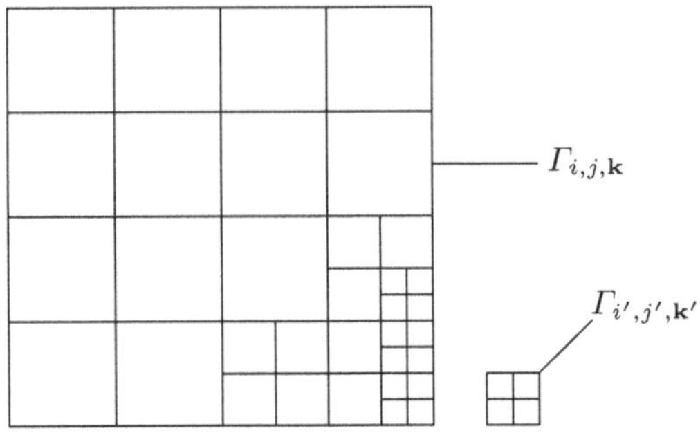

Fig. 5. Adaptive subdivision of the domains of integration

4.5 A-posteriori compression

If the entries of the compressed system matrix $\mathbf{A}_J$ have been computed, we may apply an a-posteriori compression by setting all entries to zero, which are smaller than a level dependent threshold. That way, a matrix $\widehat{\mathbf{A}}_J$ is obtained which has less nonzero entries than the matrix $\mathbf{A}_J$. Clearly, this does not accelerate the calculation of the matrix coefficients. But the requirement to the memory is reduced if the system matrix has to be stored. Especially if the linear system of equations has to be solved for several right hand sides, like for instance in shape optimization (cf. [23, 24]), the faster matrix-vector multiplication pays off. To our experience the a-posteriori compression reduces the number of nonzero coefficients by a factor 2–5.

Theorem 3 ([13, 31]). *We define the a-posteriori compression by*

$$[\widehat{\mathbf{A}}_J]_{(j,\mathbf{k}),(j',\mathbf{k}')} = \begin{cases} 0, & \text{if } \left|[\mathbf{A}_J]_{(j,\mathbf{k}),(j',\mathbf{k}')}\right| \le \varepsilon_{j,j'}, \\ [\mathbf{A}_J]_{(j,\mathbf{k}),(j',\mathbf{k}')}, & \text{if } \left|[\mathbf{A}_J]_{(j,\mathbf{k}),(j',\mathbf{k}')}\right| > \varepsilon_{j,j'}. \end{cases}$$

Herein, the level dependent threshold $\varepsilon_{j,j'}$ is chosen as in (13) with $\delta \in (d, \widetilde{d}+r)$ from (10). Then, the optimal order of convergence (12) of the Galerkin scheme is not compromised.

4.6 Wavelet preconditioning

The system matrices arising from operators of nonzero order are ill conditioned since there holds $\operatorname{cond}_{\ell^2} \mathbf{A}_J \sim 2^{2J|q|}$. According to [12, 47], the wavelet approach offers a simple diagonal preconditioner based on the norm equivalences (5).

Theorem 4. *Let the diagonal matrix $\mathbf{D}_J^r$ defined by*

$$[\mathbf{D}_J^r]_{(j,\mathbf{k}),(j',\mathbf{k}')} = 2^{rj}\delta_{j,j'}\delta_{\mathbf{k},\mathbf{k}'}, \quad \mathbf{k} \in \nabla_j, \quad \mathbf{k}' \in \nabla_{j'}, \quad j_0 \le j, j' < J. \quad (15)$$

Then, if $\widetilde{\gamma} > -q$, the diagonal matrix $\mathbf{D}_J^{2q}$ defines an asymptotically optimal preconditioner to $\mathbf{A}_J$, i.e., $\mathrm{cond}_{\ell^2}(\mathbf{D}_J^{-q}\mathbf{A}_J\mathbf{D}_J^{-q}) \sim 1$.

Remark 2. The entries on the main diagonal of $\mathbf{A}_J$ satisfy $\left(\mathcal{A}\psi_{j,\mathbf{k}}, \psi_{j,\mathbf{k}}\right)_{L^2(\Gamma)} \sim 2^{2qj}$. Therefore, the above preconditioning can be replaced by a diagonal scaling. In fact, the diagonal scaling improves and even simplifies the wavelet preconditioning.

As the numerical results in [35] confirm, this preconditioning works well in the two dimensional case. However, in three dimensions, the results are not satisfactory. Fig. 6 refers to the condition numbers of the stiffness matrices with respect to the single layer operator on a square discretized by piecewise bilinears. We employed different constructions for wavelets with four vanishing moments (spanning identical spaces, cf. [31, 36] for details). In spite of the preconditioning, the condition numbers with respect to the wavelets are not significantly better than with respect to the single-scale basis. We mention that the situation becomes even worse for operators defined on more complicated geometries.

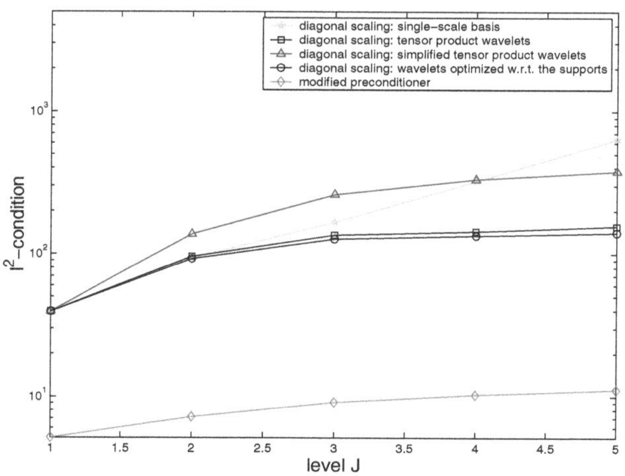

Fig. 6. ℓ^2-condition numbers for the single layer operator on the unit square

A slight modification of the wavelet preconditioner yields much better results. The simple trick is to combine the above preconditioner with the mass matrix which yields an appropriate *operator based* preconditioning, cf. [31].

Theorem 5. *Let $\mathbf{D}_J^r$ be defined as in* (15) *and $\mathbf{B}_J := (\Psi_J, \Psi_J)_{L^2(\Gamma)}$ denote the mass matrix. Then, if $\widetilde{\gamma} > -q$, the matrix $\mathbf{C}_J^{2q} = \mathbf{D}_J^q\mathbf{B}_J\mathbf{D}_J^q$ defines an asymptotically optimal preconditioner to $\mathbf{A}_J$, i.e.,*

$$\mathrm{cond}_{\ell^2}\left((\mathbf{C}_J^{2q})^{-1/2}\mathbf{A}_J(\mathbf{C}_J^{2q})^{-1/2}\right) \sim 1.$$

This preconditioner decreases the condition numbers impressively, cf. Fig. 6. Moreover, the condition depends now only on the underlying spaces but not on the chosen wavelet basis. To our experiences the condition reduces about the factor 10–100 compared to the preconditioner (15). We like to mention that, employing the fast wavelet transform, the application of this preconditioner requires only the inversion of a single-scale mass matrix, which is diagonal in case of piecewise constant and very sparse in case of piecewise bilinear ansatz functions.

4.7 Numerical results

Let Ω be the gearwheel shown in Fig. 7, represented by 504 patches. We seek the function $U \in H^1(\Omega)$ satisfying the interior Dirichlet problem

$$\Delta U = 0 \text{ in } \Omega, \qquad U = 1 \text{ on } \Gamma. \tag{16}$$

The ansatz

$$U(\mathbf{x}) = \frac{1}{4\pi}\int_\Gamma \frac{u(\mathbf{y})}{\|\mathbf{x}-\mathbf{y}\|}\mathrm{d}\sigma_\mathbf{y}, \quad \mathbf{x} \in \Omega, \tag{17}$$

yields the Fredholm boundary integral equation of the first kind $\mathcal{V}u = 1$ for the unknown density function u. Herein, $\mathcal{V} : H^{-1/2}(\Gamma) \to H^{1/2}(\Gamma)$ denotes the *single layer operator* given by

$$(\mathcal{V}u)(\mathbf{x}) := \frac{1}{4\pi}\int_\Gamma \frac{u(\mathbf{y})}{\|\mathbf{x}-\mathbf{y}\|}\mathrm{d}\sigma_\mathbf{y}, \quad \mathbf{x} \in \Gamma. \tag{18}$$

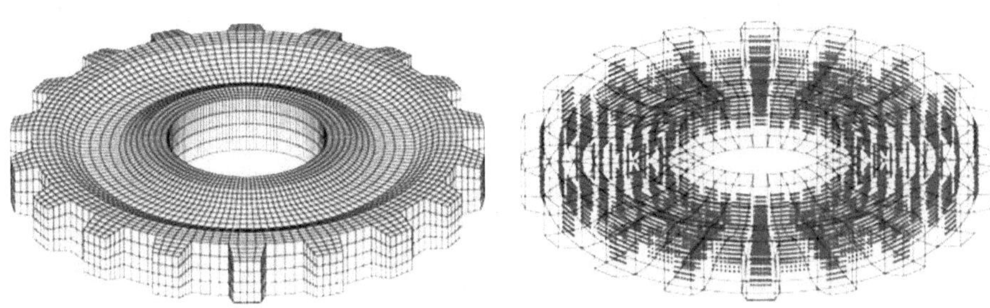

Fig. 7. The mesh on the surface Γ and the evaluation points $\mathbf{x}_i$

We discretize the given boundary integral equation by piecewise constant wavelets with three vanishing moments which is consistent to (10). We compute the discrete solution $\mathbf{U}_J := [U_J(\mathbf{x}_i)]$ according to (17) from the approximated density u_J, where the evaluation points $\mathbf{x}_i$ are specified in Fig. 7. If the density u is in $H^1(\Gamma)$ we obtain for $\mathbf{x} \in \Omega$ the pointwise estimate $|U(\mathbf{x}) - U_J(\mathbf{x})| \le c_\mathbf{x}\|u - u_J\|_{H^{-2}(\Gamma)} \lesssim 2^{-3J}\|u\|_{H^1(\Gamma)}$, cf. [54]. Unfortunately,

Table 1. Numerical results with respect to the gearwheel

J	N_J	$\|\mathbf{1} - \mathbf{U}_J\|_\infty$		a-priori	a-posteriori	cpu-time	#iterations
1	2016	1.4e-2		13%	10%	1 sec.	37
2	8086	8.8e-3	(1.6)	4.6%	3.7%	13 sec.	37
3	32256	4.2e-3	(2.1)	1.5%	1.0%	92 sec.	45
4	129024	7.6e-5	(55)	4.6e-1%	1.9e-1%	677 sec.	44
5	516096	3.9e-5	(2.0)	1.3e-1%	3.9e-2%	3582 sec.	49

we can only expect a reduced rate of convergence since $u \notin H^1(\Gamma)$ due to the presence of corner and edge singularities.

We present in Table 1 the results produced by the wavelet Galerkin scheme. The 3rd column refers to the absolute ℓ^∞-error of the point evaluations $\mathbf{U}_J$. In the 4th and 5th columns we tabulate the number of relevant coefficients. Note that on the level 5 only 650 and 200 relevant matrix coefficients per degree of freedom remain after the a-priori and a-posteriori compression, respectively. One figures out of the 6th column the over-all computing times, including compressing, assembling and solving the linear system of equations. Since the single layer operator is of order -1 preconditioning becomes an issue. Therefore, in the last column we specified the number of cg-iterations, preconditioned by the operator based preconditioner (cf. Thm. 5). The computations have been performed on a single processor of a Sun Fire V20z Server with two 2.2 MHz AMD Opteron processors and 4 GB main memory per processor.

We remark that the density u coincides with the Neumann data of the exterior analogue to (16). Therefore, the quantity $C(\Omega) = \int_\Gamma u(\mathbf{x}) d\sigma_\mathbf{x}$ is the capacity of the present domain, which we computed as $C(\Omega) = 27.02$.

5 Recent developments

5.1 Adaptivity

Wavelet matrix compression can be viewed as a non-uniform approximation of the Schwartz kernel $k(\mathbf{x}, \mathbf{y})$ with respect to the typical singularity at $\mathbf{x} = \mathbf{y}$ (cf. Defintion 1). If the domain has corners and edges, the solution itself admits singularities. In this case an adaptive scheme will reduce the number of unknowns drastically without compromising the overall accuracy. Adaptive methods for BEM have been considered by several authors, see e.g. [5, 25, 39, 40, 48] and the references therein. However, we are not aware of any paper concerning the combination of adaptive BEM with recent fast methods for integral equations like e.g. the fast multipole method.

A core ingredient of our adaptive strategy is the approximate application of (infinite dimensional) operators that ensures asymptotically optimal complexity in the following sense. If (in the given discretization framework) the unknown solution u can be approximated in the energy norm with an optimal choice of N degrees of freedom at a rate N^{-s}, then the adaptive scheme matches this rate by producing for any target accuracy ε an approximate solution u_ε such that $\|u - u_\varepsilon\|_{H^q(\Gamma)} \leq \varepsilon$ at a computational expense that stays proportionally to $\varepsilon^{-1/s}$ as ε tends to zero, see [6, 7]. Notice that $N^{-\overline{s}}$, where $\overline{s} := (d - q)/n$, is the best possible rate of convergence, gained in case of uniform refinement if $u \in H^d(\Gamma)$. Since the computation of the relevant matrix coefficients is by far the most expensive step in our algorithm, we cannot use the approach of [6, 7]. In [14] we adopted the strategy of the *best N-term approximation* by the notion of *tree approximation*, as considered in [8, 21].

The algorithm is based on an iterative method for the *continuous equation* (1), expanded with respect to the wavelet basis. To this end we assume the wavelet basis Ψ to be normalized in the energy space. Then, (1) is equivalent to the well posed problem of finding $u = \Psi\mathbf{u}$ such that the *infinite dimensional* linear system of equations

$$\mathbf{A}\mathbf{u} = \mathbf{f}, \quad \mathbf{A} = (\mathcal{A}\Psi, \Psi)_{L^2(\Gamma)}, \quad \mathbf{f} = (f, \Psi)_{L^2(\Gamma)}, \tag{19}$$

holds. The application of the operator to a function is approximated by an appropriate (finite dimensional) matrix-vector multiplication. Given a finitely supported vector $\mathbf{v}$ and a target accuracy ε, we choose *wavelet trees* τ_j according to

$$\|\mathbf{v} - \mathbf{v}|_\tau\|_{\ell^2} \leq 2^{j\overline{s}}\varepsilon, \quad j = 0, 1, \ldots, J := \left\lceil \frac{\log_2(\|\mathbf{v}\|_{\ell^2}/\varepsilon)}{\overline{s}} \right\rceil$$

and define the *layers* $\Delta_j := \tau_{j+1} \setminus \tau_j$. These layers play now the role of the levels in case of the non-adaptive scheme, i.e. we will balance the accuracy layer-dependent. We adopt the compression rules defined in [50] to define operators $\mathbf{A}_j$, having only $\mathcal{O}(2^j(1 + j)^{-2(n+1)})$ relevant coefficients per row and column while satisfying

$$\|\mathbf{A} - \mathbf{A}_j\|_{\ell^2} \leq \frac{2^{-j\overline{s}}}{(1 + j)^{2(n+1)}}.$$

Then, the approximate matrix-vector multiplication

$$\mathbf{w} := \sum_{j=0}^{J-1} \mathbf{A}_j \mathbf{v}|_{\Delta_j}$$

gives raise to the estimate $\|\mathbf{A}\mathbf{v} - \mathbf{w}\|_{\ell^2} \leq \varepsilon$. Combing this approximate matrix-vector product with an suitable iterative solver for (19) (cf. [6]) or the adaptive Galerkin type algorithm from [26] we achieve the desired goal of optimal complexity. We refer the reader to [14] for the details.

5.2 Wavelet matrix compression for complex geometries

If the geometry is given as collection $\{\pi_i\}_{i=1}^N$ of piecewise polygonal panels, which is quite common for the treatment of complex geometries, we cannot use the previous approach since the multiscale hierarchy (3) is realized by refinement. However, following [51], by agglomeration we can construct piecewise constant wavelets that are orthogonal to polynomials *in the space*

$$\int_\Gamma \psi_{j,k}(\mathbf{x})\mathbf{x}^{\boldsymbol{\alpha}} \mathrm{d}\sigma_{\mathbf{x}} = 0, \quad |\boldsymbol{\alpha}| < \widetilde{d}. \tag{20}$$

To this end, we have to introduce a hierarchical non-overlapping subdivision of the boundary Γ, called the *cluster tree* T (see Fig. 8). The cluster-tree should be a balanced 2^n tree of depth J, such that that we have approximately 2^{jn} clusters ν per level j with size $\operatorname{diam}\nu \sim 2^{-j}$. Then, starting on the finest level J with the piecewise constant ansatz functions $\Phi_J^{\pi_i} := \{\phi_i\}$, we define scaling functions $\Phi_j^\nu = \{\phi_{j,k}^\nu\}$ and wavelets $\Psi_j^\nu = \{\psi_{j,k}^\nu\}$ of a cluster ν on level j recursively via $[\Phi_j^\nu, \Psi_j^\nu] = \Phi_{j+1}^\nu [\mathbf{V}_{j,\Phi}^\nu, \mathbf{V}_{j,\Psi}^\nu]$. The coefficient matrices $\mathbf{V}_{j,\phi}^\nu$ and $\mathbf{V}_{j,\psi}^\nu$ are computed via singular value decomposition of the moment matrices

$$\mathbf{M}_j^\nu := \left[\int_\Gamma \mathbf{x}^{\boldsymbol{\alpha}} \Phi_{j+1}^\nu(\mathbf{x})\mathrm{d}\sigma\right]_{|\boldsymbol{\alpha}|<\widetilde{d}} = \mathbf{U}\boldsymbol{\Sigma}\mathbf{V}^T = \mathbf{U}[\mathbf{S},\mathbf{0}]\left[\mathbf{V}_{j,\Phi}^\nu, \mathbf{V}_{j,\Psi}^\nu\right]^T,$$

see [51] for details. Therefore, we obtain a multiscale hierarchy with respect to the spaces $V_j := \operatorname{span}\{\Phi_j^\nu : \nu \text{ is cluster of level } j\}$. The spaces $W_j := \operatorname{span}\{\Psi_j^\nu : \nu \text{ is cluster of level } j\}$ satisfy $V_{j+1} = V_j \oplus W_j$, in particular $V_j \perp W_j$ due to the orthogonality of $[\mathbf{V}_{j,\Phi}^\nu, \mathbf{V}_{j,\Psi}^\nu]$. Hence, we can define an orthonormal wavelet basis by $\Psi_N := \Phi_0^\Gamma \cup \{\Psi_j^\nu : \nu \in T\}$. In [32] the authors proved the norm equivalences (5) in a range $(-1/2, 1/2)$.

The cancellation property (20) is stronger than (6) since no smoothness of the manifold is required. Therefore, we can weaken the assumptions on the kernel. The kernel function $k(\mathbf{x},\mathbf{y})$ is supposed to be analytical in the space variables $\mathbf{x},\mathbf{y} \in \mathbb{R}^{n+1}$, apart from the singularity $\mathbf{x} = \mathbf{y}$, satisfying the decay property

$$|\partial_{\mathbf{x}}^{\boldsymbol{\alpha}} \partial_y^{\boldsymbol{\beta}} k(\mathbf{x},\mathbf{y})| \lesssim \frac{\boldsymbol{\alpha}!\boldsymbol{\beta}!}{(r\|\mathbf{x}-\mathbf{y}\|)^{n+2q+|\boldsymbol{\alpha}|+|\boldsymbol{\beta}|}}$$

for some $r > 0$ uniformly in the $(n+1)$-dimensional multi-indices $\boldsymbol{\alpha}$ and $\boldsymbol{\beta}$. Of practical interest in our considerations are first kind Fredholm integral equations for the single layer operator ($q = -1/2$) and second kind Fredholm integral equations for the double layer operator ($q = 0$).

The system matrix becomes again quasi-sparse in wavelet coordinates provided that $\widetilde{d} > d - 2q$. Since this time the wavelets are not smooth, only the first compression in (9) applies which results in $\mathcal{O}(N \log N)$ relevant matrix coefficients. In Fig. 9 one finds the system matrix (left) and its sparsity pattern (right). In case of the single layer operator the proven norm equivalences

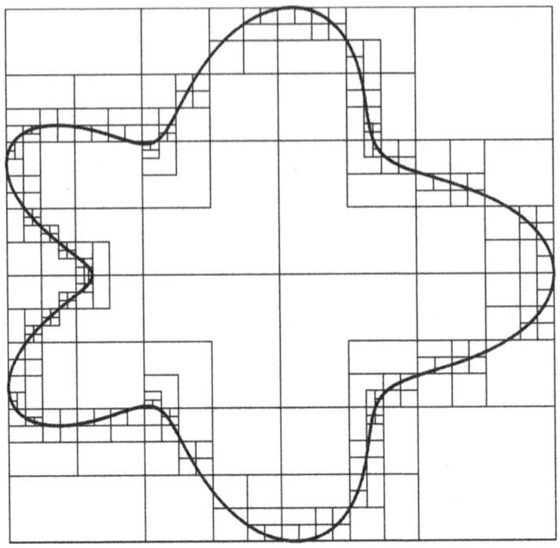

Fig. 8. The clustering of a given two dimensional boundary

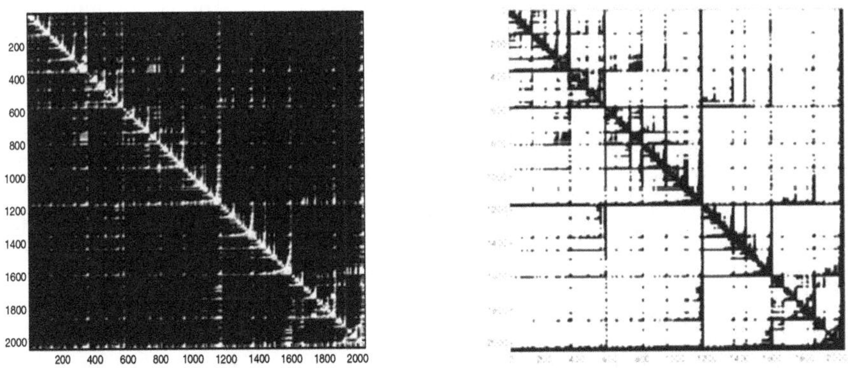

Fig. 9. The wavelet system matrix and its compression pattern

imply that the condition number of the diagonally scaled system matrix grows only polylogarithmically, cf. [42].

By using fast methods, like multipole or $\mathcal{H}$-matrices, it is possible to set up the sparse system matrix in nearly linear complexity, i.e. linear except for a polylogarithmical factor. The time consumed for basis transforms of solution and load vectors as well as for the iterative solution of the linear system of equations is nearly negligible. Numerical results presented in [32] demonstrate that we succeeded in extending the wavelet matrix compression to general geometries.

References

1. M. Bebendorf and S. Rjasanow. Adaptive low-rank approximation of collocation matrices. *Computing*, 70:1–24, 2003.
2. G. Beylkin, R. Coifman, and V. Rokhlin. The fast wavelet transform and numerical algorithms. *Comm. Pure and Appl. Math.*, 44:141–183, 1991.
3. C. Canuto, A. Tabacco, and K. Urban. The wavelet element method, part I: Construction and analysis. *Appl. Comput. Harm. Anal.*, 6:1–52, 1999.
4. J. Carnicer, W. Dahmen, and J. Peña. Local decompositions of refinable spaces. *Appl. Comp. Harm. Anal.*, 3:127–153, 1996.
5. C. Carstensen and E.P. Stephan. Adaptive boundary element methods for some first kind integral equations. *SIAM J. Numer. Anal.*, 33:2166–2183, 1996.
6. A. Cohen, W. Dahmen, and R. DeVore. Adaptive Wavelet Methods for Elliptic Operator Equations – Convergence Rates. *Math. Comp.*, 70:27–75, 2001.
7. A. Cohen, W. Dahmen, and R. DeVore. Adaptive Wavelet Methods II – Beyond the Elliptic Case. *Found. Comput. Math.*, 2:203–245, 2002.
8. A. Cohen, W. Dahmen, and R. DeVore. Adaptive wavelet schemes for nonlinear variational problems. *SIAM J. Numer. Anal.*, 41:1785–1823, 2003.
9. A. Cohen, I. Daubechies, and J.-C. Feauveau. Biorthogonal bases of compactly supported wavelets. *Pure Appl. Math.*, 45:485–560, 1992.
10. A. Cohen and R. Masson. Wavelet adaptive method for second order elliptic problems – boundary conditions and domain decomposition. *Numer. Math.*, 86:193–238, 2000.
11. M. Costabel. Boundary integral operators on Lipschitz domains: Elementary results. *SIAM J. Math. Anal.*, 19:613–626, 1988.
12. W. Dahmen. Wavelet and multiscale methods for operator equations. *Acta Numerica*, 6:55–228, 1997.
13. W. Dahmen, H. Harbrecht, and R. Schneider. Compression techniques for boundary integral equations – optimal complexity estimates. *SIAM J. Numer. Anal.*, 43:2251–2271, 2006.
14. W. Dahmen, H. Harbrecht and R. Schneider. Adaptive Methods for Boundary Integral Equations – Complexity and Convergence Estimates. *IGPM Report # 250, RWTH Aachen, Germany*, 2005. submitted to Math. Comp.
15. W. Dahmen, B. Kleemann, S. Prößdorf, and R. Schneider. A multiscale method for the double layer potential equation on a polyhedron. In H.P. Dikshit and C.A. Micchelli, editors, *Advances in Computational Mathematics*, pages 15–57, World Scientific Publ., Singapore, 1994.
16. W. Dahmen, A. Kunoth, and K. Urban. Biorthogonal spline-wavelets on the interval – stability and moment conditions. *Appl. Comp. Harm. Anal.*, 6:259–302, 1999.
17. W. Dahmen, S. Prößdorf, and R. Schneider. Wavelet approximation methods for periodic pseudodifferential equations. Part II – Matrix compression and fast solution. *Adv. Comput. Math.*, 1:259–335, 1993.
18. W. Dahmen, S. Prößdorf, and R. Schneider. Multiscale methods for pseudo-differential equations on smooth closed manifolds. In C.K. Chui, L. Montefusco, and L. Puccio, editors, *Proceedings of the International Conference on Wavelets: Theory, Algorithms, and Applications*, pages 385–424, 1994.
19. W. Dahmen and R. Schneider. Composite wavelet bases for operator equations. *Math. Comp.*, 68:1533–1567, 1999.

20. W. Dahmen and R. Schneider. Wavelets on manifolds I. Construction and domain decomposition. *Math. Anal.*, 31:184–230, 1999.
21. W. Dahmen, R. Schneider, and Y. Xu. Nonlinear functionals of wavelet expansions—adaptive reconstruction and fast evaluation. *Numer. Math.*, 86:49–101, 2000.
22. M. Duffy. Quadrature over a pyramid or cube of integrands with a singularity at the vertex. *SIAM J. Numer. Anal.*, 19:1260–1262, 1982.
23. K. Eppler and H. Harbrecht. 2nd Order Shape Optimization using Wavelet BEM. *Optim. Methods Softw.*, 21:135–153, 2006.
24. K. Eppler and H. Harbrecht. Efficient treatment of stationary free boundary problems. *WIAS-Preprint 965, WIAS Berlin, Germany*, 2004. to appear in Appl. Numer. Math.
25. B. Faermann. Local a-posteriori error indicators for the Galerkin discretization of boundary integral equations. *Numer. Math.*, 79:43–76, 1998.
26. T. Gantumur, H. Harbrecht, and R. Stevenson. An Optimal Adaptive Wavelet Method Without Coarsening. *Preprint No. 1325, Department of Mathematics, Universiteit Utrecht, The Netherlands*, 2005. to appear in Math. Comp.
27. L. Greengard and V. Rokhlin. A fast algorithm for particle simulation. *J. Comput. Phys.*, 73:325–348, 1987.
28. W. Hackbusch. *Integral equations. Theory and numerical treatment.* International Series of Numerical Mathematics, volume 120, Birkhäuser Verlag, Basel, 1995.
29. W. Hackbusch. A sparse matrix arithmetic based on $\mathcal{H}$-matrices. Part I: Introduction to $\mathcal{H}$-matrices. *Computing*, 64:89–108, 1999.
30. W. Hackbusch and Z.P. Nowak. On the fast matrix multiplication in the boundary element method by panel clustering. *Numer. Math.*, 54:463–491, 1989.
31. H. Harbrecht. Wavelet Galerkin schemes for the boundary element method in three dimensions. *PHD Thesis, Technische Universität Chemnitz, Germany*, 2001.
32. H. Harbrecht, U. Kähler, and R. Schneider. Wavelet Galerkin BEM on unstructured meshes. *Comput. Vis. Sci.*, 8:189–199, 2005.
33. H. Harbrecht, F. Paiva, C. Pérez, and R. Schneider. Biorthogonal wavelet approximation for the coupling of FEM-BEM. *Numer. Math.*, 92:325–356, 2002.
34. H. Harbrecht, F. Paiva, C. Pérez, and R. Schneider. Wavelet preconditioning for the coupling of FEM-BEM. *Num. Lin. Alg. Appl.*, 10:197–222, 2003.
35. H. Harbrecht and R. Schneider. Wavelet Galerkin Schemes for 2D-BEM. In J. Elschner, I. Gohberg and B. Silbermann, editors, *Problems and methods in mathematical physics*, Operator Theory: Advances and Applications, volume 121, pages 221–260, Birkhäuser Verlag, Basel, 2001.
36. H. Harbrecht and R. Schneider. Biorthogonal wavelet bases for the boundary element method. *Math. Nachr.*, 269–270:167–188, 2004.
37. H. Harbrecht and R. Schneider. Wavelet Galerkin Schemes for Boundary Integral Equations – Implementation and Quadrature. *SIAM J. Sci. Comput.*, 27:1347–1370, 2006.
38. H. Harbrecht and R. Stevenson. Wavelets with patchwise cancellation properties. *Preprint No. 1311, Department of Mathematics, Universiteit Utrecht, The Netherlands*, 2004. to appear in Math. Comp.
39. M. Maischak, P. Mund, and E.P. Stephan. Adaptive multilevel BEM for acoustic scattering. Symposium on Advances in Computational Mechanics,

Vol. 2, (Austin, TX, 1997). *Comput. Methods Appl. Mech. Engrg.*, 150:351–367, 1997.

40. P. Mund, and E.P. Stephan. An adaptive two-level method for hypersingular integral equations in R^3. Proceedings of the 1999 International Conference on Computational Techniques and Applications (Canberra). *ANZIAM J.*, 42:C1019–C1033, 2000.
41. J.-C. Nedelec. *Acoustic and Electromagnetic Equations — Integral Representations for Harmonic Problems.* Springer Verlag, 2001.
42. P. Oswald: Multilevel norms for $H^{-1/2}$. *Computing*, 61:235–255, 1998.
43. T. von Petersdorff, R. Schneider, and C. Schwab. Multiwavelets for second kind integral equations. *SIAM J. Num. Anal.*, 34:2212–2227, 1997.
44. T. von Petersdorff and C. Schwab. Fully discretized multiscale Galerkin BEM. In W. Dahmen, A. Kurdila, and P. Oswald, editors, *Multiscale wavelet methods for PDEs*, pages 287–346, Academic Press, San Diego, 1997.
45. S. Sauter and C. Schwab. Quadrature for the *hp*-Galerkin BEM in $\mathbb{R}^3$. *Numer. Math.*, 78:211–258, 1997.
46. S. Sauter and C. Schwab. *Randelementmethoden. Analyse, Numerik und Implementierung schneller Algorithmen.* B.G. Teubner, Stuttgart, 1997.
47. R. Schneider. *Multiskalen- und Wavelet-Matrixkompression: Analysisbasierte Methoden zur Lösung großer vollbesetzter Gleichungssysteme.* B.G. Teubner, Stuttgart, 1998.
48. H. Schulz and O. Steinbach. A new a posteriori error estimator in adaptive direct boundary element methods. The Dirichlet problem. *Calcolo*, 37:79–96, 2000.
49. C. Schwab. Variable order composite quadrature of singular and nearly singular integrals. *Computing*, 53:173–194, 1994.
50. R. Stevenson: On the compressibility of operators in wavelet coordinates. *SIAM J. Math. Anal.*, 35:1110–1132, 2004.
51. J. Tausch and J. White: Multiscale bases for the sparse representation of boundary integral operators on complex geometries. *SIAM J. Sci. Comput.*, 24:1610–1629, 2003.
52. E.E. Tyrtyshnikov. Mosaic sceleton approximation. *Calcolo*, 33:47–57, 1996.
53. L. Villemoes. Wavelet analysis of refinement equations. *SIAM J. Math. Anal.*, 25:1433–1460, 1994.
54. W.L. Wendland. On asymptotic error analysis and underlying mathematical principles for boundary element methods. In C.A. Brebbia, editor, *Boundary Element Techniques in Computer Aided Engineering, NATO ASI Series E-84*, pages 417–436, Martinus Nijhoff Publ., Dordrecht-Boston-Lancaster, 1984.

Numerical Solution of Optimal Control Problems for Parabolic Systems

Peter Benner[1], Sabine Görner[1], and Jens Saak[1]

Technische Universität Chemnitz, Fakultät für Mathematik
09107 Chemnitz, Germany
{benner,sabine.goerner,jens.saak}@mathematik.tu-chemnitz.de

1 Introduction

We consider nonlinear parabolic diffusion-convection and diffusion-reaction systems of the form

$$\frac{\partial \mathbf{x}}{\partial t} + \nabla \cdot (\mathbf{c}(\mathbf{x}) - \mathbf{k}(\nabla \mathbf{x})) + \mathbf{q}(\mathbf{x}) = \mathcal{B}\mathbf{u}(t), \qquad t \in [0, T_f], \tag{1}$$

in $\Omega \in \mathbb{R}^d, d = 1, 2, 3$, with appropriate initial and boundary conditions. Here, $\mathbf{c}$ is the convective part, $\mathbf{k}$ the diffusive part and $\mathbf{q}$ is an uncontrolled source term. The state of the system depends on $\xi \in \Omega$ and the time $t \in [0, T_f]$ and is denoted by $\mathbf{x}(\xi, t)$. The control is called $\mathbf{u}(t)$ and is assumed to depend only on the time $t \in [0, T_f]$.

A control problem is defined as

$$\min_{\mathbf{u}} J(\mathbf{x}, \mathbf{u}) \quad \text{subject to (1)}, \tag{2}$$

where $J(\mathbf{x}, \mathbf{u})$ is a performance index which will be introduced later.

There are two possibilities for the appearance of the control. If the control occurs in the boundary condition, we call this problem a *boundary control problem*. It is called *distributed control problem* if the control acts in Ω or a sub-domain $\Omega_u \subset \Omega$. The control problem as in (1) is well-suited to describe a distributed control problem while boundary control will require the specification of the boundary conditions as, for instance, given below.

The major part of this article deals with the linear version of (1),

$$\frac{\partial \mathbf{x}}{\partial t} - \nabla . (a(\xi)\nabla \mathbf{x}) + d(\xi)\nabla \mathbf{x} + r(\xi)\mathbf{x} = \mathbf{B}_V(\xi)u(t), \quad \xi \in \Omega, \ t > 0, \tag{3}$$

with initial and boundary conditions

$$\begin{aligned} \alpha(\xi)\frac{\partial \mathbf{x}(\xi,t)}{\partial n} + \gamma(\xi)\mathbf{x}(\xi,t) &= \mathbf{B}_R u(t), \qquad \xi \in \partial\Omega, \\ \mathbf{x}(\xi,0) &= \mathbf{x}_0(\xi), \quad \xi \in \Omega, \end{aligned}$$

for sufficiently smooth parameters $a, d, r, \alpha, \gamma, \mathbf{x}_0$. We assume that either $\mathbf{B}_V = 0$ (boundary control system) or $\mathbf{B}_R = 0$ (distributed control system). In addition, we include in our problem an output equation of the form

$$\mathbf{y} = \mathbf{C}\mathbf{x}, \qquad t \geq 0,$$

taking into account that in practice, often not the whole state $\mathbf{x}$ is available for measurements. Here, $\mathbf{C}$ is a linear operator which often is a restriction operator.

To solve optimal control problems (2) with a linear system (3) we interpret it as a linear quadratic regulator (LQR) problem. The theory behind the LQR ansatz has already been studied in detail, e.g., in [1–6], to name only a few.

Nonlinear control problems are still undergoing extensive research. We will apply model predictive control (MPC) here, i.e., we solve linearized problems on small time frames using a *linear-quadratic Gaussian* (LQG) design. This idea is presented by Ito and Kunisch in [7]. We will briefly sketch the main ideas of this approach at the end of this article.

There exists a rich variety of other approaches to solve linear and nonlinear optimal control problems for partial differential equations. We can only refer to a selection of ideas, see e.g. [4, 8–12].

This article is divided into four parts. In the remainder of this section we will give a short introduction to linear control problems and we present the model problem used in this article. Theoretical results which justify the numerical implementation of the LQR problem will be pointed out in Sect. 2. The third section deals with computational methods for the model problem. There we go into algorithmic and implementation details and present some numerical results. Finally we give an insight into a method for nonlinear parabolic systems in Sect. 4.

1.1 Linear problems

In this section we will formulate the linear quadratic regulator (LQR) problem. We assume that $\mathcal{X}, \mathcal{Y}, \mathcal{U}$ are separable Hilbert spaces where $\mathcal{X}$ is called the state space, $\mathcal{Y}$ the observation space and $\mathcal{U}$ the control space.

Furthermore the linear operators

$$\begin{aligned} \mathbf{A} &: \operatorname{dom}(\mathbf{A}) \subset \mathcal{X} \to \mathcal{X}, \\ \mathbf{B} &: \mathcal{U} \to \mathcal{X}, \\ \mathbf{C} &: \mathcal{X} \to \mathcal{Y} \end{aligned}$$

are given. Such an abstract system can now be understood as a Cauchy problem for a linear evolution equation of the form

$$\dot{\mathbf{x}} = \mathbf{A}\mathbf{x} + \mathbf{B}\mathbf{u}, \qquad \mathbf{x}(.,0) = \mathbf{x}_0 \in \mathcal{X}. \tag{4}$$

Since in many applications the state $\mathbf{x}$ of a system can not be observed completely we consider the observation equation

$$\mathbf{y} = \mathbf{C}\mathbf{x}, \tag{5}$$

which describes the map between the states and the outputs of the system.

The abstract LQR problem is now given as the minimization problem

$$\min_{\mathbf{u}\in L^2(0,T_f;\mathcal{U})} \frac{1}{2}\int_0^{T_f} \langle \mathbf{y}, \mathbf{Q}\mathbf{y}\rangle_{\mathcal{Y}} + \langle \mathbf{u}, \mathbf{R}\mathbf{u}\rangle_{\mathcal{U}} \, \mathrm{d}t \tag{6}$$

with self-adjoint, positive definite, linear, bounded operators $\mathbf{Q}$ and $\mathbf{R}$ on $\mathcal{Y}$ and $\mathcal{U}$, respectively. Recall that if (4) is an ordinary differential equation with $\mathcal{X} = \mathbb{R}^n$, $\mathcal{Y} = \mathbb{R}^p$ and $\mathcal{U} = \mathbb{R}^m$, equipped with the standard scalar product, then we obtain an LQR problem for a *finite-dimensional system* [13]. For partial differential equations we have to choose the function spaces $\mathcal{X}, \mathcal{Y}, \mathcal{U}$ appropriately and we get an LQR system for an *infinite-dimensional system* [14,15].

Many optimal control problems for instationary linear partial differential equations can be described using the abstract LQR problem above. Additionally, many control, stabilization and parameter identification problems can be reduced to the LQR problem, see [1–3,15,16].

The infinite time case

In the infinite time case we assume that $T_f = \infty$. Then the minimization problem subject to (4) is given by

$$\min_{\mathbf{u}\in L^2(0,\infty;\mathcal{U})} \frac{1}{2}\int_0^{\infty} \langle \mathbf{y}, \mathbf{Q}\mathbf{y}\rangle_{\mathcal{Y}} + \langle \mathbf{u}, \mathbf{R}\mathbf{u}\rangle_{\mathcal{U}} \, \mathrm{d}t. \tag{7}$$

If the standard assumptions that

- $\mathbf{A}$ is the infinitesimal generator of a $\mathcal{C}^0$-semigroup $\mathbf{T}(t)$,
- $\mathbf{B}, \mathbf{C}$ are linear bounded operators and
- for every initial value there exists an admissible control $\mathbf{u} \in L^2(0,\infty;\mathcal{U})$

hold then the solution of the abstract LQR problem can be obtained analogously to the finite-dimensional case (see [14,17,18]). We then have to consider the algebraic operator Riccati equation

$$0 = \Re(\mathbf{P}) = \mathbf{C}^*\mathbf{Q}\mathbf{C} + \mathbf{A}^*\mathbf{P} + \mathbf{P}\mathbf{A} - \mathbf{P}\mathbf{B}\mathbf{R}^{-1}\mathbf{B}^*\mathbf{P}, \tag{8}$$

where the linear operator $\mathbf{P}$ will be the solution of (8) if $\mathbf{P} : \operatorname{dom} \mathbf{A} \to \operatorname{dom} \mathbf{A}^*$ and $\langle \hat{\mathbf{x}}, \Re(\mathbf{P})\mathbf{x} \rangle = 0$ for all $\mathbf{x}, \hat{\mathbf{x}} \in \operatorname{dom}(\mathbf{A})$. The optimal control is then given as the *feedback control*

$$\mathbf{u}_*(t) = -\mathbf{R}^{-1}\mathbf{B}^*\mathbf{P}_\infty \mathbf{x}_*(t), \tag{9}$$

which has the form of a regulator or closed-loop control. Here, $\mathbf{P}_\infty$ is the minimal nonnegative self-adjoint solution of (8), $\mathbf{x}_*(t) = \mathbf{S}(t)\mathbf{x}_0(t)$, and $\mathbf{S}(t)$ is the $\mathcal{C}^0$-semigroup generated by $\mathbf{A} - \mathbf{B}\mathbf{R}^{-1}\mathbf{B}^*\mathbf{P}_\infty$. Using further standard assumptions it can be shown, see e.g. [5], that $\mathbf{P}_\infty$ is the unique nonnegative solution of (8). Most of the required conditions, particularly the restrictive assumption that $\mathbf{B}$ is bounded, can be weakened [1,2].

The finite time case

The finite time case arises if $T_f < \infty$ in (6). Then the numerical solution is more complicated since we have to solve the operator differential Riccati equation

$$\dot{\mathbf{P}}(t) = -(\mathbf{C}^*\mathbf{Q}\mathbf{C} + \mathbf{A}^*\mathbf{P}(t) + \mathbf{P}(t)\mathbf{A} - \mathbf{P}(t)\mathbf{B}\mathbf{R}^{-1}\mathbf{B}^*\mathbf{P}(t)). \tag{10}$$

The optimal control is obtained as

$$\mathbf{u}_*(t) = -\mathbf{R}^{-1}\mathbf{B}^*\mathbf{P}_*(t)\mathbf{x}_*(t),$$

where $\mathbf{P}_*(t)$ is the unique solution of (10) in complete analogy to the infinite time case in (9).

1.2 Discretization

For the discretization of an optimal control problem we can follow different strategies. In the literature the following two alternatives are often used:

- "Optimize–then–discretize"
  That is, we compute the optimal control with methods of optimization first and discretize afterwards.
- "Discretize–then–optimize"
  Here, we discretize at first and optimize the discrete problem.

The literature provides a large amount of approaches which are based on non-smooth Newton's methods or sequential quadratic programming (SQP) methods, see, e.g., [11,12,19].

In contrast to the strategies above we want to examine the strategy "semidiscretize–optimize–semidiscretize". If we semidiscretize in space first, for instance by using a Galerkin ansatz with finite element ansatz functions, we obtain a linear finite-dimensional system. The structure and solution of the resulting system are analogous to those of the infinite dimensional system.

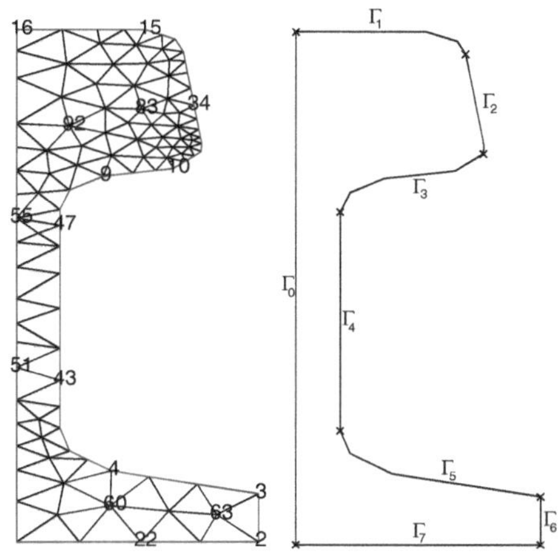

Fig. 1. Initial mesh with observation nodes (left) and partitioning of the boundary (right)

Thus, we can formulate the following general strategy for solving LQR problems, where, of course, the applicability of the described *Riccati approach* has to be tested for every situation, in particular the conditions on the regularity of the boundary and the solution play a decisive role.

1. Find a spatial discretization for the partial differential equation using a Galerkin projection of $\mathcal{X}$ on a finite-dimensional subspace $\mathcal{X}^N$ with matrix representations A^N, B^N, C^N, Q^N of the corresponding linear operators $\mathbf{A}, \mathbf{B}, \mathbf{C}, \mathbf{Q}$.
2. Solve the finite-dimensional LQR problem.
3. Apply the finite-dimensional feedback law to the infinite-dimensional system.
4. If necessary, refine the discretization.

1.3 The model problem

The control of the cooling process for a rail profile in a rolling mill serves as a benchmark problem for our approach. The model has been discussed in detail in the literature in the context of optimization by Tröltzsch and others (see [20–22] and references therein). First results concerning the LQR design for this problem can be found in [23–26].

As in [20–22, 27] the steel profile is assumed to stretch infinitely into the z-direction. This admits the assumption of a stationary heat distribution in z-direction. That means we can restrict ourselves to a 2-dimensional heat diffusion process on cross-sections of the profile $\Omega \subset \mathbb{R}^2$ as shown in Fig. 1. Measurements for defining the geometry of the cross-section are taken from

[20]. As one can see in Fig. 1 the domain exploits the symmetry of the profile introducing an artificial boundary Γ_0 on the symmetry axis.

We will concentrate on the linearized version of the state equation introduced in [20–22]. The linearization is derived by taking means of the material parameters ρ, λ and c. This is admissible as long as we work in temperature regimes above approximately 700-800°C (depending on the kind of steel used) where changes of ρ, λ and c are small and we do not have multiple phases and phase transitions in the material. We partition the non-artificial boundary into 7 parts, each of them located between two neighboring corners of the domain (see Fig. 1 for details). The control u is assumed to be constant with respect to the spatial variable ξ on each part Γ_i of the boundary. Thus we obtain the following model:

$$\begin{aligned} c\rho\frac{\partial x(\xi,t)}{\partial t} &= \nabla.(\lambda\nabla x(\xi,t)) && \text{in } \Omega\times(0,T), \\ -\lambda\frac{\partial x(\xi,t)}{\partial n} &= g_i(t,x,u) && \text{on } \Gamma_i \text{ for } i=0,\dots,7, \\ x(\xi,0) &= x_0(\xi) && \text{in } \Omega. \end{aligned} \tag{11}$$

We now have to describe the heat transfer across the surface of the material, i.e. the boundary conditions. The boundary condition according to Newton's cooling law is given as

$$-\frac{\partial x(\xi,t)}{\partial n} = \frac{\kappa_k}{\lambda}(x(\xi,t) - x_{ext,k}(t)). \tag{12}$$

Note that $x_{ext,k}(t)$ is assumed to be constant on Γ_k and therefore does not depend explicitly on ξ. For a more detailed derivation of this condition see [26].

Here, we will take the external temperature as the control. The mathematical advantage of this choice is that the multiplication of control and state which would lead to a bilinear control system in case of the heat transfer coefficient as control is avoided.

2 Theoretical results

2.1 Approximation of abstract cauchy problems

The theoretical fundament for our approach was set by Gibson [18]. The ideas and proofs used for the boundary control problem considered here closely follow the extension of Gibson's method proposed by Banks and Kunisch [28] for distributed control systems arising from parabolic equations. Similar approaches can be found in [2,26]. Common to all those approaches is to formulate the control system for a parabolic system as an abstract Cauchy problem in an appropriate Hilbert space setting. For numerical approaches this Hilbert space $\mathcal{X}$ is approximated by a sequence of finite-dimensional spaces $(\mathcal{X}^N)_{N\in\mathbb{N}}$, e.g., by spatial finite element approximations, leading to large sparse systems of ordinary differential equations in $\mathbb{R}^n$. Following the theory in [28] those approximations do not even have to be subspaces of the Hilbert space of solutions.

Before stating the main theoretical result we will first collect some approximation prerequisites we will need for the theorem. We call them (BK1) and (BK2) for they were already formulated in [28] (and called H1 and H2 there). In the following Π^N is the canonical projection operator mapping from the infinite-dimensional space $\mathcal{X}$ to its finite-dimensional approximation $\mathcal{X}^N$. The first and natural prerequisite is:

For each N and $x_0 \in \mathcal{X}^N$ there exists an admissible control $u^N \in L^2(0,\infty;\mathcal{U})$ and any admissible control drives the states to 0 asymptotically. (BK1)

Additionally one needs the following properties for the approximation as $N \to \infty$. Assume that for each N, A^N is the infinitesimal generator of a $\mathcal{C}^0$-semigroup $T^N(t)$, then we require:

(BK2)

(i) For all $\varphi \in \mathcal{X}$ we have uniform convergence $T^N(t)\Pi^N\varphi \to \mathbf{T}(t)\varphi$ on any bounded subinterval of $[0,\infty)$.

(ii) For all $\phi \in \mathcal{X}$ we have uniform convergence $T^N(t)^*\Pi^N\phi \to \mathbf{T}(t)^*\phi$ on any bounded subinterval of $[0,\infty)$.

(iii) For all $v \in \mathcal{U}$ we have $B^N v \to \mathbf{B}v$ and for all $\varphi \in \mathcal{X}$ we have ${B^N}^*\varphi \to \mathbf{B}^*\varphi$.

(iv) For all $\varphi \in \mathcal{X}$ we have $Q^N\Pi^N\varphi \to \mathbf{Q}\varphi$.

With these we can now formulate the main result.

Theorem 1 (Convergence of the finite-dimensional approximations). *Let* (BK1) *and* (BK2) *hold. Moreover, assume* $\mathbf{R} > 0$, $\mathbf{Q} \geq 0$ *and* $Q^N \geq 0$. *Further, let* P^N *be the solutions of the AREs for the finite-dimensional systems and let the minimal nonnegative self-adjoint solution* $\mathbf{P}$ *of* (8) *for (4), (5) and (7) exist. Moreover, let* $\mathbf{S}(t)$ *and* $S^N(t)$ *be the operator semigroups generated by* $\mathbf{A}-\mathbf{B}\mathbf{R}^{-1}\mathbf{B}^*\mathbf{P}$ *on* $\mathcal{X}$ *and* $A^N - B^N\mathbf{R}^{-1}{B^N}^*P^N$ *on* $\mathcal{X}^N$, *respectively, with* $\|\mathbf{S}(t)\varphi\| \to 0$ *as* $t \to \infty$ *for all* $\varphi \in \mathcal{X}$.

If there exist positive constants M_1, M_2 *and* ω *independent of* N *and* t, *such that*

$$\begin{aligned} \|S^N(t)\|_{\mathcal{X}^N} &\leq M_1 e^{-\omega t}, \\ \|P^N\|_{\mathcal{X}^N} &\leq M_2, \end{aligned} \tag{13}$$

then

$$P^N \Pi^N \varphi \rightarrow \mathbf{P}\varphi \qquad \textit{for all } \varphi \in \mathcal{X},$$
$$S^N(t)\Pi^N \varphi \rightarrow \mathbf{S}(t)\varphi \quad \textit{for all } \varphi \in \mathcal{X}, \tag{14}$$

converge uniformly in t on bounded subintervals of $[0, \infty)$ as $N \rightarrow \infty$ and

$$\|\mathbf{S}(t)\| \leq M_1 e^{-\omega t} \textit{ for } t \geq 0. \tag{15}$$

Theorem 1 gives the theoretical justification for the numerical method used for the linear problems described in this paper. It shows that the finite-dimensional closed-loop system obtained from optimizing the semidiscretized control problem indeed converges to the infinite-dimensional closed-loop system. Deriving a similar result for the nonlinear problem is an open problem.

The proof of Theorem 1 is given in [26]. It very closely follows that of [28, Theorem 2.2]. The only difference is the definition of the sesquilinear form on which the mechanism of the proof is based. It has an additional term in the boundary control case discussed here, but one can check that this term does not destroy the required properties of the sesquilinear form.

2.2 Tracking control

In contrast to stabilization problems, where one searches for a stabilizing feedback K (i.e. a feedback such that the closed loop operator $A - BK$ is stable), we are searching for a feedback which drives the state to a given reference trajectory asymptotically. Thus the tracking problem is in fact a stabilization problem for the deviation of the current state from the desired state. We will show in this section, that for linear operators A and B tracking can easily be incorporated into an existing solver for the stabilization problem with only a small computational overhead.

A common trick (see, e.g., [29]) to handle inhomogeneities in system theory for ODEs is the following. Given

$$\dot{x} = Ax + Bu + f,$$

let $\hat{x}$ be a solution of the uncontrolled system $\dot{x} = Ax + f$, such that $f = \dot{\hat{x}} - A\hat{x}$. Then

$$\dot{x} - \dot{\hat{x}} = Ax + Bu - A\hat{x}$$

from which we get a homogenous linear system

$$\dot{z} = Az + Bu \quad \text{where} \quad z = x - \hat{x}.$$

We want to apply this to the abstract Cauchy problem. Assume $(\tilde{\mathbf{x}}, \tilde{\mathbf{u}})$ is a reference pair solving

$$\dot{\tilde{\mathbf{x}}} = \mathbf{A}\tilde{\mathbf{x}} + \mathbf{B}\tilde{\mathbf{u}}.$$

We rewrite the tracking type control system as a stabilization problem for the difference $\mathbf{z} = \mathbf{x} - \tilde{\mathbf{x}}$ as follows:

$$\dot{\mathbf{z}} = \mathbf{A}\mathbf{z} + \mathbf{B}\mathbf{v} \quad \overset{\mathbf{v}=-\mathbf{K}\mathbf{z}}{\Leftrightarrow} \quad \dot{\mathbf{x}} = \mathbf{A}\mathbf{x} - \mathbf{B}\mathbf{K}\mathbf{x} + \dot{\tilde{\mathbf{x}}} - \mathbf{A}\tilde{\mathbf{x}} + \mathbf{B}\mathbf{K}\tilde{\mathbf{x}}. \tag{16}$$

So the only difference between the tracking type and stabilization problems is the *known* inhomogeneity $\mathbf{f} := \dot{\tilde{\mathbf{x}}} - (\mathbf{A} - \mathbf{B}\mathbf{K})\tilde{\mathbf{x}}$. Note that the operators do not change at all. That means we have to solve the same Riccati equation (8) in both cases, thus one only has to add the inhomogeneity $\mathbf{f}$ (which can be computed once and in advance directly after the feedback operator is obtained) to the solver for the closed loop system in the tracking type case provided that in the cost function (7) $\mathbf{y} = \mathbf{C}\mathbf{x}$ has been replaced by $\mathbf{C}(\mathbf{x} - \tilde{\mathbf{x}})$.

3 Computational methods and results

In this section we will discuss the computational methods used to achieve an efficient implementation for the numerical solution of the model problem. In Subsect. 3.1 we will first explain the algorithms used to solve the problem. There we will especially review the algorithm employed to solve the large sparse Riccati equation. We will focus on the case of infinite final time, where we have to deal with an algebraic Riccati equation (ARE), but we will also sketch the method used for a differential Riccati equation (DRE) in the finite final time case. After that we will briefly explain the concrete implementation in Subsection 3.2 and give an overview of the problems which may be solved with our implementation at the current stage. In the closing subsection we will present selected numerical results of our computations for tracking type control systems. Numerical experiments for the stabilization problem have already been published in [23–26].

3.1 Algorithmic details

The approach we present here admits two different implementations which can be seen as implementations of the well known horizontal and vertical methods of lines from numerical methods for partial differential equations. In the case of the vertical method of lines we use a finite element semidiscretization in space to set up the approximate finite-dimensional problems. This approximation is then used to formulate an LQR system for an ordinary differential equation. The LQR system for this ordinary differential equation is then solved by computing the feedback, retrieving the closed loop system and applying an ODE solver to the closed loop system. The case of the horizontal method of lines is very similar to the algorithm used when solving the PDE forward problem. We only have to introduce a step computing the feedback operator and a step which updates the boundary conditions according to this operator for the boundary control system.

So in both cases we need to compute the feedback operator for the approximate finite-dimensional systems. As we use finite element approximations here we have to deal with matrices of dimension larger than 1000 which makes it infeasible to use classical methods for the solution of the Riccati equations as these are of cubic complexity. In the late 90's Li and Penzl [30, 31] independently proposed a method for the efficient solution of large sparse Lyapunov equations. These methods are based on earlier work of Wachspress [32] on the application of an ADI-like method exploiting sparsity and the oftenly encountered very low numerical rank of their solutions.

The method developed by Li and Penzl can also be used to solve the large sparse Riccati equations appearing in this approach, since the Fréchet derivative of the Riccati operator is a Lyapunov operator. Thus we can apply Newton's method to the nonlinear matrix Riccati equations and in each step solve the Lyapunov equations efficiently by the ADI approach.

For the finite final time, it is shown in [33] that backward differentiation (BDF) methods can be combined with the above Newton-ADI-method to solve the differential Riccati equation (10) efficiently.

Low rank Cholesky factor ADI Newton method

We will consider a system of the form (4), (5) here to characterize the low rank Cholesky factor ADI Newton method. It is sufficient to consider this case, because a finite-dimensional system of the form

$$\begin{aligned} M\dot{x}(t) &= Ax(t) + Bu(t) \\ y(t) &= Cx(t) \\ x(0) &= x_0 \end{aligned} \tag{17}$$

can easily be transformed into the representation (4), (5) by the following procedure. First split the matrix M into $M = M_L M_U$ (where $M_L = M_U^T$ in the symmetric positive semidefinite case) and define

$$z(t) := M_U x(t).$$

Then

$$\dot{z}(t) = \dot{M_U} x(t) + M_U \dot{x}(t) = M_U \dot{x}(t)$$

and by defining

$$\begin{aligned} \tilde{A} &:= M_L^{-1} A M_U^{-1}, \\ \tilde{B} &:= M_L^{-1} B, \\ \tilde{C} &:= C M_U^{-1}, \end{aligned}$$

we can rewrite the system in the form (4), (5). The mass matrix from the finite element semidiscretization of the heat equation is always symmetric and positive definite and thus we can always apply the above procedure to the finite-dimensional systems. There also exists a method which avoids decompositions of M by rewriting the linear systems of equation arising inside the

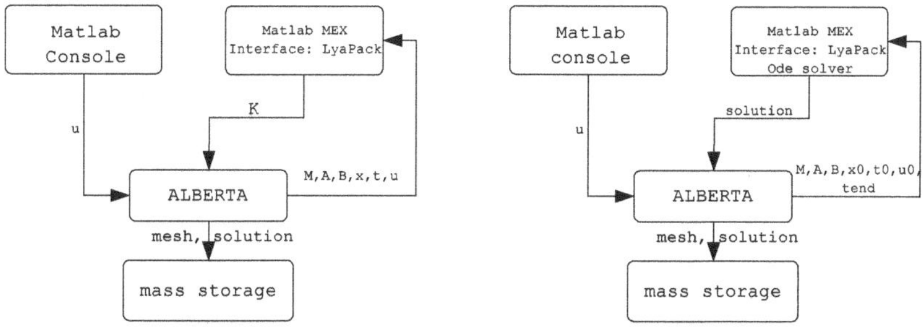

Fig. 2. Data flow in the horizontal (left) and vertical (right) method of lines implementations

ADI method instead of the control system (see [34] for details). We discussed the above method in detail here, because it is used in the LyaPack software package used to solve the Riccati equations in our implementation.

The low rank Cholesky factor ADI Newton method implemented in LyaPack can be seen as a modification of the classical Newton-like method for algebraic Riccati equations (Kleinman iteration, [35]), where the Lyapunov subproblems in each Newton step are solved by the low rank Cholesky factor ADI method. The most important feature of this method is that it does not work with the generally full $n \times n$ iterates X_i but with low rank Cholesky factors thereof ($Z_i Z_i^T = X_i$). The rank of Z_i is generally full but it has $r_i \ll n$ columns which drastically reduces memory consumption and computational complexity. LyaPack also provides functions iterating directly on the rectangular (number of rows = number of system inputs $\ll n$) feedback matrix possibly reducing the complexity even further.

3.2 Implementation details

For the implementation we combine the software packages LyaPack[1] (see [36]) and ALBERTA[2] (see [37]). LyaPack is a collection of MATLAB-routines for solving large-scale Lyapunov equations, LQR problems, and model reduction tasks. Therefore we used MATLAB to initialize the computation. That means we setup the initial control parameters and the time measurement routines at a MATLAB-console. After that the finite element discretization is generated by a mex call to a C-function utilizing the finite element method (fem) library ALBERTA. Inside this routine the system matrices M, A and B are assembled. After system assembly the program returns to the MATLAB prompt providing the matrices and problem data like current time and temperature profile. Now the matrices C, Q, and R are initialized and LyaPack is used to compute the feedback matrix.

[1]available from: `http://www.tu-chemnitz.de/`

[2]available from: `http://www.alberta-fem.de`

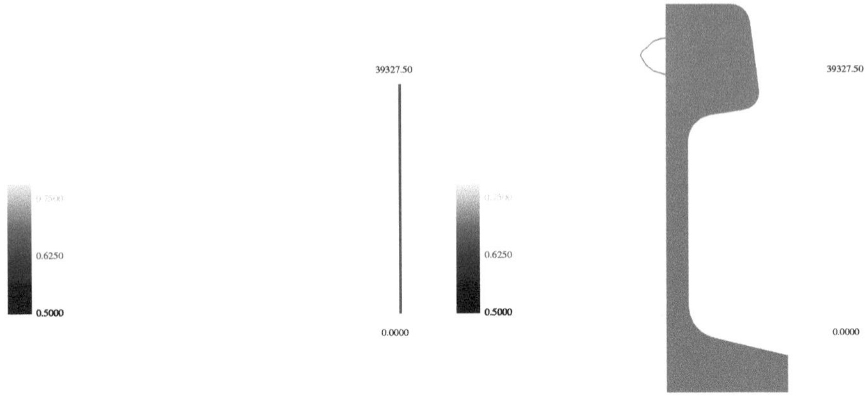

Fig. 3. Initial (left) and final (right) temperature distribution for a tracking type system steering to constant 700°C

In case of the horizontal method of lines the program now returns to the ALBERTA subroutine providing the feedback and with it the new control parameters for the boundary conditions. With these it continues the standard forward computation updating the boundary conditions by use of the feedback matrix.

In case of the vertical method of lines the feedback is used to generate the closed-loop system. This is then solved with a standard ODE-solver using MATLAB. After that the program uses the ALBERTA function again to store the solution on a mass storage device for visualization and post processing tasks in the same format used in the above case.

Both implementations have their advantages. The horizontal method of lines implementation can use operator information for the selection of (time) stepsizes and can easily be generalized to what might be called an adaptive LQR system, where the idea is to stabilize a system with nonlinear PDE constraints by systems for local (in time) linearizations of the constraints. This method is similar to receding horizon- or model predictive control techniques which will be addressed in Sect. 4. On the other hand the vertical method of lines implementation is easier to generalize to tracking-type control systems (see Sect. 2.2).

To close this section on the implementation details we give a table showing what our current implementation is capable of computing.

	$x_{ext,k}$ as control in (12)	$\frac{\kappa_k}{\lambda}$ as control in (12)	tracking
vertical m.o.l.	X	+	X
linear horiz. m.o.l.	X	+	–
nonlin. horiz. m.o.l.	O	+	–

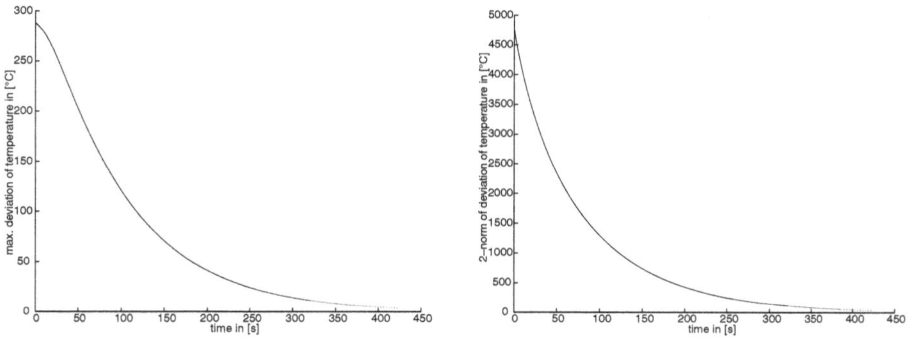

Fig. 4. Deviation of temperature from the reference (700°C) in maximum norm (left) and Euclidian norm (right)

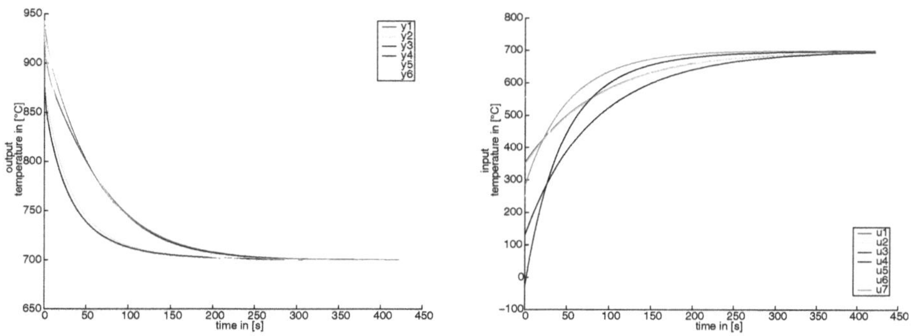

Fig. 5. Evolution of temperature at the outputs (left) and control inputs (right)

In the tabular an X denotes a fully supported feature. That means the feature is implemented and there exists a rigorous theory for this approach. An O denotes a feature which is fully implemented but the theoretical backing is not complete. The features marked '+' already give promising results although they are not covered by the theory. So under slight changes in the implementation (e.g. a posteriori error estimates) they might become fully valid and theoretically confirmed in future research.

3.3 Numerical results

We will now present an example of a tracking control system. We want to control the state (temperature distribution) to constant 700°C. For the particular problem considered here one also knows the reference control which has to be applied to stay at this state. From (12) it is easy to see that we have to introduce an exterior temperature (cooling temperature) of 700°C, because (12) then becomes an isolation boundary condition.

It is in general not necessary to know $\tilde{u}$ for this approach. We have seen in Sect. 2.2 that we only need to know that there exists such a control. We do not have to know the reference control itself for the computations, because to calculate the inhomogeneity f we only use $\tilde{x}$ and its derivative with respect to time. On the other hand we might need $\tilde{u}$ to regain the real control u from the artificial control v.

We start the calculation with the same initial temperature (see Fig. 3 on the left) distribution we already used e.g. in [26]. The computational time horizon is equivalent to approximately 7 minutes of real time. The time-bars in Fig. 3 have to be scaled down by a factor of 100 (see [26] for details on the scalings) to read real time in seconds. The temperatures are scaled such that 1.0 is equivalent to 1000°C. The isolines in Fig. 3 are plotted at a distance of 15°C. Thus from Fig. 3 we can conclude that the maximum deviation of temperatures from 700°C is smaller than 15°C. In fact it is only 3.84°C after approximately 7 minutes. The average deviation at that time is already at about 0.186°C (also compare Fig. 4). Figure 5 shows the evolution of the control parameters (i.e., temperatures of the cooling fluid) on the right. The plotted values represent the real control temperatures one would have to apply. In comparison to $v = u - \tilde{u}$ in (16) we already added the reference control $\tilde{u}$ of 700°C to the computed values of v.

4 Nonlinear parabolic systems

Nonlinear problems will arise if the system equation or the boundary conditions are nonlinear. In the following we consider the minimization problem

$$\min_{\mathbf{u}\in L^2(0,T_f;\mathcal{U})} \int_0^{T_f} \mathbf{g}(\mathbf{y}(t), \mathbf{u}(t))\, dt, \quad 0 < T_f \leq \infty, \tag{18}$$

subject to the semilinear equation

$$\dot{\mathbf{x}} = \mathbf{f}(\mathbf{x}) + \mathbf{B}\mathbf{u}(t), \qquad t \in [0, T_f], \quad \mathbf{x}(0) = \mathbf{x}_0. \tag{19}$$

The idea of receding horizon control (RHC) or model predictive control (MPC) is to decompose the time interval $[0, T_f]$ in (19) in smaller subintervals $[T_i, T_{i+1}]$ with

$$0 = T_0 < T_1 < T_2 < \ldots < T_{\ell-1} < T_\ell = T_f$$

and

$$T \geq \max\{T_{i+1} - T_i \mid i = 0, 1, \ldots, \ell - 1\}$$

for given T. Now we have to solve optimal control problems on the time frames $[T_i, T_i + T]$ successively, that is we replace (18) and (19) by

$$\min_{\mathbf{u}\in L^2(T_i,T_i+T;\mathcal{U})} \int_{T_i}^{T_i+T} \mathbf{g}(\mathbf{y}(t), \mathbf{u}(t))\, dt + \mathbf{g}_f(\mathbf{x}(T_i + T)) \tag{20}$$

subject to the dynamical semilinear system

$$\dot{\mathbf{x}}(t) = \mathbf{f}(\mathbf{x}(t)) + \mathbf{B}\mathbf{u}(t) \quad \text{for } t \geq T_i, \quad \mathbf{u}(t) \in \mathcal{U}, \tag{21}$$

and the initial condition

$$\mathbf{x}(T_i) = \mathbf{x}_*(T_i). \tag{22}$$

Here $\mathbf{u}_*$ is the optimal control and $\mathbf{x}_*$ the optimal trajectory for the optimal control problem on $[T_{i-1}, T_{i-1} + T]$. The second term $\mathbf{g}_f(\mathbf{x}(T_i + T))$ in the cost functional (20) is called *terminal cost* and penalizes the states at the end of the finite horizon. It is required to establish the asymptotic stabilization property of the MPC scheme. To obtain the approximated optimal control on $[0, T_f]$ we have to compose the optimal controls on the subintervals $[T_i, T_{i+1}]$.

The strategy of MPC/RHC is used successfully in particular for control problems with ordinary differential equations, e.g. [38,39]. The literature also provides research into partial differential equations, see [10,40–42], where different techniques are used for solving the subproblems (20)–(22).

We want to present a tracking approach which was introduced in [7,43,44]. In [7] the authors use a linear quadratic Gaussian (LQG) design which allows to include noise and observers into the model. Therefor we consider equation (19) as a nonlinear stochastic system

$$\dot{\mathbf{x}} = \mathbf{f}(\mathbf{x}) + \mathbf{B}\mathbf{u}(t) + \mathbf{d}(t), \qquad t \in [0, T_f], \quad \mathbf{x}(0) = \mathbf{x}_0, \tag{23}$$

where $\mathbf{d}(t)$ is an unknown Gaussian disturbance process. The observation process

$$\mathbf{y}(t) = \mathbf{C}\mathbf{x}(t) + \mathbf{n}(t)$$

provides partial observations of the state $\mathbf{x}(t)$, where $\mathbf{n}(t)$ is a measurement noise process.

Now we consider the time frame $[T_i, T_i + T]$ and define an operating point, for example

$$\bar{\mathbf{x}} = \frac{1}{T} \int_{T_i}^{T_i+T} \bar{\mathbf{x}}^*(t) dt \quad \text{or} \quad \bar{\mathbf{x}} = \bar{\mathbf{x}}^*(T_i + T), \tag{24}$$

where $(\bar{\mathbf{u}}^*, \bar{\mathbf{x}}^*)$ is the reference pair on $[T_i, T_i + T]$ which is known from applications or results from an open-loop computation. If we linearize $\mathbf{f}(\mathbf{x})$ at $\bar{\mathbf{x}}$, we will obtain the following linear optimal control problem on $[T_i, T_i + T]$:

$$\min_{\tilde{\mathbf{u}} \in L^2(T_i, T_i+T; \mathcal{U})} \frac{1}{2} \int_{T_i}^{T_i+T} \langle \mathbf{z}, \mathbf{Q}\mathbf{z} \rangle_{\mathcal{Y}} + \langle \tilde{\mathbf{u}}, \mathbf{R}\tilde{\mathbf{u}} \rangle_{\mathcal{U}} \, dt + \mathbf{g}_f(\mathbf{x}(T_i + T))$$

subject to

$$\begin{aligned} \dot{\mathbf{z}}(t) &= \mathbf{A}\mathbf{z}(t) + \mathbf{B}\tilde{\mathbf{u}}(t) + \mathbf{d}(t), \quad \mathbf{z}(0) = \eta_0, \\ \tilde{\mathbf{y}}(t) &= \mathbf{C}\mathbf{z}(t) + \mathbf{n}(t) \end{aligned}$$

where

$$\mathbf{z}(t) := \mathbf{x}(t) - \bar{\mathbf{x}}^*(t), \qquad \tilde{\mathbf{u}}(t) := \mathbf{u}(t) - \bar{\mathbf{u}}^*(t)$$

and

$$\mathbf{A} := \frac{d}{d\mathbf{x}}\mathbf{f}(\bar{\mathbf{x}}).$$

It can be shown, see e.g. [39], that if the terminal cost $\mathbf{g}_f$ bounds the infinite horizon cost for the nonlinear system (starting from $T_i + T$), the cost function to be minimized is an upper bound for

$$\min_{\tilde{\mathbf{u}} \in L^2(T_i,\infty;\mathcal{U})} \frac{1}{2} \int_{T_i}^{\infty} \langle \mathbf{z}, \mathbf{Q}\mathbf{z} \rangle_{\mathcal{Y}} + \langle \tilde{\mathbf{u}}, \mathbf{R}\tilde{\mathbf{u}} \rangle_{\mathcal{U}} \, dt. \tag{25}$$

So we can consider the infinite time case on every time frame. After application of the minimum principle we have to solve the algebraic Riccati equation (8), which corresponds to the LQR problem, as well as the dual equation

$$\tilde{\Re}(\mathbf{W}) := \mathbf{A}\mathbf{W} + \mathbf{W}\mathbf{A}^* - \mathbf{W}\mathbf{C}^*\mathbf{C}\mathbf{W} + \mathbf{S} = 0 \tag{26}$$

for the state estimation by using a Kalman filter, where $\mathbf{S}$ is an appropriate positive semidefinite operator. The best estimate $\hat{\mathbf{x}}$ can be obtained by solving the so called compensator equation

$$\begin{aligned}
\dot{\hat{\mathbf{x}}}(t) &= \mathbf{A}(\bar{\mathbf{x}})(\hat{\mathbf{x}}(t) - \bar{\mathbf{x}}^*(t)) + \mathbf{f}(\bar{\mathbf{x}}^*(t)) + \mathbf{B}\mathbf{u}(t) + \mathbf{W}_*\mathbf{C}^*(y(t) - \mathbf{C}\hat{\mathbf{x}}(t)), \\
\hat{\mathbf{x}}(0) &= \mathbf{x}_0 + \eta_0,
\end{aligned}$$

where $\mathbf{W}_*$ is the positive semidefinite solution to (26). The associated feedback law is now given as

$$\mathbf{u}(t) = \mathbf{u}^*(t) - \mathbf{R}^{-1}\mathbf{B}^*\mathbf{P}_*(\hat{\mathbf{x}}(t) - \bar{\mathbf{x}}^*(t))$$

where $\mathbf{P}_*$ is the positive semidefinite solution to (8).

Now we have computed the solution on the time frame $[T_i, T_{i+1}]$. In the next step we determine the solution on $[T_{i+1}, T_{i+2}]$ by repeating the procedure above, that is linearization, solving the two dual algebraic Riccati equations and determination of the optimal control. Since the current horizon is moving forward this strategy is called *receding horizon control* or *moving horizon control*.

The numerical implementation for such problems is similar to that described in Sect. 3. We need efficient algorithms to solve the two Riccati equations and the ordinary differential equation in every time frame. Details can be found in Sects. 3.1 and 3.2.

It is also possible to linearize along a given (time-dependent) reference trajectory instead of using a constant operating point as in (24). Then we have to solve two differential Riccati equations which have the following form

$$\dot{\mathbf{P}}(t) + \mathbf{A}(t)^*\mathbf{P}(t) + \mathbf{P}(t)\mathbf{A}(t) - \mathbf{P}(t)\mathbf{B}\mathbf{R}^{-1}\mathbf{B}^*\mathbf{P}(t) + \mathbf{C}^*\mathbf{Q}\mathbf{C} = 0, \quad (27)$$
$$\dot{\mathbf{W}}(t) + \mathbf{A}(t)\mathbf{W}(t) + \mathbf{W}(t)\mathbf{A}(t)^* - \mathbf{W}(t)\mathbf{C}^*\mathbf{C}\mathbf{W}(t) + \mathbf{S} = 0, \quad (28)$$

where $\mathbf{A}(t) \equiv \mathbf{A}(\bar{\mathbf{x}}^*(t))$. For the numerical solution of the differential Riccati equation we refer to [33].

The numerical implementation of this approach for nonlinear parabolic-type problems such as semilinear and quasilinear heat, convection-diffusion, and reaction-diffusion equations is under current investigation. For solving the sub-problems on each time frame, we make intensive use of the algorithms developed for the linear case discussed in Sect. 3.

Acknowledgments

The authors wish to thank Hermann Mena from Escuela Politécnica Nacional Quito (Ecuador) for his contributions in the finite final time case.

References

1. I. Lasiecka and R. Triggiani. *Differential and Algebraic Riccati Equations with Application to Boundary/Point Control Problems: Continuous Theory and Approximation Theory.* Number 164 in Lecture Notes in Control and Information Sciences. Springer-Verlag, Berlin, 1991.
2. I. Lasiecka and R. Triggiani. *Control Theory for Partial Differential Equations: Continuous and Approximation Theories I. Abstract Parabolic Systems.* Cambridge University Press, Cambridge, UK, 2000.
3. I. Lasiecka and R. Triggiani. Control theory for partial differential equations: Continuous and approximation theories II. Abstract hyperbolic-like systems over a finite time horizon. In *Encyclopedia of Mathematics and its Applications*, volume 75, pages 645–1067. Cambridge University Press, Cambridge, 2000.
4. J.L. Lions. *Optimal Control of Systems Governed by Partial Differential Equations.* Springer-Verlag, Berlin, FRG, 1971.
5. A. Bensoussan, G. Da Prato, M.C. Delfour, and S.K. Mitter. *Representation and Control of Infinite Dimensional Systems, Volume II.* Systems & Control: Foundations & Applications. Birkäuser, Boston, Basel, Berlin, 1992.
6. A.V. Balakrishnan. Boundary control of parabolic equations: L-Q-R theory. In *Theory of nonlinear operators, Proc. 5th int. Summer Sch.*, number 6N in Abh. Akad. Wiss. DDR 1978, pages 11–23, Berlin, 1977. Akademie-Verlag.
7. K. Ito and K. Kunisch. Receding horizon control with incomplete observations. Preprint, October 2003.
8. F. Tröltzsch. *Optimale Steuerung partieller Differentialgleichungen - Theorie, Verfahren und Anwendungen.* Vieweg, Wiesbaden, 2005. In German.
9. A. Bensoussan, G. Da Prato, M.C. Delfour, and S.K. Mitter. *Representation and Control of Infinite Dimensional Systems, Volume I.* Systems & Control: Foundations & Applications. Birkäuser, Boston, Basel, Berlin, 1992.
10. M. Hinze. *Optimal and instantaneous control of the instationary Navier-Stokes equations.* Habilitationsschrift, TU Berlin, Fachbereich Mathematik, 2002.

11. K.-H. Hoffmann, I. Lasiecka, G. Leugering, J. Sprekels, and F. Tröltzsch, editors. *Optimal control of complex structures. Proceedings of the international conference, Oberwolfach, Germany, June 4–10, 2000*, volume 139 of *ISNM, International Series of Numerical Mathematics*. Birkhäuser, Basel, Switzerland, 2002.
12. K.-H. Hoffmann, G. Leugering, and F. Tröltzsch, editors. *Optimal control of partial differential equations. Proceedings of the IFIP WG 7.2 international conference, Chemnitz, Germany, April 20–25, 1998*, volume 133 of *ISNM, International Series of Numerical Mathematics*. Birkhäuser, Basel, Switzerland, 1999.
13. E.D. Sontag. *Mathematical Control Theory*. Springer-Verlag, New York, NY, 2nd edition, 1998.
14. R.F. Curtain and T. Pritchard. *Infinite Dimensional Linear System Theory*, volume 8 of *Lecture Notes in Control and Information Sciences*. Springer-Verlag, New York, 1978.
15. R.F. Curtain and H. Zwart. *An Introduction to Infinite-Dimensional Linear Systems Theory*, volume 21 of *Texts in Applied Mathematics*. Springer-Verlag, New York, 1995.
16. H.T. Banks, editor. *Control and Estimation in Distributed Parameter Systems*, volume 11 of *Frontiers in Applied Mathematics*. SIAM, Philadelphia, PA, 1992.
17. J. Zabczyk. Remarks on the algebraic Riccati equation. *Appl. Math. Optim.*, 2:251–258, 1976.
18. J.S. Gibson. The Riccati integral equation for optimal control problems in Hilbert spaces. *SIAM J. Cont. Optim.*, 17:537–565, 1979.
19. L.T. Biegler, O. Ghattas, M. Heinkenschloß and B. van Bloemen Waanders, editors. *Large-Scale PDE-Constrained Optimization*, volume 30 of *Lecture Notes in Computational Science and Engineering*. Springer-Verlag, Berlin, 2003.
20. F. Tröltzsch and A. Unger. Fast solution of optimal control problems in the selective cooling of steel. *Z. Angew. Math. Mech.*, 81:447–456, 2001.
21. K. Eppler and F. Tröltzsch. Discrete and continuous optimal control strategies in the selective cooling of steel profiles. *Z. Angew. Math. Mech.*, 81(2):247–248, 2001.
22. R. Krengel, R. Standke, F. Tröltzsch, and H. Wehage. Mathematisches Modell einer optimal gesteuerten Abkühlung von Profilstählen in Kühlstrecken. Preprint 98-6, Fakultät für Mathematik TU Chemnitz, November 1997.
23. J. Saak. Effiziente numerische Lösung eines Optimalsteuerungsproblems für die Abkühlung von Stahlprofilen. Diplomarbeit, Fachbereich 3/Mathematik und Informatik, Universität Bremen, D-28334 Bremen, September 2003.
24. P. Benner and J. Saak. Efficient numerical solution of the LQR-problem for the heat equation. *Proc. Appl. Math. Mech.*, 4(1):648–649, 2004.
25. P. Benner and J. Saak. A semi-discretized heat transfer model for optimal cooling of steel profiles. In P. Benner, V. Mehrmann, and D. Sorensen, editors, *Dimension Reduction of Large-Scale Systems*, Lecture Notes in Computational Science and Engineering, pages 353–356. Springer-Verlag, Berlin/Heidelberg, Germany, 2005.
26. P. Benner and J. Saak. Linear-quadratic regulator design for optimal cooling of steel profiles. Technical Report SFB393/05-05, Sonderforschungsbereich 393 *Parallele Numerische Simulation für Physik und Kontinuumsmechanik*, TU Chemnitz, D-09107 Chemnitz (Germany), 2005. Available from `http://www.tu-chemnitz.de/sfb393/sfb05pr.html`.

27. K. Eppler and F. Tröltzsch. Discrete and continuous optimal control strategies in the selective cooling of steel profiles. Preprint 01-3, DFG Schwerpunktprogramm *Echtzeit-Optimierung groSSer Systeme*, 2001. Available from `http://www.zib.de/dfg-echtzeit/Publikationen/Preprints/Preprint-01-3.html`.
28. H.T. Banks and K. Kunisch. The linear regulator problem for parabolic systems. *SIAM J. Cont. Optim.*, 22:684–698, 1984.
29. S.K. Godunov. *Ordinary Differential Equations with Constant Coefficient.*, volume 169 of *Translations of Mathematical Monographs*. American Mathematical Society, 1997.
30. P. Benner, J.-R. Li, and T. Penzl. Numerical solution of large Lyapunov equations, Riccati equations, and linear-quadratic control problems. Unpublished manuscript, 2000.
31. T. Penzl. A cyclic low rank Smith method for large sparse Lyapunov equations. *SIAM J. Sci. Comput.*, 21(4):1401–1418, 2000.
32. E.L. Wachspress. Iterative solution of the Lyapunov matrix equation. *Appl. Math. Letters*, 107:87–90, 1988.
33. P. Benner and H. Mena. BDF methods for large-scale differential Riccati equations. In B. De Moor, B. Motmans, J. Willems, P. Van Dooren, and V. Blondel, editors, *Proc. of Mathematical Theory of Network and Systems, MTNS 2004*, 2004.
34. P. Benner. Solving large-scale control problems. *IEEE Control Systems Magazine*, 14(1):44–59, 2004.
35. D.L. Kleinman. On an iterative technique for Riccati equation computations. *IEEE Trans. Automat. Control*, AC-13:114–115, 1968.
36. T. Penzl. LYAPACK Users Guide. Technical Report SFB393/00-33, Sonderforschungsbereich 393 *Numerische Simulation auf massiv parallelen Rechnern*, TU Chemnitz, 09107 Chemnitz, FRG, 2000. Available from `http://www.tu-chemnitz.de/sfb393/sfb00pr.html`.
37. A. Schmidt and K. Siebert. *Design of Adaptive Finite Element Software; The Finite Element Toolbox ALBERTA*, volume 42 of *Lecture Notes in Computational Science and Engineering*. Springer-Verlag, Berlin/Heidelberg, 2005.
38. F. Allgöwer, T.Badgwell, J. Qin, J. Rawlings, and S. Wright. Nonlinear predictive control and moving horizon estimation – an introductory overview. In P. Frank, editor, *Advances in Control*, pages 391–449. Springer-Verlag, Berlin/Heidelberg, 1999.
39. C.-H. Chen. A quasi-infinite horizon nonlinear model predictive control scheme with guaranteed stability. *Automatica*, 34:1205–1217, 1998.
40. M. Böhm, M.A. Demetriou, S. Reich, and I.G. Rosen. Model reference adaptive control of distributed parameter systems. *SIAM J. Cont. Optim.*, 36(1):33–81, 1998.
41. H. Choi, M. Hinze, and K. Kunisch. Instantaneous control of backward-facing step flows. *Appl. Numer. Math.*, 31(2):133–158, 1999.
42. M. Hinze and S. Volkwein. Analysis of instantaneous control for the Burgers equation. *Optim. Methods Softw.*, 18(3):299–315, 2002.
43. K. Ito and K. Kunisch. On asymptotic properties of receding horizon optimal control. *SIAM J. Cont. Optim.*, 40:1455–1472, 2001.
44. K. Ito and K. Kunisch. Receding horizon optimal control for infinite dimensional systems. *ESAIM: Control Optim. Calc. Var.*, 8:741–760, 2002.

Part III

Applications

Parallel Simulations of Phase Transitions in Disordered Many-Particle Systems

Thomas Vojta

Department of Physics, University of Missouri–Rolla
Rolla, MO 65401, USA
vojtat@umr.edu

1 Introduction

Phase transitions in classical and quantum many-particle systems are one of the most fascinating topics in contemporary condensed matter physics. The strong fluctuations associated with a phase transition or critical point can lead to unusual behavior and to novel, exotic phases in its vicinity, with consequences for problems such as quantum magnetism, unconventional superconductivity, non-Fermi liquid physics, and glassy behavior in doped semiconductors. In many realistic systems, impurities, dislocations, and other forms of quenched disorder play an important role. They often modify or even enhance the effects of the critical fluctuations. An interesting, if intricate, aspect of phase transitions with quenched disorder are the rare regions. These are large spatial regions that, due to a strong disorder fluctuation, are either devoid of impurities or have stronger interactions than the bulk system. They can be locally in one of the phases even though the bulk system may be in another phase. The slow fluctuations of these rare regions can dominate the behavior of the entire system.

Conventional theoretical approaches to many-particle systems such as diagrammatic perturbation theory and the perturbative renormalization group are not particularly well suited for strongly disordered systems. In particular, they cannot easily account for the rare regions which are nonperturbative degrees of freedom because their probability is exponentially small in their volume. For this reason, much of our understanding in this area has been achieved by large-scale computer simulations. However, computational studies of disordered many-particle systems come with their own challenges. In addition to the intrinsic problem of dealing with a macroscopic number of interacting particles, the presence of impurities and defects requires studying many samples to determine averages and distribution functions of observable quantities.

Here, we describe large-scale parallel computer simulation studies of several classical, quantum, and nonequilibrium phase transitions with quenched disorder. Examples include classical Ising models with extended defects, disordered quantum magnets, and phase transitions in nonequilibrium spreading processes. We demonstrate that disorder can have drastic effects on these transitions ranging from Griffiths singularities in the disordered phase of a classical magnet to a complete destruction of the phase transition by smearing if the disorder is sufficiently correlated in space or (imaginary) time. We classify these phase transition scenarios based on the effective dimensionality of the rare regions. This chapter is organized as follows. In Sect. 2, we introduce the main concepts and develop a classification of disorder effects. Computational results for several phase transitions are presented Sects. 3, 4, and 5. Sect. 6 is devoted to details of our computational approach and specifics of our implementations. We conclude in Sect. 7.

2 Phase transitions, disorder, and rare regions

In this section, we give a brief introduction into the effects of impurities and defects on phase transitions and critical points. For definiteness, we restrict ourselves to simple order-disorder transitions between conventional phases such as the transition between the paramagnetic (magnetically disordered) and ferromagnetic (magnetically ordered) states of a magnetic material.[1] We consider impurities and defects which introduce spatial variations in the coupling strength (i.e., spatial variations in the tendency towards the ordered phase) but no frustration or random external fields. This type of impurities is sometimes referred to as weak disorder, random-T_c disorder, or, from analogy with quantum field theory, random-mass disorder. The two main questions to be addressed are: (i) Will the phase transition remain sharp in the presence of disorder? and (ii) If a sharp transition survives, will the critical behavior change in response to the disorder?

2.1 Average disorder and Harris criterion

The question of how quenched disorder influences phase transitions has a long history. Initially it was suspected that disorder destroys any critical point because in the presence of defects, the system divides itself up into spatial regions which independently undergo the phase transition at different temperatures (see Ref. [22] and references therein). However, subsequently it became clear that generically a phase transition remains sharp in the presence of defects, at least for classical systems with short-range disorder correlations.

The fate of a particular clean critical point under the influence of impurities is controlled by the Harris criterion [24]: If the correlation length critical

---

[1]Note that *disorder* has two meanings in this field: On the one hand, in *disordered phase*, it denotes the non-symmetry-broken (paramagnetic) phase even in the absence of impurities. On the other hand, in *quenched disorder* it refers to impurities and defects. Unfortunately, this double meaning is well established in the literature.

exponent ν fulfills the inequality $\nu \geq 2/d$ where d is the spatial dimensionality, the critical point is (perturbatively) stable, and quenched disorder should not affect its critical behavior. At these transitions, the disorder strength *decreases* under coarse graining, and the system becomes asymptotically homogeneous at large length scales. Consequently, the critical behavior of the dirty system is identical to that of the clean system. Technically, this means the disorder is renormalization group irrelevant, and the clean renormalization group fixed point is stable. In this class of systems, the macroscopic observables are self-averaging at the critical point, i.e., the relative width of their probability distributions vanishes in the thermodynamic limit [1, 65]. A prototypical example in this class is the three-dimensional classical Heisenberg model whose clean correlation length exponent is $\nu \approx 0.698$ (see, e.g., [29]), fulfilling the Harris criterion.

If the Harris criterion is violated, the clean critical point is destabilized, and the behavior must change. Nonetheless, a sharp transition can still exist. Depending on the behavior of the average disorder strength under coarse graining, one can distinguish two classes. In the first class, the system remains inhomogeneous at all length scales with the relative strength of the inhomogeneities approaching a finite value for large length scales. The resulting critical point still displays conventional power-law scaling but with new critical exponents which differ from those of the clean system (and fulfill the Harris criterion). These transitions are controlled by renormalization group fixed points with a nonzero value of the disorder strength. Macroscopic observables are not self-averaging, but the relative width of their probability distributions approaches a size-independent constant [1, 65]. An example in this class is the classical three-dimensional Ising model. Its clean correlation length exponent, $\nu \approx 0.627$ (see, e.g. [18]) does not fulfill the Harris criterion. Introduction of quenched disorder, e.g., via dilution, thus leads to a new critical point with an exponent of $\nu \approx 0.684$ [3].

At critical points in the last class, the relative magnitude of the inhomogeneities *increases* without limit under coarse graining. The corresponding renormalization group fixed points are characterized by infinite disorder strength. At these infinite-randomness critical points, the power-law scaling is replaced by activated (exponential) scaling. The probability distributions of macroscopic variables become very broad (even on a logarithmic scale) with the width diverging with system size. Consequently, averages are often dominated by rare events, e.g., spatial regions with atypical disorder configurations. This type of behavior was first found in the McCoy-Wu model, a two-dimensional Ising model with bond disorder perfectly correlated in one dimension [42]. However, it was fully understood only when Fisher [15, 17] solved the one-dimensional random transverse field Ising model by a version of the Ma-Dasgupta-Hu real space renormalization group [37]. Since then, several infinite-randomness critical points have been identified, mainly at quantum phase transitions since the disorder, being perfectly correlated in (imaginary) time, has a stronger effect on quantum phase transitions than on thermal ones. Examples include one-dimensional random quantum spin chains as well as one-dimensional and two-dimensional random quantum Ising models [5,16,39,46,67].

2.2 Rare regions and Griffiths effects

In the last subsection we have discussed scaling scenarios for phase transitions with quenched disorder based on the *global, i.e., average,* behavior of the disorder strength under coarse graining. In this subsection, we focus on the effects of *rare* strong spatial disorder fluctuations. Such fluctuations can lead to very interesting non-perturbative effects not only directly at the phase transition but also in its vicinity.

In a quenched disordered system, the presence of random impurities and defects leads to a spatial variation of the coupling strength. As a result, there can be rare large spatial regions that are locally in the ordered phase even though the bulk system is in the disordered phase. The dynamics of such rare regions is very slow because flipping them requires a coherent change of the order parameter over a large volume. Griffiths [21] was the first to show that these rare regions can lead to singularities in the free energy in an entire parameter region near the critical point which is now known as the Griffiths region or the Griffiths phase [47].

Recently, a general classification of rare region effects at order-disorder phase transitions in systems with weak, random mass type, disorder [63] has been suggested. This classification can be understood as follows. The probability w for finding a large spatial region of linear size L devoid of impurities is exponentially small in its volume, $w \sim \exp(-cL^d)$. The importance of the rare regions now depends on how rapidly the contribution of a *single* region to observables increases with its size. Three cases can be distinguished.

(i) If the effective dimensionality $d_{\rm RR}$ of the rare regions is *below* the lower critical dimensionality d_c^- of the problem, their energy gap depends on their size via a power law, $\epsilon_L \sim L^{-\psi}$. Thus, the contribution of a rare region to observables can at most grow as a power of its size. As a result, the low-energy density of states due to the rare regions is exponentially small. This leads to exponentially weak rare region effects characterized by an essential singularity in the free energy [4,21]. The leading scaling behavior *at* the dirty critical point is of conventional power-law type. To the best of our knowledge, these exponentially weak thermodynamic Griffiths singularities have not yet been observed in experiments. In contrast, the long-time *dynamics* is dominated by the rare regions. Inside the Griffiths phase, the spin autocorrelation function $C(t)$ decays as $\ln C(t) \sim -(\ln t)^{d/(d-1)}$ for Ising systems [6,8,13,47] and as $\ln C(t) \sim -t^{1/2}$ for Heisenberg systems [7,8]. Examples for the case of exponentially weak Griffiths effects can be found in generic classical equilibrium systems (where the rare regions are finite in all directions and thus effectively zero-dimensional).

Table 1. Classification of rare region (RR) effects at critical points in the presence of weak quenched disorder according to the effective dimensionality d_{RR} of the rare regions

RR dimension	Griffiths effects	Dirty critical point	Critical scaling
$d_{\mathrm{RR}} < d_c^-$	weak exponential	conventional	power law
$d_{\mathrm{RR}} = d_c^-$	strong power-law	infinite randomness	activated
$d_{\mathrm{RR}} > d_c^-$	RR become static	smeared transition	no scaling

(ii) In the second class, the rare regions are exactly *at* the lower critical dimension, and their energy gap shows an exponential dependence on their volume L^d. In this case, the contribution of a single rare region to observables can grow exponentially with its size. The resulting power-law low-energy density of states of the rare regions leads to a power-law Griffiths singularity with a nonuniversal continuously varying exponent. The scaling behavior at the dirty critical point itself turns out to be of exotic activated (exponential) type instead of conventional power-law scaling. This second case is realized, e.g., in classical Ising models with perfectly correlated disorder in one direction (linear defects) [34, 42] and random quantum Ising models (where the disorder correlations are in imaginary time direction) [15, 17, 39, 46, 67]. In these systems, several thermodynamic observables including the average susceptibility actually diverge in a finite region of the disordered phase rather than just at the critical point. Similar phenomena have also been found in quantum Ising spin glasses [19, 48, 56].

(iii) Finally, in the third class, the rare regions can undergo the phase transition independently from the bulk system, i.e., they are *above* the lower critical dimension. In this case, the dynamics of the locally ordered rare regions completely freezes [40, 41], and they develop a truly static order parameter. As a result, the global phase transition is destroyed by smearing [60]. This happens, e.g., for itinerant quantum Ising magnets. In the tail of the smeared transition, the order parameter is extremely inhomogeneous, with statically ordered islands or droplets coexisting with the disordered bulk of the system.

A summary of the general classification of rare region effects is given in table 1. It is expected to apply to all continuous order-disorder transitions (which can be described by a Landau-Ginzburg-Wilson theory) with short-range interactions. (Long-range spatial interactions will modify the rare region effects but it is not yet fully understood how [12]). In the remainder of this chapter we will present computational results for several such transitions and discuss them on the basis of the above classification.

3 Classical Ising magnet with planar defects

3.1 The model

Our first example is a classical equilibrium system, viz. a three-dimensional classical Ising magnet. The disorder is perfectly correlated in $d_C = 2$ dimensions but uncorrelated in $d_\perp = d - d_C = 1$ dimensions, i.e., the system has uncorrelated planar defects. The critical behavior of such systems has been investigated in some detail in the literature, but the consistent picture has been slow to emerge. Early renormalization group analysis [33] based on a single expansion in $\epsilon = 4 - d$ did not produce a critical fixed point, leading to the conclusion that the phase transition is either smeared or of first order. Later work [2] which included an expansion in the number of correlated dimensions d_C lead to a fixed point with conventional power law scaling. Notice, however, that the perturbative renormalization group calculations missed all effects coming form the rare regions. Recently, a theory based on extremal statistics arguments [61] has predicted that in this system rare region effects completely destroy the sharp phase transition by smearing. The predictions of this theory were confirmed in simulations of mean-field type models [61].

Here we report results of large-scale Monte-Carlo simulations aimed at testing the theoretical predictions for a realistic model with short-range interactions [53]. Our starting point is a three-dimensional Ising model with planar defects. Classical Ising spins $S_{ijk} = \pm 1$ reside on a cubic lattice. They interact via nearest-neighbor interactions. In the clean system all interactions are identical and have the value J. The defects are modeled via 'weak' bonds randomly distributed in one dimension (uncorrelated direction). The bonds in the remaining two dimensions (correlated directions) remain equal to J. The system effectively consists of blocks separated by parallel planes of weak bonds. Thus, $d_\perp = 1$ and $d_C = 2$. The Hamiltonian of the system is given by:

$$\begin{aligned} H = & - \sum_{\substack{i=1,\ldots,L_\perp \\ j,k=1,\ldots,L_C}} J_i S_{i,j,k} S_{i+1,j,k} \\ & - \sum_{\substack{i=1,\ldots,L_\perp \\ j,k=1,\ldots,L_C}} J (S_{i,j,k} S_{i,j+1,k} + S_{i,j,k} S_{i,j,k+1}), \end{aligned} \tag{1}$$

where $L_\perp (L_C)$ is the length in the uncorrelated (correlated) direction, i, j and k are integers counting the sites of the cubic lattice, J is the coupling constant in the correlated directions and J_i is the random coupling constant in the uncorrelated direction. The J_i are drawn from a binary distribution:

$$J_i = \begin{cases} cJ & \text{with probability} \quad p \\ J & \text{with probability} \quad 1-p \end{cases} \tag{2}$$

characterized by the concentration p and the relative strength c of the weak bonds $(0 < c \leq 1)$. The fact that one can independently vary concentration and strength of the defects in an easy way is the main advantage of this binary disorder distribution. The order parameter of the magnetic phase transition is the total magnetization:

$$m = \frac{1}{V} \sum_{i,j,k} \langle S_{i,j,k} \rangle, \tag{3}$$

where $V = L_\perp L_C^2$ is the volume of the system, and $\langle \cdot \rangle$ is the thermodynamic average.

3.2 Numerical method

We have performed large scale Monte-Carlo simulations of the Hamiltonian (1) employing the the Wolff cluster algorithm [66]. It can be used because the disorder in our system is not frustrated. As discussed above, the expected smearing of the transition is a result of exponentially rare events. Therefore sufficiently large system sizes are required in order to observe it. We have simulated system sizes ranging from $L_\perp = 50$ to $L_\perp = 200$ in the uncorrelated direction and from $L_C = 50$ to $L_C = 400$ in the remaining two correlated directions, with the largest system simulated having a total of 32 million spins. We have chosen $J = 1$ and $c = 0.1$ in the eq. (2), i.e., the strength of a 'weak' bond is 10% of the strength of a strong bond. The simulations have been performed for various disorder concentrations $p = \{0.2, 0.25, 0.3\}$. The values for concentration p and strength c of the weak bonds have been chosen in order to observe the desired behavior over a sufficiently broad interval of temperatures. The temperature range has been $T = 4.325$ to $T = 4.525$, close to the critical temperature of the clean three-dimensional Ising model $T_c^0 = 4.511$.

To achieve optimal performance of the simulations, one must carefully choose the number N_S of disorder realizations (i.e., samples) and the number N_I of measurements during the simulation of each sample. Assuming full statistical independence between different measurements (quite possible with a cluster update), the variance σ_T^2 of the final result (thermodynamically and disorder averaged) for a particular observable is given by [3]

$$\sigma_T^2 = (\sigma_S^2 + \sigma_I^2/N_I)/N_S \tag{4}$$

where σ_S is the disorder-induced variance between samples and σ_I is the variance of measurements within each sample. Since the computational effort is roughly proportional to $N_I N_S$ (neglecting equilibration for the moment), it is then clear that the optimum value of N_I is very small. One might even be tempted to measure only once per sample. On the other hand, with too short measurement runs most computer time would be spent on equilibration.

In order to balance these requirements we have used a large number N_S of disorder realizations, ranging from 30 to 780, depending on the system size and

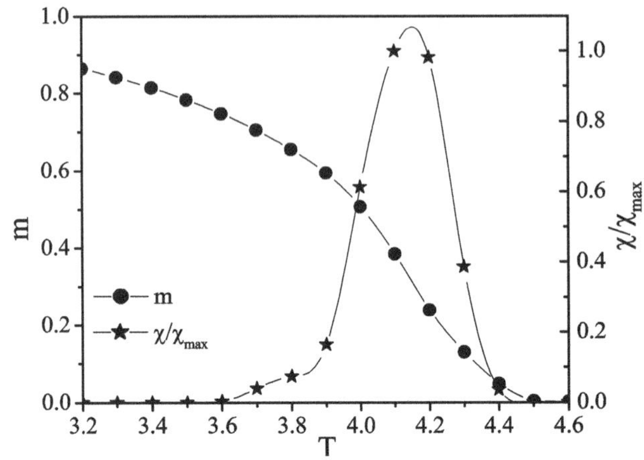

Fig. 1. Average magnetization m and susceptibility χ (spline fit) as functions of T for $L_\perp = 100$, $L_C = 200$ and $p = 0.2$ averaged over 200 disorder realizations (from [53])

rather short runs of 100 Monte-Carlo sweeps, with measurements taken after every sweep. (A sweep is defined by a number of cluster flips so that the total number of flipped spins is equal to the number of sites, i.e., on the average each spin is flipped once per sweep.) The length of the equilibration period for each sample is also 100 Monte-Carlo sweeps. The actual equilibration times have typically been of the order of 10-20 sweeps at maximum. Thus, an equilibration period of 100 sweeps is more than sufficient.

3.3 Results

In this subsection we summarize the results obtained by our simulations, more details are presented in Ref. [53]. Fig. 1 gives an overview of total magnetization and susceptibility as functions of temperature averaged over 200 samples of size $L_\perp = 100$ and $L_C = 200$ with an impurity concentration $p = 0.2$. We note that at the first glance the transition looks like a sharp phase transition with a critical temperature between $T = 4.3$ and $T = 4.4$, rounded by conventional finite size effects. In order to distinguish this conventional scenario from the disorder induced smearing of the phase transition, we have performed a detailed analysis of the system in a temperature range in the immediate vicinity of the clean critical temperature $T_c^0 = 4.511$.

In Fig. 2, we plot the logarithm of the total magnetization vs. $|T_c^0 - T|^{-\nu}$ averaged over 240 samples for system size $L = 200$, $L_C = 280$ and three disorder concentrations $p = \{0.2, 0.25, 0.3\}$. Here, ν is the three-dimensional clean critical Ising exponent. For all three concentrations the data follow a stretched exponential

$$m(T) \sim e^{-B|T-T_c^0|^{-\nu}} \qquad (\text{for } T \to T_c^0-), \tag{5}$$

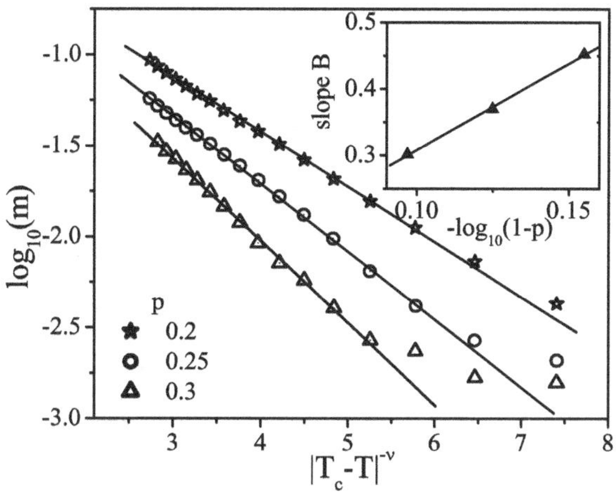

Fig. 2. Logarithm of the total magnetization m as a function of $|T_c^0 - T|^{-\nu}$ ($\nu = 0.627$) for several impurity concentrations $p = 0.2, 0.25, 0.3$, averaged over 240 disorder realizations. System size $L_\perp = 200$, $L_C = 280$. The statistical errors are smaller than a symbol size for all $\log_{10}(m) > -2.5$. Inset: Decay slope B as a function of $-\log(1-p)$ (from [53])

over more than an order of magnitude in m with the exponent for the clean Ising model $\nu = 0.627$. The deviation from the straight line for small m is due to the conventional finite size effects. In the inset we show that the decay constant B depends linearly on $-\log(1-p)$. This stretched exponential dependence of the magnetization on the distance $|T - T_c^0|$ from the clean critical point can be easily understood from rare region arguments [61]. The probability w for finding a large region of linear size $L_\perp$ containing only strong bonds is, up to pre-exponential factors:

$$w \sim (1-p)^{L_\perp} = e^{\log(1-p)L_\perp}. \tag{6}$$

Such a rare region is equivalent to a two-dimensional Ising model in slab geometry. It undergoes a phase transition, i.e., it develops static long-range (ferromagnetic) order at some temperature $T_c(L_\perp)$ below the clean critical temperature T_c^0. The value of $T_c(L_\perp)$ varies with the length of the rare region; the longest islands will develop long-rage order closest to the clean critical point. A rare region is equivalent to a slab of the clean system, we can thus use finite size scaling to obtain:

$$T_c^0 - T_c(L) = |t_c(L)| = AL^{-\phi}, \tag{7}$$

where ϕ is the finite-size scaling shift exponent of the clean system and A is the amplitude for the crossover from three dimensions to a slab geometry infinite in two (correlated) dimension but with finite length in the third (uncorrelated)

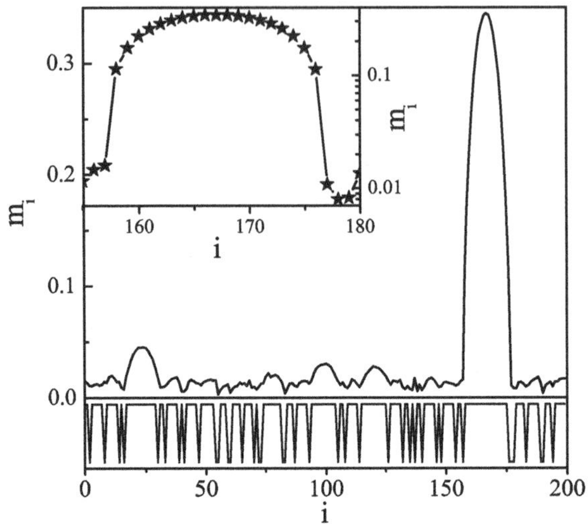

Fig. 3. Local magnetization m_i as function of position i in the uncorrelated direction (system size $L = 200$, $L_C = 200$ and temperature $T = 4.425$, one particular disorder realization). The statistical error is approximately $5 \cdot 10^{-3}$. Lower panel: The coupling constant J_i in the uncorrelated direction as a function of position i. Inset: Log-linear plot of the region in the vicinity of the largest ordered island ([53])

direction. The reduced temperature $t = T - T_c^0$ measures the distance form the *clean* critical point. Since the clean 3d Ising model is below its upper critical dimension ($d_c^+ = 4$), hyperscaling is valid and the finite-size shift exponent $\phi = 1/\nu$. Combining (6) and (7) we get the probability for finding an island of length $L_\perp$ which becomes critical at some t_c as:

$$w(t_c) \sim e^{-B|t_c|^{-\nu}} \quad (\text{for } t_c \to 0-) \tag{8}$$

with the constant $B = -\log(1-p)A^\nu$. The total (average) magnetization m at some reduced temperature t is obtained by integrating over all rare regions which have $t_c > t$. Since the functional dependence on t of the local magnetization on the island is of power-law type it does not enter the leading exponentials but only pre-exponential factors, leading directly to the desired stretched exponential (5).

Because different rare regions undergo the phase transitions at different temperatures, the magnetization is spatially very inhomogeneous in the tail of the smeared phase transition. Close to the clean critical point the system contains a few ordered islands (rare regions devoid of impurities) typically far apart in space. The remaining bulk system is essentially still in the disordered phase. Fig. 3 illustrates such a situation. It displays the local magnetization m_i of a particular disorder realization as a function of the position i in the uncorrelated direction for the size $L_\perp = 200$, $L_C = 200$ at a temperature $T = 4.425$ in the tail of the smeared transition. The lower panel shows the local coupling constant J_i as a function of i. The figure shows that a sizable magnetization has developed on the longest island only (around position $i = 160$). One can also observe that order starts to emerge on the next longest island located close to $i = 25$. Far form these islands the system is still in its disordered phase. In the thermodynamic limit, the local magnetization away from the ordered rare regions should be exponentially small. However, in the simulations of a finite size system the local magnetization has a lower cut-off which is produced by finite-size fluctuations of the order parameter. These fluctuations are governed by the central limit theorem and can be estimated as $m_{\text{bulk}} \approx 1/\sqrt{N_{\text{cor}}} \approx \sqrt{L_{\text{cl}}^2/L_C^2} \approx 5 \cdot 10^{-3}$ in agreement with the typical off-island value in Fig. 3. Here, N_{cor} is the number of correlated volumes per slab as determined by the size off the Wolff cluster. L_{cl} is a typical linear size of a Wolff cluster which is, at $T = 4.425$, $L_{\text{cl}} \approx 10$. In the inset of Fig. 3 we zoom in on the region around the largest island. The local magnetization, plotted on the logarithmic scale, exhibits a rapid drop-off with the distance from the ordered island. This drop-off suggests a relatively small (a few lattice spacings) bulk correlation length ξ_0 in this parameter region.

In summary, our simulations have shown that the phase transition in an Ising model with planar defects is destroyed by smearing. This result is in agreement with the general classification put forward in Sect. 2. The dimensionality of the rare regions is $d_{\text{RR}} = 2$ which is larger than the lower critical dimension of the Ising model which is $d_c^- = 1$. The behavior in the tail of the resulting smeared transition agrees well with the predictions of extremal statistics theory [61].

4 Diluted bilayer quantum Heisenberg antiferromagnet

4.1 Model and Method

In this section we discuss the zero-temperature quantum phase transitions in a diluted bilayer quantum antiferromagnet. The spins in each two-dimensional layer interact via nearest neighbor exchange $J_\parallel$, and the interplane coupling is $J_\perp$. The clean version of this model has been studied extensively [26,38,51]. For $J_\perp \gg J_\parallel$, neighboring spins from the two layers form singlets, and the ground state is paramagnetic. In contrast, for $J_\parallel \gg J_\perp$ the system develops antiferromagnetic (Néel) order. Both phases are separated by a quantum phase transition at $J_\perp/J_\parallel \approx 2.525$. Random disorder is introduced by removing *pairs* (dimers) of adjacent spins, one from each layer. The Hamiltonian of the model with dimer dilution is:

$$H = J_\parallel \sum_{\substack{\langle i,j \rangle \\ a=1,2}} \epsilon_i \epsilon_j \hat{\mathbf{S}}_{i,a} \cdot \hat{\mathbf{S}}_{j,a} + J_\perp \sum_i \epsilon_i \hat{\mathbf{S}}_{i,1} \cdot \hat{\mathbf{S}}_{i,2}, \tag{9}$$

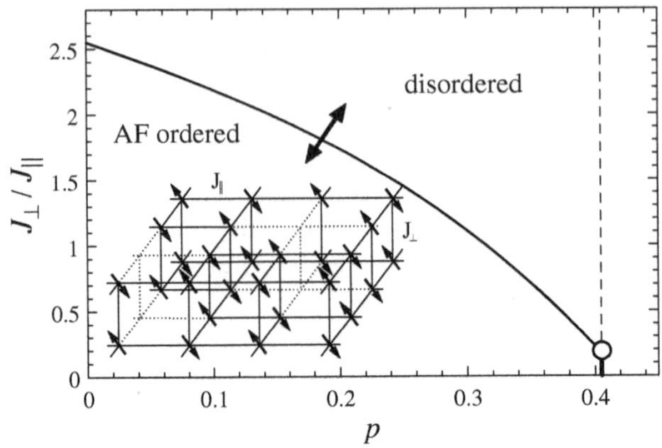

Fig. 4. Phase diagram [59] of the diluted bilayer Heisenberg antiferromagnet, as function of $J_\perp/J_\parallel$ and dilution p. The dashed line is the percolation threshold, the open dot is the multicritical point of Refs. [50, 59]. The arrow indicates the QPT studied here. Inset: The model: Quantum spins (arrows) reside on the two parallel square lattices. The spins in each plane interact with the coupling strength $J_\parallel$. Interplane coupling is $J_\perp$. Dilution is done by removing dimers (from [54])

and ϵ_i=0 (ϵ_i=1) with probability p $(1-p)$.

The phase diagram of the dimer-diluted bilayer Heisenberg model has been studied by Sandvik [50] and Vajk and Greven [59], see Fig. 4. For small $J_\perp$, magnetic order survives up to the percolation threshold $p_p \approx 0.4072$, and a multicritical point exists at $p = p_p$ and $J_\perp/J_\parallel \approx 0.16$.

We have performed extensive simulations [54] of the critical behavior at the generic transition occurring for $0 < p < p_p$ and driven by $J_\perp$. To this end we have first mapped the quantum Hamiltonian (9) onto a classical model. The low-energy properties of bilayer quantum antiferromagnets are represented by a (2+1)-dimensional O(3) quantum rotor model [10] with the rotor coordinate $\hat{\mathbf{n}}_i$ corresponding to $\hat{\mathbf{S}}_{i,1} - \hat{\mathbf{S}}_{i,2}$ and the angular momentum $\hat{\mathbf{L}}_i$ representing $\hat{\mathbf{S}}_{i,1} + \hat{\mathbf{S}}_{i,2}$ (see, e.g., Chap. 5 of [49]). This quantum rotor model in turn is equivalent to a three-dimensional classical Heisenberg model with the disorder perfectly correlated in imaginary time direction, as can be easily seen from a path integral representation of the partition function. Thus, our classical Hamiltonian reads:

$$H = K \sum_{\langle i,j\rangle,\tau} \epsilon_i \epsilon_j \mathbf{n}_{i,\tau} \cdot \mathbf{n}_{j,\tau} + K \sum_{i,\tau} \epsilon_i \mathbf{n}_{i,\tau} \cdot \mathbf{n}_{i,\tau+1}, \quad (10)$$

where $\mathbf{n}_{i,\tau}$ is an O(3) unit vector. The coupling constant βK of the classical model is related to the ratio $J_\parallel/J_\perp$ of the quantum model. Here, $\beta \equiv 1/T$ where T is an effective "classical" temperature, not equal to the real temperature which is zero. We set $K = 1$ and drive the classical system through the transition by tuning the classical temperature T.

We also note that dimer dilution in the bilayer antiferromagnet (9) does not introduce random Berry phases because the Berry phase contributions from the two spins of each unit cell cancel [10,49]. In contrast, for site dilution, the physics changes completely: The random Berry phases (which have no classical analogue) are equivalent to impurity-induced moments [52], and those become weakly coupled via bulk excitations. Thus, for all $p < p_p$ the ground state shows long-range order, independent of $J_\perp/J_\parallel$! This effect is absent for dimer dilution, and both phases of the clean system survive for small p.

The classical model (10) is studied by Monte-Carlo simulations using the efficient Wolff cluster algorithm [66]. We investigate linear sizes up to $L = 120$ in space direction and $L_\tau = 384$ in imaginary time, for impurity concentrations $p = \frac{1}{8}, \frac{1}{5}, \frac{2}{7}$ and $\frac{1}{3}$. The results are averaged over $10^3 - 10^4$ disorder realizations. Each sample is equilibrated using 100 Monte-Carlo sweeps (spin-flips per site). For large dilutions, $p = \frac{2}{7}$ and $\frac{1}{3}$ we perform both Wolff and Metropolis sweeps to equilibrate small dangling clusters. During the measurement period of another 100-200 sweeps we calculate magnetization, susceptibility, specific heat and correlation functions.

4.2 Results

A quantity particularly suitable to locate the critical point and to study the critical behavior is the Binder ratio:

$$g_{\mathrm{av}} = \left[1 - \frac{\langle|\mathbf{M}|^4\rangle}{3\langle|\mathbf{M}|^2\rangle^2}\right]_{\mathrm{av}}, \tag{11}$$

where $\mathbf{M} = \sum_{i,\tau} \mathbf{n}_{i,\tau}$, $[\ldots]_{\mathrm{av}}$ denotes the disorder average and $\langle\ldots\rangle$ denotes the Monte-Carlo average for each sample. It also allows one to distinguish between power law dynamical scaling (correlation time behaves like a power of correlation length, $\xi_\tau \sim \xi^z$) and activated dynamical scaling ($\ln \xi_\tau \sim \xi^\psi$). Because the Binder ratio has scale dimension 0, its finite-size scaling form is given by

$$g_{\mathrm{av}} = \tilde{g}_C(tL^{1/\nu}, L_\tau/L^z) \qquad \text{or} \tag{12}$$

$$g_{\mathrm{av}} = \tilde{g}_A(tL^{1/\nu}, \log(L_\tau)/L^\mu) \tag{13}$$

for conventional scaling or for activated scaling, respectively. Two important characteristics follow: (i) For fixed L, g_{av} has a peak as a function of L_τ. The peak position $L_\tau^{\max}$ marks the *optimal* sample shape, where the ratio L_τ/L roughly behaves like the corresponding ratio of the correlation lengths in time and space directions, ξ_τ/ξ. At the critical temperature T_c, the peak value $g_{\mathrm{av}}^{\max}$ is independent of L. Thus, for power law scaling, plotting g_{av} vs. $L_\tau/L_\tau^{\max}$ at T_c should collapse the data, without the need for a value of z. In contrast, for activated scaling the g_{av} data should collapse when plotted as a function of $\log(L_\tau)/\log(L_\tau^{\max})$. (ii) For samples of the optimal shape ($L_\tau = L_\tau^{\max}$), plots of g_{av} vs. temperature for different L cross at T_c. Based on these two

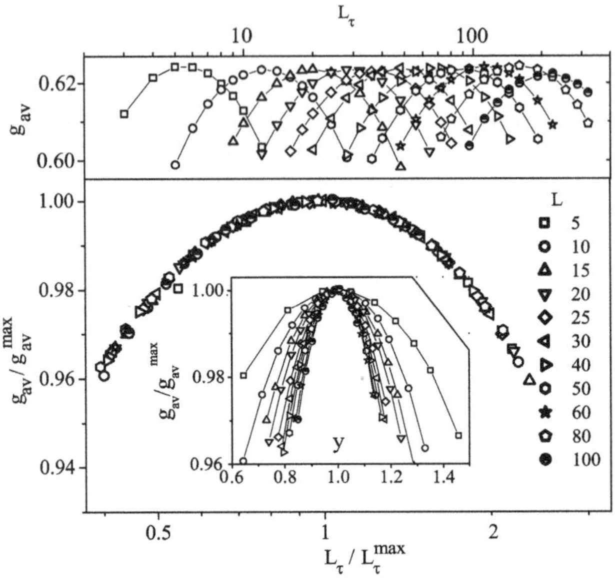

Fig. 5. Upper panel: Binder ratio $g_{\rm av}$ as a function of L_τ for various L ($p = \frac{1}{5}$). Lower panel: Power-law scaling plot $g_{\rm av}/g_{\rm av}^{\rm max}$ vs. $L_\tau/L_\tau^{\rm max}$. Inset: Activated scaling plot $g_{\rm av}/g_{\rm av}^{\rm max}$ vs. $y = \log(L_\tau)/\log(L_\tau^{\rm max})$ (from [54])

characteristics, we use a simple iterative procedure to determine both the optimal shapes and the location of the critical point.

We now turn to our results. To distinguish between activated and power-law dynamical scaling we perform a series of calculations at the critical temperature. The upper panel of Fig. 5 shows the Binder ratio $g_{\rm av}$ as a function of L_τ for various $L = 5 \ldots 100$ and dilution $p = \frac{1}{5}$ at $T = T_c = 1.1955$. The statistical error of $g_{\rm av}$ is below 0.1% for the smaller sizes and not more than 0.2% for the largest systems. As expected at T_c, the maximum Binder ratio for each of the curves does not depend on L. To test the conventional power-law scaling form, eq. (12), we plot $g_{\rm av}/g_{\rm av}^{\rm max}$ as a function of $L_\tau/L_\tau^{\rm max}$ in the lower panel of Fig. 5. The data scale extremely well, giving statistical errors of $L_\tau^{\rm max}$ in the range between 0.3% and 1%. For comparison, the inset shows a plot of $g_{\rm av}$ as a function of $\log(L_\tau)/\log(L_\tau^{\rm max})$ corresponding to eq. (13). The data clearly do not scale which rules out the activated scaling scenario. The results for the other impurity concentrations $p = \frac{1}{8}, \frac{2}{7}, \frac{1}{3}$ are completely analogous.

Having established conventional power-law dynamical scaling, we proceed to determine the dynamical exponent z. In Fig. 6, we plot $L_\tau^{\rm max}$ vs. L for all four dilutions p. The curves show significant deviations from pure power-law behavior which can be attributed to corrections to scaling due to irrelevant operators. In such a situation, a direct power-law fit of the data will only yield *effective* exponents. To find the true *asymptotic* exponents we take the leading correction to scaling into account by using the ansatz $L_\tau^{\rm max}(L) =$

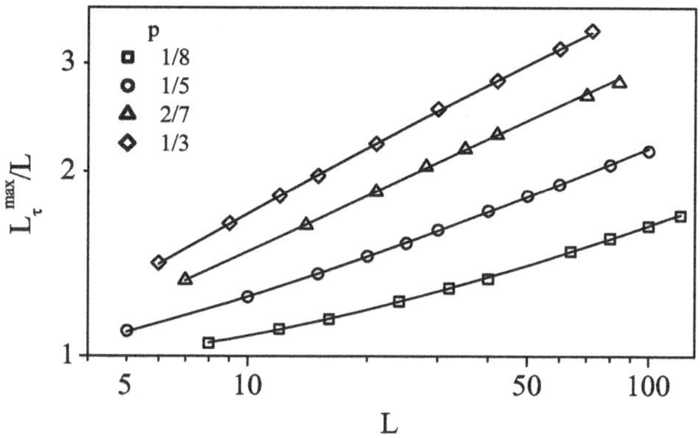

Fig. 6. $L_\tau^{\max}/L$ vs. L for four disorder concentrations $p = \frac{1}{8}, \frac{1}{5}, \frac{2}{7}$ and $\frac{1}{3}$. Solid lines: Fit to $L_\tau^{\max} = aL^z(1 + bL^{-\omega_1})$ with $z = 1.310(6)$ and $\omega_1 = 0.48(3)$ (from [54])

$aL^z(1+bL^{-\omega_1})$ with universal (dilution-independent) exponents z and ω_1 but dilution-dependent a and b. A combined fit of all four curves gives $z = 1.310(6)$ and $\omega_1 = 0.48(3)$ where the number in brackets is the standard deviation of the last given digit. The fit is of high quality with $\chi^2 \approx 0.7$ per degree of freedom indicating that the data follow our ansatz without systematic deviations. The fit is also robust against removing complete data sets or removing points form the lower or upper end of each set. We thus conclude that the asymptotic dynamical exponent z is indeed universal. Note that the leading corrections to scaling vanish very close to $p = \frac{2}{7}$; the curvature of the $L_\tau^{\max}(L)$ curves in Fig. 6 is opposite above and below this concentration.

To find the correlation length exponent ν, we perform simulations in the vicinity of T_c for samples with the optimal shape ($L_\tau = L_\tau^{\max}$) to keep the second argument of the scaling function (12) constant. The exponent ν can be determined from the L-dependence of the scale-factor x_L necessary to collapse the data. A combined fit to the ansatz $x_L = cL^{1/\nu}(1 + dL^{-\omega_2})$ where ν and ω_2 are universal, gives $\nu = 1.16(3)$ and $\omega_2 = 0.5(1)$. As above, the fit is robust and of high quality ($\chi^2 \approx 1.2$). Importantly, as expected for the true asymptotic exponent, ν fulfills the Harris criterion [24], $\nu > 2/d{=}1$. Note that both irrelevant exponents ω_1 and ω_2 agree within their error bars, suggesting that the same irrelevant operator controls the leading corrections to scaling for both z and ν. We have also calculated total magnetization and susceptibility. The corresponding exponents $\beta/\nu = 0.56(5)$ and $\gamma/\nu = 2.15(10)$ have slightly larger error bars than z and ν. Nonetheless, they fulfill the hyperscaling relation $2\beta + \gamma = (d + z)\nu$ which is another argument for our results being asymptotic rather than effective exponents.

In summary, our computer simulations have shown that the zero-temperature quantum phase transition in the diluted bilayer Heisenberg quantum antiferromagnet is characterized by a conventional critical point with power-law dynamical scaling and universal critical exponents that fulfill the Harris criterion. These results are in agreement with the general classification suggested in Sect. 2. The effective dimensionality of the rare regions (including the imaginary time direction which is important for quantum phase transitions) is $d_{\mathrm{RR}} = 1$. The lower critical dimension for an $O(3)$ Heisenberg model is $d_c^- = 2$. Thus, $d_{\mathrm{RR}} < d_c^-$, and the rare region effects are exponentially small.

5 Nonequilibrium phase transitions in the disordered contact process

5.1 The contact process

The examples discussed in Sects. 3 and 4 concerned systems in thermal equilibrium. However, nonequilibrium systems can also undergo "phase" transitions between different nonequilibrium steady states. These transitions are characterized by large scale fluctuations and collective behavior over large distances and times very similar to the behavior at equilibrium critical points. Examples of such nonequilibrium transitions can be found in population dynamics and epidemics, chemical reactions, growing surfaces, and in granular flow and traffic jams (for recent reviews see, e.g., Refs. [9, 27, 35, 44, 55, 57])

A prominent class of nonequilibrium phase transitions separates active fluctuating states from inactive, absorbing states where fluctuations cease entirely. Recently, much effort has been devoted to classifying possible universality classes of these absorbing state phase transitions [27, 44]. The generic universality class is directed percolation (DP) [23]. According to a conjecture by Janssen and Grassberger [20, 30], all absorbing state transitions with a scalar order parameter, short-range interactions, and no extra symmetries or conservation laws belong to this class.

The contact process [25] is a prototypical system in the directed percolation universality class. It can be interpreted, e.g., as a model for the spreading of a disease. The contact process is defined on a d-dimensional hypercubic lattice. Each lattice site $\mathbf{r}$ can be active (occupied by a particle) or inactive (empty). In the course of the time evolution, active sites can infect their neighbors, or they can spontaneously become inactive. Specifically, the dynamics is given by a continuous-time Markov process during which particles are created at empty sites at a rate $\lambda n/(2d)$ where n is the number of active nearest neighbor sites. Particles are annihilated at rate μ (which is often set to unity without loss of generality). The ratio of the two rates controls the behavior of the system.

For small birth rate λ, annihilation dominates, and the absorbing state without any particles is the only steady state (inactive phase). For large birth rate λ, there is a steady state with nonzero particle density (active phase). The two phases are separated by a nonequilibrium phase transition in the directed percolation universality class at $\lambda = \lambda_c^0$. The central quantity in the contact process is the average density of active sites at time t

$$\rho(t) = \frac{1}{L^d} \sum_{\mathbf{r}} \langle n_{\mathbf{r}}(t) \rangle \tag{14}$$

where $n_{\mathbf{r}}(t)$ is the particle number at site $\mathbf{r}$ and time t, L is the linear system size, and $\langle \ldots \rangle$ denotes the average over all realizations of the Markov process. The longtime limit of this density (i.e., the steady state density)

$$\rho_{\mathrm{st}} = \lim_{t \to \infty} \rho(t) \tag{15}$$

is the order parameter of the nonequilibrium phase transition.

5.2 Contact process with point defects

Quenched spatial disorder can be introduced by making the birth rate λ a random function of the lattice site $\mathbf{r}$. We assume the disorder to be spatially uncorrelated; and we use a binary probability distribution

$$P[\lambda(\mathbf{r})] = (1 - p)\,\delta[\lambda(\mathbf{r}) - \lambda] + p\,\delta[\lambda(\mathbf{r}) - c\lambda] \tag{16}$$

where p and c are constants between 0 and 1. This distribution allows us to independently vary spatial density p of the impurities and their relative strength c. The impurities locally *reduce* the birth rate, therefore, the nonequilibrium transition will occur at a value λ_c that is larger than the clean critical birth rate λ_c^0.

The investigation of disorder effects on the directed percolation transition actually has a long history, but a coherent picture has emerged only recently. The directed percolation universality class violates the Harris criterion $d\nu > 2$ in all dimensions $d < 4$, because the exponent values are $\nu \approx 1.097$ (1D), 0.73 (2D), and 0.58 (3D) [27]. A field-theoretic renormalization group study [31] confirmed the instability of the DP critical fixed point. Moreover, no new critical fixed point was found. Instead the renormalization group displays runaway flow towards large disorder, indicating unconventional behavior. Early Monte-Carlo simulations [43] showed significant changes in the critical exponents while later studies [36] of the two-dimensional contact process with dilution found logarithmically slow dynamics in violation of power-law scaling. In addition, rare region effects similar to Griffiths singularities were found to lead to slow dynamics in a whole parameter region in the vicinity of the phase transition. Recently, an important step towards understanding spatial disorder effects on the DP transition has been made by Hooyberghs et al. [28]. These authors used a version of the Ma-Dasgupta-Hu strong-disorder renormalization group [37] and showed that the transition is controlled by an infinite-randomness critical point, at least for sufficiently strong disorder.

Numerical method

Here, we report the results of large scale Monte-Carlo simulations of the one-dimensional disordered contact process [58]. There is a number of different ways to actually implement the contact process on the computer (all equivalent with respect to the universal behavior). We follow the widely used algorithm described, e.g., by Dickman [11]. Runs start at time $t = 0$ from some configuration of occupied and empty sites. Each event consists of randomly selecting an occupied site $\mathbf{r}$ from a list of all N_p occupied sites, selecting a process: creation with probability $\lambda(\mathbf{r})/[1 + \lambda(\mathbf{r})]$ or annihilation with probability $1/[1 + \lambda(\mathbf{r})]$ and, for creation, selecting one of the neighboring sites of $\mathbf{r}$. The creation succeeds, if this neighbor is empty. The time increment associated with this event is $1/N_p$. Note that in this implementation of the disordered contact process both the creation rate and the annihilation rate vary from site to site in such a way that their sum is constant (and equal to one).

Using this algorithm, we have performed simulations for system sizes between $L = 1000$ and $L = 10^7$. We have studied impurity concentrations $p = 0.2, 0.3, 0.4, 0.5, 0.6$ and 0.7 as well as relative impurity strengths of $c = 0.2, 0.4, 0.6$ and 0.8. To explore the extremely slow dynamics associated with the predicted infinite-randomness critical point, we have simulated very long times up to $t = 10^9$ which is, to the best of our knowledge, at least three orders of magnitude in t longer than previous simulations of the disordered contact process. In all cases we have averaged over a large number (at least 480) of different disorder realizations.

Results

A first set of calculations starts from a full lattice and follows the time evolution of the average density. This means, at time $t = 0$, all sites are active and $\rho(0) = 1$. Figure 7 gives an overview of the time evolution of the density for a system of 10^6 sites with $p = 0.3, c = 0.2$, covering the λ range from the conventional inactive phase, $\lambda < \lambda_c^0$ all the way to the active phase, $\lambda > \lambda_c$. For birth rates below and at the clean critical point $\lambda_c^0 \approx 3.298$, the density decay is very fast, clearly faster then a power law. Above λ_c^0, the decay becomes slower and asymptotically seems to follow a power-law. For even larger birth rates the decay seems to be slower than a power law while the largest birth rates give rise to a nonzero steady state density, i.e., the system is in the active phase.

To test the strong-disorder renormalization group theory of Hooyberghs et al. [28], we plot the time dependence of the density according to the predicted activated scaling law

$$\rho(t) \sim [\ln(t)]^{-\bar{\delta}}, \tag{17}$$

with $\bar{\delta} = 0.38197$. Fig. 8 shows our data for a system of 10^4 sites with $p = 0.3$ and $c = 0.2$. As before, the data are averages over 480 runs, each with a different disorder realization. The evolution of the density at $\lambda = 5.24$ follows eq. (17) over almost six orders of magnitude in time. Therefore, we conclude that the critical point of the disordered contact process is indeed of infinite-randomness type [28] with $\lambda = \lambda_c = 5.24$ being the critical birthrate for

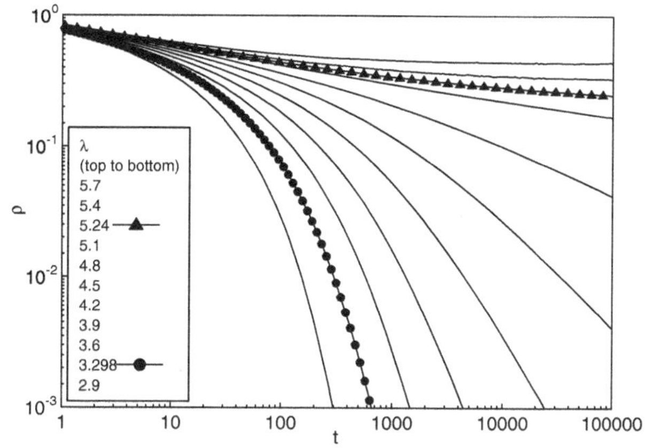

Fig. 7. Overview of the time evolution of the density for a system of 10^6 sites with $p = 0.3$ and $c = 0.2$. The clean critical point $\lambda_c^0 \approx 3.298$ and the dirty critical point $\lambda_c \approx 5.24$ are specially marked (from [58])

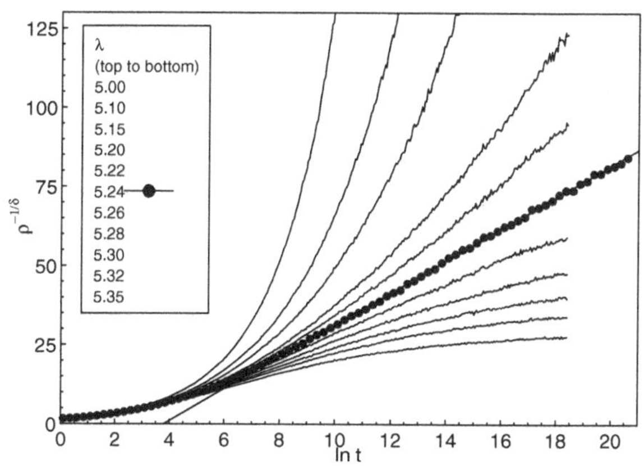

Fig. 8. $\rho^{-1/\bar{\delta}}$ vs. $\ln(t)$ for a system of 10^4 sites with $p = 0.3$ and $c = 0.2$. The filled circles mark the critical birth rate $\lambda_c = 5.24$, and the straight line is a fit of the long-time behavior to eq. (17) (from [58])

$p = 0.3, c = 0.2$. We have performed analogous calculation for two different sets of parameters. In the first, we kept $p = 0.3$ but varied c from 0.2 to 0.8; in the second we kept $c = 0.2$ and varied p from 0.2 to 0.5. We found that the critical point is characterized by the logarithmic density decay (17) with a universal exponent $\bar{\delta} = 0.38197$ for all parameter sets including the case of weak disorder.

In addition to the critical point, we have also studied the Griffiths region between the clean critical birthrate, $\lambda_c^0 = 3.298$ and the dirty critical birthrate

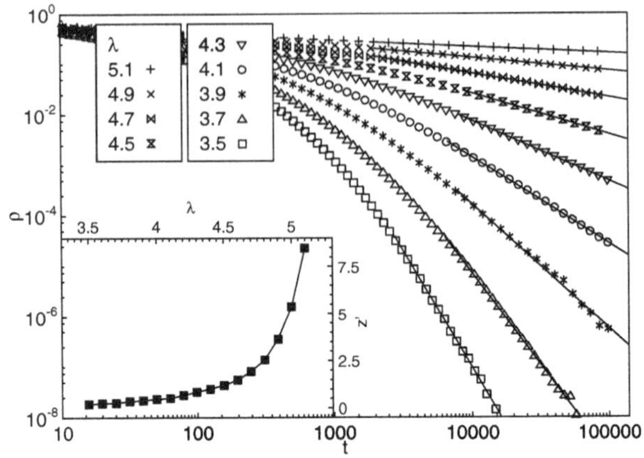

Fig. 9. Log-log plot of the density time evolution in the Griffiths region for systems with $p = 0.3, c = 0.2$ and several birth rates λ. The system sizes are 10^7 sites for $\lambda = 3.5, 3.7$ and 10^6 sites for the other λ values. The straight lines are fits to the power law $\rho(t) \sim t^{-1/z'}$ predicted in eq. (18). Inset: Dynamical exponent z' vs. birth rate λ (from [58])

$\lambda_c = 5.24$. According to an early prediction by Noest [43], the long-time decay of the density in the Griffiths region should asymptotically follow a power-law,

$$\rho(t) \sim t^{-d/z'} \tag{18}$$

where z' is a customarily used nonuniversal dynamical exponent.

Figure 9 shows a double-logarithmic plot of the density time evolution for birth rates $\lambda = 3.5 \ldots 5.1$ and $p = 0.3, c = 0.2$. The system sizes are between 10^6 and 10^7 lattice sites. For all birth rates λ shown, $\rho(t)$ follows (18) over several orders of magnitude in ρ (except for the largest λ where we could observe the power law only over a smaller range in ρ because the decay is too slow). The nonuniversal dynamical exponent z' can be obtained by fitting the long-time asymptotics of the curves in Fig. 9 to eq. (18). The inset of Fig. 9 shows z' as a function of the birth rate λ. As predicted, z' increases with increasing λ throughout the Griffiths region with an apparent divergence around $\lambda = \lambda_c = 5.24$.

In summary, our large-scale simulations of the contact process with point defects have provided strong evidence that the critical point is of infinite-randomness type with universal critical exponents (even for weak bare disorder). The critical point is accompanied by strong Griffiths singularities characterized by non-universal power-law decay of the density. These results are in excellent agreement with the general classification of dirty phase transitions suggested in Sect. 2. The dimensionality of the rare regions in the contact process with point defects is $d_{\mathrm{RR}} = 0$ (they are of finite size). However, the zero-dimensional contact process is right at its lower critical dimension d_c^-,

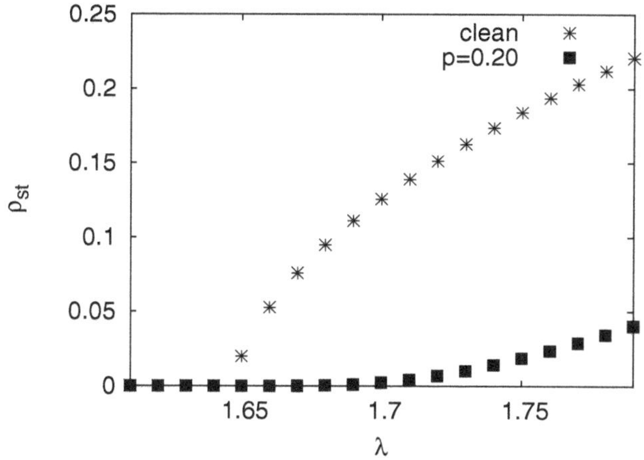

Fig. 10. Stationary density ρ_{st} as a function of birth rate λ for a clean system and a system with impurity concentration $p = 0.2$. System size is $L = 1000$ (from [14])

because the life time of a cluster depends exponentially on its size. Thus, $d_{\text{RR}} = d_c^-$, and the classification predicts power-law Griffiths effects and activated scaling at the dirty critical point.

5.3 Contact process with extended defects

In this subsection we consider the contact process in the presence of spatially extended (linear or planar) defects. From the general arguments in Sect. 2, we expect the disorder correlations to enhance the impurity effects. Indeed, using optimal fluctuation theory, it has recently been predicted [62], that extended defects destroy the phase transition in the contact process by smearing.

Here, we report Monte-Carlo simulations of a contact process on a square lattice [14]. The disorder consists of linear defects, i.e., the local birthrate $\lambda(\mathbf{r})$ with $\mathbf{r} = (x, y)$ depends only on x (in y-direction, it is perfectly correlated). Except for this difference, the simulations proceed analogously to those described in the last subsection. We have investigated linear system sizes up to $L = 3000$ and impurity concentrations $p = 0.2, 0.25, 0.3, 0.35$ and 0.4. The relative strength of the birth rate on the impurities was $c = 0.2$ for all simulations. The data presented below represent averages of 200 disorder realizations.

Let us first focus our attention on the stationary state of our contact process. Fig. 10 shows a comparison of the stationary density ρ_{st} as a function of λ between the clean system and a dirty system with $p = 0.2$. The clean system ($p = 0$) has a sharp phase transition with a power-law singularity of the density, $\rho_{\text{st}} \sim (\lambda - \lambda_c^0)^\beta$ at the clean critical point $\lambda_c^0 \approx 1.65$ with $\beta \approx 0.58$ in agreement with the literature [36]. In contrast, in the dirty system, the

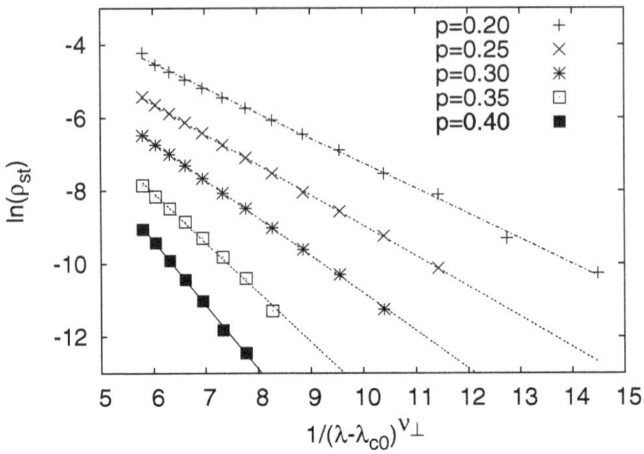

Fig. 11. Left: Logarithm of the stationary density $\rho_{\rm st}$ as a function of $(\lambda - \lambda_c^0)^{-\nu_\perp} = (\lambda - \lambda_c^0)^{-0.734}$ for several impurity concentrations p and $L = 3000$. The straight lines are fits to eq. (19) (from [14])

density increases much more slowly with λ after crossing the clean critical point. This suggests either a critical point with a very large exponent β or exponential behavior.

In the tail of a smeared phase transition, the stationary density is expected to behave as

$$\rho_{\rm st}(\lambda) \sim \exp\left(-B(\lambda - \lambda_c^0)^{-\nu_\perp}\right) . \tag{19}$$

This can be obtained from an extremal statistics theory [62] for rare regions with strong local infection rate similar to the theory sketched in section 3. Here, $\nu_\perp$ is the spatial correlation length exponent of the clean system. To test this prediction, in Fig. 11, we plot $\ln \rho_{\rm st}$ as a function of $(\lambda - \lambda_c^0)^{\nu_\perp}$ for several impurity concentrations p. The data show that the density tail is indeed exponential, following the exponential law (19) over at least two orders of magnitude in $\rho_{\rm st}$. The clean two-dimensional spatial correlation length exponent is $\nu_\perp = 0.734$ [64]. Fits of the data to eq. (19) can be used to determine the decay constants B. As predicted by extremal statistics theory, the decay constants depend linearly on $\tilde{p} = -\ln(1-p)$.

In addition to the stationary density, we have also studied its time evolution in the tail of the smeared transition. According to extremal statistics theory [62], the density decay right at the clean critical point should follow a stretched exponential

$$\ln \rho(t) \sim -t^{1/(1+z)} . \tag{20}$$

where $z = 1.76$ is the dynamical exponent of the clean two-dimensional contact process [64]. This prediction is tested in Fig. 12 for our two-dimensional contact process with linear defects. The figure shows that the data follow the

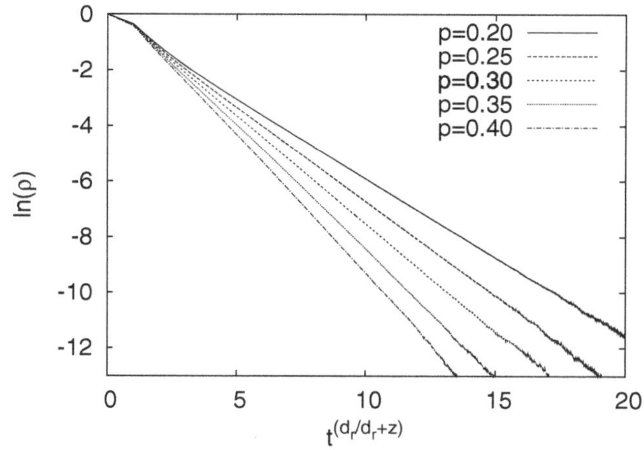

Fig. 12. Logarithm of the density at the clean critical point λ_c^0 as a function of $t^{1/(1+z)} = t^{0.362}$ for several impurity concentrations ($p = 0.2, \ldots, 0.4$ from top to bottom) and $L = 3000$. The long-time behavior follows a stretched exponential $\ln \rho = -Et^{0.362}$ (from [14])

predicted stretched exponential behavior over more than three orders of magnitude in ρ. The very slight deviation of the curves from a straight line can be attributed to the pre-exponential factors neglected in the extremal statistics theory.

We have also simulated the dynamics above the clean critical point, $\lambda > \lambda_c^0$. In this parameter region, the stationary density is nonzero as shown in Fig. 11. The approach of the density to this nonzero stationary value follows a power-law with a nonuniversal exponent [14, 62].

In summary, our large-scale computer simulations have established that the phase transition in the two-dimensional contact process with linear defects is smeared because rare regions can undergo the phase transition independently from the rest of the system. This result agrees with the general classification of dirty phase transitions suggested in Sect. 2. The dimensionality of the rare regions in the contact process with linear defects is $d_{\mathrm{RR}} = 1$ which is larger than the lower critical dimension $d_c^- = 0$ of the contact process. Thus, the global transition must be smeared.

6 Computational implementation

Computational studies of disordered many-particle systems generally require a very high computational effort. In addition to dealing with a large number of degrees of freedom, the presence of impurities and defects requires studying large numbers (from 100 to several 10000) of samples or disorder realizations to explore the averages or even distribution functions of macroscopic observables.

Fortunately, this very complication also makes disordered many-particle systems ideal candidates for massive parallel simulations. In the simplest case, one can distribute (farm out) the N_{samp} disorder realizations over the N_{CPU} available processors at the beginning such that each processor is assigned N_{samp}/N_{CPU} samples. This can be done, e.g., by using different random number seeds in the function that generates the impurities or defects. The simulations then proceed independently for the different samples, minimizing the communication requirements. After each processor has carried out the simulation for all the samples assigned to it, the data are collected and processed to obtain the desired averages or distribution functions. In this way, one achieves linear speedup (total time inversely proportional to the number of available processors) as long as the number of simulated samples is larger than the number of processors.

The only major drawback of this scheme lies in the fact that depending on the algorithm, different samples of a strongly disordered system may require vastly different simulation times. If the number of samples per processor is fixed from the outset, some idle time is thus unavoidable. To overcome this problem we have implemented an algorithm that dynamically assigns the samples to available processors. In the beginning, a master process hands each of the simulation processes a specific sample (i.e., a random number seed). After a process finishes the simulation, it sends the results to the master process and is handed a new sample. This is repeated until all samples have been simulated. With this modification of the simple farming process, no cycles are wasted, and the speedup of our simulations is nearly perfect. In all other aspects, the simulations use established procedures for parallel computing, in particular, communication is via MPI.

Monte-Carlo simulations of large many-particle systems with quenched disorder require huge numbers of (pseudo) random numbers. The quality of the random number generator (long period, low correlations) is therefore of particular importance. Our main "work horse" generator has been the combined linear feedback shift register generator LFSR113 suggested by P. L'Ecuyer [32] which is very fast and has a long period of about 2^{113}. We have also used other random number generators for comparison and testing purposes including the popular RAN2 from Ref. [45].

7 Summary and conclusions

In this chapter we have discussed the results of large-scale parallel Monte-Carlo simulations of quantum and classical phase transitions with quenched disorder. We have paid particular attention to the effects of rare strong spatial disorder fluctuations, the so-called rare regions. Our results show that disorder correlations, either in space or in imaginary time (at quantum phase transitions) strongly enhance the disorder effects.

We have focused on what is arguably the simplest type of phase transition in the presence of impurities and defects, viz., order-disorder transitions with random-T_c (random mass) type disorder. In this scenario, both phases are conventional (i.e., disorder is irrelevant at the stable fixed points corresponding to the bulk phases). The only disorder effect is a local variation of the tendency towards the ordered phase. For such transitions (with short-range spatial interactions), a general classification of rare region effects has been suggested, based on the effective dimensionality d_{RR} of the rare regions [63]. Three cases can be distinguished.

(i) If d_{RR} is below the lower critical dimension d_c^- of the problem, the rare region effects are exponentially small because the probability of a rare region decreases exponentially with its volume but the contribution of each region to observables increases only as a power law. In this case, the critical point is of conventional power-law type.

(ii) In the second class, with $d_{\mathrm{RR}} = d_c^-$, the Griffiths effects are of power-law type because the exponentially rarity of the rare regions in L_r is overcome by an exponential increase of each region's contribution. In this class, the critical point is controlled by an infinite-randomness fixed point with activated scaling.

(iii) Finally, for $d_{\mathrm{RR}} > d_c^-$, the rare regions can undergo the phase transition independently from the bulk system. This leads to a destruction of the sharp phase transition by smearing.

All the examples discussed in this chapter are in agreement with this classification. The diluted bilayer quantum Heisenberg antiferromagnet falls into the first class, because $d_{\mathrm{RR}} = 1$ (disorder correlations in imaginary time direction) but $d_c^- = 2$ for Heisenberg symmetry. The contact process with point defects belongs to class (ii) since $d_{\mathrm{RR}} = d_c^- = 0$. Finally, the classical Ising model with plane defects and the contact process with line defects have a smeared transition, case (iii). In the former system, the plane defects lead to $d_{\mathrm{RR}} = 2$ while $d_c^- = 1$ for Ising symmetry; in the latter system $d_{\mathrm{RR}} = 1$ but $d_c^- = 0$. Thus, all our simulation results provide support for the classification put forward in Ref. [63].

In conclusion, rare regions, i.e. rare strong spatial disorder fluctuations, can have pronounced effects on phase transitions in systems with quenched disorder. These effects range from classical Griffiths phenomena to the much stronger quantum Griffiths singularities, and to a complete destruction of the sharp phase transition by smearing. The simulation results summarized here have helped clarifying these rare region effects for order-disorder transitions between conventional phases. In the future, it will be interesting to see whether additional new rare region effects [63] can be found for transitions where the phases themselves are unconventional, such as spin glass or random singlet phases.

Acknowledgements

The research presented here would have been impossible without the contributions and suggestions of many friends and colleagues including D. Belitz, T.R. Kirkpatrick, J. Schmalian, M. Schreiber, R. Sknepnek, and M. Vojta.

This work was supported in part by SFB 393 of the German Research Foundation, by the National Science Foundation under grant no. DMR-0339147 and by the University of Missouri Research Board. Thomas Vojta is a Cottrell Scholar of Research Corporation.

The early simulations for this work have been performed on the CLIC cluster at Chemnitz University of Technology, the later calculations have been carried out on the PEGASUS cluster in the Physics Department at the University of Missouri-Rolla. The author is also grateful for the hospitality of the Aspen Center for Physics and the Kavli Institute for Theoretical Physics, Santa Barbara (supported by the NSF via grant no. PHY99-07949), where parts of the work have been performed.

References

1. Aharony, A., Harris, A.B.: Absence of Self-averaging and universal fluctuations in random systems near critical points. *Phys. Rev. Lett.*, 77:3700, 1996.
2. Boyanovsky D., Cardy, J.L.: Critical behavior of m-component magnets with correlated impurities. *Phys. Rev. B* 26:154, 1982.
3. Ballesteros, H.G., Fernández, L.A., Martín-Mayor, V., Muñoz Sudupe, A.: Critical exponents of the three-dimensional diluted Ising model. *Phys. Rev. B*, 58:2740, 1998.
4. Bray, A.J., Huifang, D.: Griffiths singularities in random magnets: Results for a soluble model. *Phys. Rev. B*, 40:6980, 1989.
5. Bhatt, R.N., Lee, P.A.: Scaling Studies of Highly Disordered Spin-1/2 Antiferromagnetic Systems. *Phys. Rev. Lett.*, 48:344, 1982.
6. Bray A.J., Rodgers, G.J.: Dynamics of random ising ferromagnets in the Griffiths phase. *Phys. Rev. B*, 38:9252, 1988.
7. Bray, A.J.: Nature of the Griffiths phase. Phys. Rev. Lett., 59:586, 1987.
8. Bray, A.J.: Dynamics of dilute magnets above T_c. *Phys. Rev. Lett.*, 60:720, 1988.
9. Chopard B., Droz, M.: Cellular Automaton Modeling of Physical Systems. Cambridge University Press, Cambridge, 1998.
10. Chakravarty, S., Halperin, B.I., Nelson, D.R.: Two-dimensional quantum Heisenberg antiferromagnet at low temperatures. *Phys. Rev. B*, 39:2344, 1989.
11. Dickman, R.: Reweighting in nonequilibrium simulations. *Phys. Rev. E*, 60:R2441, 1999.
12. Dobrosavljevic V., Miranda, E.: Absence of Conventional Quantum Phase Transitions in Itinerant Systems with Disorder. *Phys. Rev. Lett.*, 94:187203, 2005.
13. Dhar, D., Randeria, M., Sethna, J.: Griffiths singularities in the dynamics of disordered Ising models. *Europhys. Lett.*, 5:485, 1988.
14. Dickison, M, Vojta, T.: Monte Carlo simulations of the smeared phase transition in a contact process with extended defects. *J. Phys. A*, 38:1199, 2005.

15. Fisher, D.S.: Random transverse-field Ising spin chains. *Phys. Rev.*, 69:534, 1992.
16. Fisher, D.S.: Random antiferromagnetic quantum spin chains. *Phys. Rev. B*, 50:3799, 1994.
17. Fisher, D.S.: Critical behavior of random transverse-field Ising spin chains. *Phys. Rev. B*, 51:6411, 1995.
18. Ferrenberg A.M., Landau, D.P.: Critical behavior of the three-dimensional Ising model: A high-resolution Monte Carlo study. *Phys. Rev. B*, 44:5081, 1991.
19. Guo, M., Bhatt R., Huse, D.: Quantum Griffiths singularities in the transverse-field Ising spin glass. *Phys. Rev. B*, 54:3336, 1996.
20. Grassberger, P.: On phase transitions in Schlögl's second model. *Z. Phys. B*, 47:365, 1982.
21. Griffiths, R.B.: Nonanalytic behavior above the critical point in a random Ising ferromagnet. *Phys. Rev. Lett.*, 23:17, 1969.
22. Grinstein, G.: Phases and phase transitions of quenched disordered systems. In: Cohen, E.G.D. (ed) Fundamental Problems in Statistical Mechanics VI. Elsevier, New York (1985) p.147
23. Grassberger, P., de la Torre, A.: Reggeon field theory (Schlögl's first model) on a lattice: Monte Carlo calculations of critical behaviour. *Ann. Phys. (NY)*, 122:373 , 1979.
24. Harris, A.B.: Effect of random defects on the critical behaviour of Ising models. *J. Phys. C*, 7:1671, 1974.
25. Harris, T.E.: Contact interactions on a lattice. *Ann. Prob.*, 2:969, 1974.
26. Hida, K.: Low temperature properties of the double layer quantum Heisenberg antiferromagnet-modified spin wave method. *J. Phys. Soc. Jpn.*, 59:2230, 1990.
27. Hinrichsen, H.: Non-equilibrium critical phenomena and phase transitions into absorbing states. *Adv. Phys.*, 49:815, 2000.
28. Hooyberghs, J., Igloi, F., Vanderzande, C.: Strong Disorder Fixed Point in Absorbing-State Phase Transitions. *Phys. Rev. Lett.*, 90:100601, 2003.
29. Holm C., Janke, W.: Critical exponents of the classical three-dimensional Heisenberg model: A single-cluster Monte Carlo study. *Phys. Rev. B*, 48:936, 1993.
30. Janssen, H.K.: On the nonequilibrium phase transition in reaction-diffusion systems with an absorbing stationary state. *Z. Phys. B*, 42:151, 1981.
31. Janssen, H.K.: Renormalized field theory of the Gribov process with quenched disorder. *Phys. Rev. E*, 55:6253, 1997.
32. L'Ecuyer, P.: Tables of maximally-equidistributed combined LFSR generators. *Mathematics of Computation,*, 68:225, 261, 1999.
33. Lubensky, T.C: Critical properties of random-spin models from the epsilon expansion. *Phys. Rev. B*, 11:3573, 1975.
34. McCoy, B.M.: Incompleteness of the Critical Exponent Description for Ferromagnetic Systems Containing Random Impurities. *Phys. Rev. Lett.*, 23:383, 1969.
35. Marro, J., Dickman, R.: Nonequilibrium Phase Transitions in Lattice Models. Cambridge University Press, Cambridge (1996)
36. Moreira, A.G., Dickman, R.: Critical dynamics of the contact process with quenched disorder. *Phys. Rev. E*, 54:R3090 , 1996.
37. Ma, S.K., Dasgupta, C., Hu, C.-K.: Random antiferromagnetic chain. *Phys. Rev. Lett.*, 43:1434 , 1979.
38. Millis, A.J., Monien, H.: Spin Gaps and Spin Dynamics in $La_{2-x}Sr_xCuO_4$ and $YBa_2Cu_3O_{7-\delta}$. *Phys. Rev. Lett.*, 70:2810, 1993.

39. Motrunich, O., Mau, S.-C., Huse, D.A., Fisher, D.S.: Infinite-randomness quantum Ising critical fixed points. *Phys. Rev. B*, 61:1160, 2000.
40. Millis, A.J., Morr, D.K., Schmalian, J.: Local Defect in Metallic Quantum Critical Systems. *Phys. Rev. Lett.*, 87:167202, 2001.
41. Millis, A.J., Morr, D.K., Schmalian, J.: Quantum Griffiths effects in metallic systems. *Phys. Rev. B*, 66:174433, 2002.
42. McCoy B.M., Wu, T.T.: Theory of a Two-Dimensional Ising Model with Random Impurities. I. Thermodynamics. *Phys. Rev.*, 176:631, 1968.
43. Noest, A.J.: New universality for spatially disordered cellular automata and directed percolation. *Phys. Rev. Lett.*, 57:90, 1986.
44. Odor G.: Universality classes in nonequilibrium lattice systems. *Rev. Mod. Phys.*, 76:663, 2004.
45. Press, W.H., Teukolsky, S.A., Vetterling, W.T., Flannary, B.P.: Numerical Recipes in Fortran. Cambridge University Press, Cambridge (1992)
46. Pich, C., Young, A.P., Rieger, H., Kawashima, N.: Critical behavior and Griffiths-McCoy singularities in the two-dimensional random quantum Ising ferromagnet. *Phys. Rev. Lett.*, 81:5916, 1998.
47. Randeria, M., Sethna, J., Palmer, R.G.: Low-Frequency Relaxation in Ising Spin-Glasses. *Phys. Rev. Lett.*, 54:1321, 1985.
48. Rieger, H., Young, A.P.: Griffiths Singularities in the Disordered Phase of a Quantum Ising Spin Glass. *Phys. Rev. B*, 54:3328, 1996.
49. Sachdev, S.: Quantum Phase Transitions. Cambridge University Press, Cambridge (1999)
50. Sandvik, A.W.: Multicritical Point in a Diluted Bilayer Heisenberg Quantum Antiferromagnet. *Phys. Rev. Lett.*, 89:177201, 2002.
51. Sandvik, A.W., Scalapino, D.J.: Order-disorder transition in a two-layer quantum antiferromagnet. *Phys. Rev. Lett.*, 72:2777, 1994.
52. S. Sachdev and M. Vojta, Non-magnetic impurities as probes of insulating and doped Mott insulators in two dimensions. In: Fokas, A. et al (Eds.): Proceedings of the XIII International Congress on Mathematical Physics. International Press, Boston (2001).
53. Sknepnek, R., Vojta, T.: Smeared phase transition in a three-dimensional Ising model with planar defects: Monte-Carlo simulations. *Phys. Rev. B*, 69:174410, 2004.
54. Sknepnek, R., Vojta, T., Vojta, M.: Exotic vs. conventional scaling and universality in a disordered bilayer quantum Heisenberg antiferromagnet. *Phys. Rev. Lett.*, 93:097201, 2004.
55. Schmittmann B., Zia, R.K.P.: Statistical mechanics of driven diffusive systems. In: Domb, C., Lebowitz, J.L.: Phase transitions and critical phenomena, Vol. 17. Academic Press, New York (1995)
56. Thill M., Huse D.: Equilibrium behaviour of quantum Ising spin glass. *Physica A* , 214:321, 1995.
57. Täuber, U.C., Howard, M., Vollmayr-Lee, B.P.: Applications of field-theoretic renormalization group methods to reaction-diffusion problems. *J. Phys. A*, 38:R79, 2005.
58. Vojta, T., Dickison, M.: Critical behavior and Griffiths effects in the disodered contact process. *Phys. Rev. E*, 72:036126, 2005.
59. Vajk, O.P., Greven, M.: Quantum Versus Geometric Disorder in a Two-Dimensional Heisenberg Antiferromagnet. *Phys. Rev. Lett.*, 89:177202, 2002.

60. Vojta. T.: Disorder induced rounding of certain quantum phase transitions. *Phys. Rev. Lett.*, 90:107202, 2003.
61. Vojta, T.: Smearing of the phase transition in Ising systems with planar defects. *J. Phys. A*, 36:10921, 2003.
62. Vojta, T.: Broadening of a nonequilibrium phase transition by extended structural defects. *Phys. Rev. E*, 70:026108, 1994.
63. Vojta, T., Schmalian, S.: Quantum Griffiths effects in itinerant Heisenberg magnets. *Phys. Rev. B*, 72:045438, 2005.
64. Voigt, C.A., Ziff, R.M.: Epidemic analysis of the second-order transition in the Ziff-Gulari-Barshad surface-reaction model. *Phys. Rev. E*, 56:R6241, 1997.
65. Wiseman S., Domany E.: Finite-size scaling and Lack of self-averaging in critical disordered systems. *Phys. Rev. Lett.*, 81:22, 1998.
66. Wolff, U.: Collective Monte Carlo updating for spin systems. *Phys. Rev. Lett.*, 62:361, 1989.
67. Young, A.P., Rieger, H.: Numerical study of the random transverse-field Ising spin chain. *Phys. Rev. B*, 53:8486, 1996.

Localization of Electronic States in Amorphous Materials: Recursive Green's Function Method and the Metal-Insulator Transition at $E \neq 0$

Alexander Croy[1], Rudolf A. Römer[2], and Michael Schreiber[1]

[1] Technische Universität Chemnitz, Institut für Physik
09107 Chemnitz, Germany
alexander.croy@s2000.tu-chemnitz.de, schreiber@physik.tu-chemnitz.de
[2] Centre for Scientific Computing and Department of Physics
University of Warwick, Coventry, CV4 7AL, United Kingdom
r.roemer@warwick.ac.uk

1 Introduction

Traditionally, condensed matter physics has focused on the investigation of perfect crystals. However, real materials usually contain impurities, dislocations or other defects, which distort the crystal. If the deviations from the perfect crystalline structure are large enough, one speaks of *disordered systems*. The Anderson model [1] is widely used to investigate the phenomenon of localisation of electronic states in disordered materials and electronic transport properties in mesoscopic devices in general. Especially the occurrence of a quantum phase transition driven by disorder from an insulating phase, where all states are localised, to a metallic phase with extended states, has led to extensive analytical and numerical investigations of the critical properties of this metal-insulator transition (MIT) [2–4]. The investigation of the behaviour close to the MIT is supported by the one-parameter scaling hypothesis [5,6]. This scaling theory originally formulated for the conductance plays a crucial role in understanding the MIT [7]. It is based on an ansatz interpolating between metallic and insulating regimes [8]. So far, scaling has been demonstrated to an astonishing degree of accuracy by numerical studies of the Anderson model [9–13]. However, most studies focused on scaling of the localisation length and the conductivity at the disorder-driven MIT in the vicinity of the band centre [9,14,15]. Assuming a power-law form for the d.c. conductivity, as it is expected from the one-parameter scaling theory, Villagonzalo et al. [6] have used the Chester-Thellung-Kubo-Greenwood formalism to calculate the temperature dependence of the thermoelectric properties numerically and showed that all thermoelectric quantities follow single-parameter scaling laws [16,17].

In this chapter we will investigate whether the scaling assumptions made in previous studies for the transition at energies outside the band centre can be reconfirmed in numerical calculations, and in particular whether the conductivity σ follows a power law close to the critical energy E_c. For this purpose we will use the recursive Green's function method [18, 19] to calculate the four-terminal conductance of a disordered system for fixed disorder strength at temperature $T = 0$. Applying the finite-size scaling analysis we will compute the critical exponent and determine the mobility edge, i.e. the MIT outside the band centre. A complementary investigation into the statistics of the energy spectrum and the states close to the MIT can be found in Chap. [20]. An analysis of the mathematical properties of the so-called binary-alloy or Bernoulli-Anderson model is done in Chap. [21].

2 The Anderson model of localisation and its metal-insulator transition

The Anderson model [1, 2] is widely used to investigate the phenomenon of localisation of electronic states in disordered materials. It is based upon a tight-binding Hamiltonian in site representation

$$\mathcal{H} = \sum_i \epsilon_i |i\rangle\langle i| + \sum_{i \neq j} t_{ij} \, |i\rangle\langle j| \,, \tag{1}$$

where $|i\rangle$ is a localized state at site i and t_{ij} are the hopping parameters, which are usually restricted to nearest neighbours. The on-site potentials ϵ_i are random numbers, chosen according to some distribution $P(\epsilon)$ [22, 23]. In what follows we take $P(\epsilon)$ to be a box distribution over the interval $[-W/2, W/2]$, thus W determines the strength of the disorder in the system. Other distributions have also been considered [2, 3, 24].

For strong enough disorder, $W > W_c(E = 0)$, all states are exponentially localized and the respective wave functions $\Psi(\mathbf{r})$ are proportional to $\mathrm{e}^{-|\mathbf{r}-\mathbf{r}_0|/\xi}$ for large distances $|\mathbf{r} - \mathbf{r}_0|$. Thus, Ψ is confined to a region of some finite size, which may be described by the so-called localisation length ξ. In this language extended states are characterised by $\xi \longrightarrow \infty$. Comparing ξ with the size L of the system one can distinguish between *strong* and *weak localisation*, for $\xi \ll L$ and $L < \xi$, respectively[3]. Here we also assume that the phase-relaxation length $\ell_\Phi \gg L$. Otherwise, the effective system size is determined by ℓ_Φ.

It turns out that the value of the critical disorder strength W_c depends on the distribution function $P(\varepsilon)$ and the dimension d of the system. In absence

[3] We note that the phrase *weak localization* in the context of the scaling theory is often used with a specific meaning, namely the onset of localization in large 2-dimensional samples, where the conductance decreases logarithmically with scale [2, 7].

of a magnetic field and for $d \leq 2$ all states are localized[4], i.e. $W_c = 0$ [7,8]. For systems with $d = 3$ the value of W_c additionally depends on the Fermi energy E and the curve $W_c(E)$ separates localized states ($W > W_c(E)$) from extended states ($W < W_c(E)$) in the phase diagram [22,23,26]. If instead of E the disorder strength is taken as a parameter, there will be a critical energy $E_c(W)$ — also called the mobility edge — and states with $|E| < E_c$ are extended and those with $|E| > E_c$ localized yielding the same phase boundary in the (E, W)-plane. At the mobility edge, states are multifractals [27]. The separation of localized and extended states is illustrated in Fig. 1, which shows a schematic density of states (DOS) of a three-dimensional (3D) Anderson model. Since for $T = 0$ localized states cannot carry any electric current, the

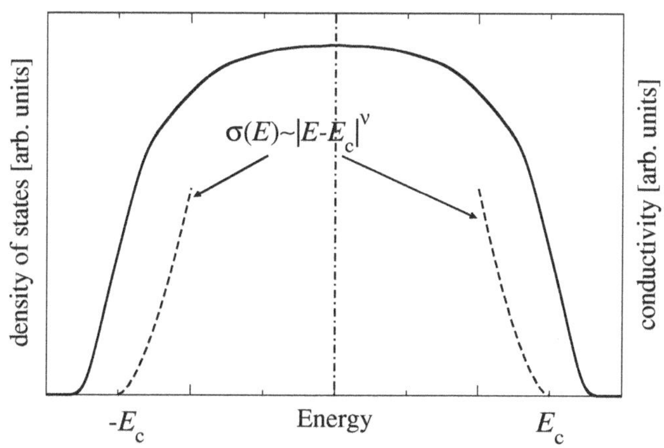

Fig. 1. Typical DOS of a 3D Anderson model for fixed $W < W_c$. The states in the grey regions are localized, otherwise they are extended. The mobility edges are indicated at $\pm E_c$. Also indicated is the power-law behaviour of $\sigma(E)$ (dashed lines) close to $\pm E_c$ according to (2)

system shows insulating behaviour, i.e. the electric conductivity σ vanishes for $|E| > E_c$ or $W > W_c$. Otherwise the system is metallic. Therefore, the transition at the critical point is called a *disorder-driven* MIT.

For the MIT in $d = 3$ it was found that σ is described by a power law at the critical point [2],

$$\sigma(E) = \begin{cases} \sigma_0 \left|1 - \frac{E}{E_c}\right|^\nu, & |E| < E_c \\ 0, & |E| > E_c \end{cases} \tag{2}$$

with ν being the universal critical exponent of the phase transition and σ_0 a constant. The value of ν has been computed numerically by various methods

[4]Strictly speaking, this is only true if $\mathcal{H}$ belongs to the Gaussian orthogonal ensemble [25].

[2,9–11] and was also derived from experiments [28,29]. The results range from 1 to 1.6, depending on the distribution $P(\epsilon)$ and the computational method [3] used.

Moreover, Wegner [30] was able to show that for non-interacting electrons the d.c. conductivity σ obeys a general scaling form close to the MIT,

$$\sigma(\varepsilon,\omega) = b^{2-d}\,\sigma(b^{1/\nu}\varepsilon, b^{z}\omega)\ . \tag{3}$$

Here ε denotes the dimensionless distance from the critical point, ω is an external parameter such as the frequency or the temperature, b is a scaling parameter and z is the dynamical exponent. For non-interacting electrons $z = d$ [31]. Assuming a finite conductivity for $\omega = 0$, one obtains from (3)

$$\sigma(\varepsilon, 0) \propto \varepsilon^{\nu(d-2)}, \tag{4}$$

where $\varepsilon = |1 - E/E_{\mathrm{c}}|$. With $d = 3$ this gives (2).

3 Computational method

An approach to calculate the d.c. conductivity from the Anderson tight-binding Hamiltonian (1) is the recursive Green's function method [18, 19, 23]. It yields a recursion scheme for the d.c. conductivity tensor starting from the Kubo-Greenwood formula [32]. Moreover, this method allows to compute the density of states and the localization length as well as the full set of thermoelectric kinetic coefficients [33]. Parallel implementations of the method are advantageous [34,35]. The method is therefore a companion to the more widely used transfer-matrix method [36, 37] or iterative diagonalisation schemes [38, 39].

3.1 Recursive Green's function method

Let $\mathcal{H} = \sum_{ij} H_{ij}|i\rangle\langle j|$ denote our hermitian tight-binding Hamiltonian. The single particle Green's function $\mathcal{G}^{\pm}(z)$ is defined as [40] $(z^{\pm} - \mathcal{H})\mathcal{G}^{\pm} = \mathbb{1}$ where $z = E \pm \mathrm{i}\gamma$ is the *complex energy* and the sign of the small imaginary part γ distinguishes between advanced and retarded Green's functions, $\mathcal{G}^{-}(E - \mathrm{i}0)$ and $\mathcal{G}^{+}(E + \mathrm{i}0)$, respectively [40]. Equivalently, $\mathcal{G}^{\pm}$ can be represented in the basis of the functions $|i\rangle$,

$$(z^{\pm}\delta_{ij} - H_{ij})G^{\pm}_{ij} = \delta_{ij}\ , \tag{5}$$

where $G^{\pm}_{ij}$ is the matrix element $\langle i|\mathcal{G}^{\pm}|j\rangle$. We note that for a hermitian Hamiltonian $G^{-}_{ij} = (G^{+}_{ji})^{*}$.

If $\mathcal{H}$ contains only nearest-neighbour hopping matrix elements, (5) can be simplified using a block matrix notation. This is equivalent to considering the system as being built up of slices or strips for 3D or 2D, respectively, along

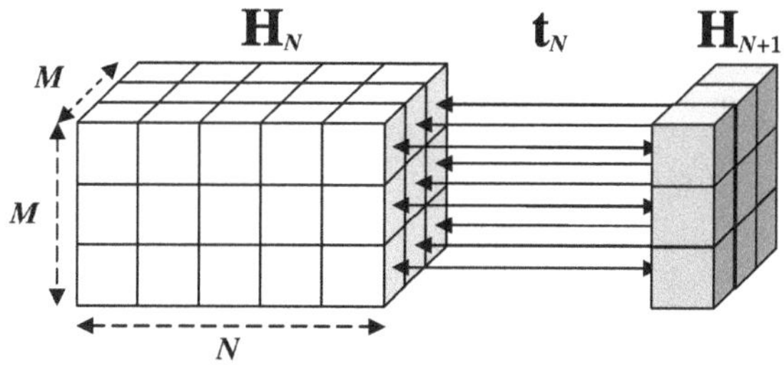

Fig. 2. Scheme of the recursive Green's function method for a 3D system. The new Green's function $\mathbf{G}^{(N+1)}$ can be calculated from the old Hamiltonian $\mathbf{H}_N$ (light grey), the new slice Hamiltonian $\mathbf{H}_{N+1}$ (dark grey) and the coupling $\mathbf{t}_N$ (solid arrows)

one lattice direction. In what follows all quantities written in bold capitals are matrices acting in the subspace of such a slice or strip. For 2D and 3D these are matrices of size $M \times M$ and $M^2 \times M^2$, respectively, where M is the lateral extension of the system (cf. Fig. 2). The left hand side of (5) is then given as

$$\begin{pmatrix} \ddots & \ddots & \ddots & 0 & 0 & \cdots \\ 0 & -\mathbf{H}_{i,i-1} & (z^{\pm}\mathbf{I} - \mathbf{H}_{ii}) & -\mathbf{H}_{i,i+1} & 0 & \cdots \\ \cdots & 0 & -\mathbf{H}_{i+1,i} & (z^{\pm}\mathbf{I} - \mathbf{H}_{i+1,i+1}) & -\mathbf{H}_{i+1,i+2} & 0 \\ \cdots & 0 & 0 & \ddots & \ddots & \ddots \end{pmatrix} \times \begin{pmatrix} \ddots & \vdots & \ddots \\ \cdots & \mathbf{G}^{\pm}_{i-1,j} & \cdots \\ \cdots & \mathbf{G}^{\pm}_{ij} & \cdots \\ \cdots & \mathbf{G}^{\pm}_{i+1,j} & \cdots \\ \ddots & \vdots & \ddots \end{pmatrix} , \tag{6}$$

where i and j now label the slices or strips. From this expression one can easily see that (5) is equivalent to

$$(z^{\pm}\mathbf{I} - \mathbf{H}_{ii})\mathbf{G}^{\pm}_{ij} - \mathbf{H}_{i,i-1}\mathbf{G}^{\pm}_{i-1,j} - \mathbf{H}_{i,i+1}\mathbf{G}^{\pm}_{i+1,j} = \mathbf{I}\delta_{ij} \,. \tag{7}$$

Using the hermiticity of $\mathcal{H}$ we define the hopping matrix $\mathbf{t}_i \equiv \mathbf{H}_{i,i+1}$ (and hence $\mathbf{t}^{\dagger}_i = \mathbf{H}_{i,i-1}$) connecting the ith and the $(i+1)$st slice. Now, we consider adding an additional slice to a system consisting of N slices. The Hamiltonian of this larger system can be written as [19]

$$\mathcal{H}^{(N+1)} \longrightarrow \mathbf{H}_{ij} + \mathbf{t}_N + \mathbf{t}^{\dagger}_N + \mathbf{H}_{N+1,N+1} \quad (i,j \le N). \tag{8}$$

The first and the last terms describe the uncoupled N-slice and the additional 1-slice system. Using $\mathbf{t}_N$ as an "interaction" the Green's function $\mathbf{G}^{(N+1)}$ of the coupled system can be calculated via Dyson's equation [19, 40],

$$\mathbf{G}_{ij}^{(N+1)} = \mathbf{G}_{ij}^{(N)} + \mathbf{G}_{iN}^{(N)} \mathbf{t}_N \mathbf{G}_{Nj}^{(N+1)} \quad (i, j \leq N). \tag{9}$$

In particular, we have

$$\mathbf{G}_{N+1,N+1}^{(N+1)} = \left[z^{\pm}\mathbf{I} - \mathbf{H}_{N+1,N+1} - \mathbf{t}_N^{\dagger} \mathbf{G}_{NN}^{(N)} \mathbf{t}_N \right]^{-1} \tag{10a}$$

$$\mathbf{G}_{ij}^{(N+1)} = \mathbf{G}_{ij}^{(N)} + \mathbf{G}_{iN}^{(N)} \mathbf{t}_N \mathbf{G}_{N+1,N+1}^{(N+1)} \mathbf{t}_N^{\dagger} \mathbf{G}_{Nj}^{(N)} \quad (i, j \leq N) \tag{10b}$$

$$\mathbf{G}_{i,N+1}^{(N+1)} = \mathbf{G}_{iN}^{(N)} \mathbf{t}_N \mathbf{G}_{N+1,N+1}^{(N+1)} \quad (i \leq N) \tag{10c}$$

$$\mathbf{G}_{N+1,j}^{(N+1)} = \mathbf{G}_{N+1,N+1}^{(N+1)} \mathbf{t}_N^{\dagger} \mathbf{G}_{Nj}^{(N)} \quad (j \leq N)\,. \tag{10d}$$

With (10) the Green's function can be obtained iteratively. Additionally, there are two kinds of boundary conditions which must be considered: across each slice and at the beginning and the end of the stack. The first kind does not present any difficulty and usually hard wall or periodic boundary conditions are employed. The second kind of boundary is connected to some subtleties with attached leads which will be addressed in Sect. 6.

3.2 Density of states and d.c. conductivity

The DOS is given in terms of Green's function by [40]

$$\rho(E) = -\frac{1}{\pi \Omega} \mathrm{Im}\,\mathrm{Tr}\,\mathcal{G}^{+} = -\frac{1}{\pi N M^2} \mathrm{Im} \sum_{i=1}^{N} \mathrm{Tr}\,\mathbf{G}_{ii}^{+} \tag{11}$$

and the d.c. conductivity σ is

$$\sigma = \frac{2e^2\hbar}{\pi \Omega m^2} \mathrm{Tr}\,\left[p\,\mathrm{Im}\,\mathcal{G}^{+}\, p\,\mathrm{Im}\,\mathcal{G}^{+} \right]\,. \tag{12}$$

Here, Ω denotes the volume of the system and m the electron mass. Using for the momentum the relation $p = \frac{\mathrm{i}m}{\hbar}[\mathcal{H}, x]$ one can rewrite (12) in position representation

$$\sigma = \frac{e^2 4}{h N M^2} \mathrm{Tr} \left\{ \gamma^2 \sum_{i,j}^{N} \mathbf{G}_{ij}^{+}\, x_j\, \mathbf{G}_{ji}^{-}\, x_i - \mathrm{i}\frac{\gamma}{2} \sum_{i}^{N} (\mathbf{G}_{ii}^{+} - \mathbf{G}_{ii}^{-})\, x_i^2 \right\}\,, \tag{13}$$

where x_i is the position of the ith slice.

Starting from these relations and using the iteration scheme (10) one can derive recursion formulæ to calculate the properties for the $(N+1)$-slice system. The results are expressed in terms of the following auxiliary matrices

$$\mathbf{R}_N = \mathbf{G}^+_{N,N} , \tag{14a}$$

$$\mathbf{B}_N = \gamma \mathbf{t}^\dagger_N \left[\sum_{ij}^{N} \mathbf{G}^{+(N)}_{Nj} x_j (2\gamma \mathbf{G}^{-(N)}_{ji} - \mathrm{i}\mathbf{I}\delta_{ij}) x_i \mathbf{G}^{+(N)}_{iN} \right] \mathbf{t}_N , \tag{14b}$$

$$\mathbf{C}^+_N = \gamma \mathbf{t}^\dagger_N \left[\sum_{i=1}^{N} \mathbf{G}^{+(N)}_{Ni} x_i \mathbf{G}^{-(N)}_{iN} \right] \mathbf{t}_N = (\mathbf{C}^+_N)^\dagger , \tag{14c}$$

$$\mathbf{C}^-_N = \gamma \mathbf{t}^\dagger_N \left[\sum_{i=1}^{N} \mathbf{G}^{-(N)}_{Ni} x_i \mathbf{G}^{+(N)}_{iN} \right] \mathbf{t}_N = (\mathbf{C}^-_N)^\dagger , \tag{14d}$$

$$\mathbf{F}_N = \mathbf{t}^\dagger_N \left[\sum_{i=1}^{N} \mathbf{G}^{+(N)}_{Ni} \mathbf{G}^{+(N)}_{iN} \right] \mathbf{t}_N . \tag{14e}$$

The derivation can be simplified assuming the new slice to be at $x_{N+1} = 0$. This leads, however, to corrections for the matrices $\mathbf{B}_N$ and $\mathbf{C}^\pm_N$ because the origin of x_i has to be shifted to the position of the current slice in each iteration step. The corrections are

$$\mathbf{B}'_N = \mathbf{B}_N + \mathrm{i}\mathbf{C}^+_N + \mathrm{i}\mathbf{C}^-_N + \frac{1}{2}\mathbf{t}^\dagger_N (\mathbf{R}_N - \mathbf{R}^\dagger_N)\mathbf{t}_N , \tag{15a}$$

$$\mathbf{C}'^\pm_N = \mathbf{C}^\pm_N - \mathrm{i}\frac{1}{2}\mathbf{t}^\dagger_N (\mathbf{R}_N - \mathbf{R}^\dagger_N)\mathbf{t}_N . \tag{15b}$$

Here we have used the identity

$$\gamma \sum_{i=1}^{N} \mathbf{G}^+_{Ni} \mathbf{G}^-_{iN} = \mathrm{i}\frac{1}{2}(\mathbf{G}^+_{NN} - \mathbf{G}^-_{NN}) = \mathrm{i}\frac{1}{2}(\mathbf{R}_N - \mathbf{R}^\dagger_N) = -\mathrm{Im}\,\mathbf{R}_N . \tag{16}$$

The derivation of the recursion relations is given in Refs. [18, 19, 23, 41], it yields the following expressions

$$s^{(N+1)}_\rho = s^{(N)}_\rho + \mathrm{Tr}\,\{\mathbf{R}_{N+1}(\mathbf{F}_N + \mathbf{I})\} , \tag{17a}$$

$$s^{(N+1)}_\sigma = s^{(N)}_\sigma + \mathrm{Tr}\,\{\mathrm{Re}\,(\mathbf{B}_N \mathbf{R}_{N+1}) + \mathbf{C}^+_N \mathbf{R}^\dagger_{N+1} \mathbf{C}^-_N \mathbf{R}_{N+1}\} , \tag{17b}$$

$$\mathbf{R}_{N+1} = \left[z^\pm \mathbf{I} - \mathbf{H}_{N+1,N+1} - \mathbf{t}^\dagger_N \mathbf{R}_N \mathbf{t}_N \right]^{-1} , \tag{17c}$$

$$\mathbf{B}_{N+1} = \mathbf{t}^\dagger_{N+1} \mathbf{R}_{N+1} \left[\mathbf{B}_N + 2\mathbf{C}^+_N \mathbf{R}^\dagger_{N+1} \mathbf{C}^-_N \right] \mathbf{R}_{N+1} \mathbf{t}_{N+1} , \tag{17d}$$

$$\mathbf{C}^+_{N+1} = \mathbf{t}^\dagger_{N+1} \mathbf{R}_{N+1} \mathbf{C}^+_N \mathbf{R}^\dagger_{N+1} \mathbf{t}_{N+1} , \tag{17e}$$

$$\mathbf{C}^-_{N+1} = \mathbf{t}^\dagger_{N+1} \mathbf{R}^\dagger_{N+1} \mathbf{C}^-_N \mathbf{R}_{N+1} \mathbf{t}_{N+1} , \tag{17f}$$

$$\mathbf{F}_{N+1} = \mathbf{t}^\dagger_{N+1} \mathbf{R}_{N+1} (\mathbf{F}_N + \mathbf{I}) \mathbf{R}_{N+1} \mathbf{t}_{N+1} . \tag{17g}$$

The DOS and the d.c. conductivity are then given as

$$\rho^{(N+1)}(E) = -\frac{1}{\pi(N+1)M^2} s_\rho^{(N+1)} , \tag{18}$$

$$\sigma^{(N+1)}(E) = \frac{e^2}{h} \frac{4}{(N+1)M^2} s_\sigma^{(N+1)} . \tag{19}$$

For a comparison with the scaling arguments, we convert the conductivity into the *two-terminal* conductance as

$$g_2 = \sigma \frac{M^2}{L} \tag{20}$$

with $L = N+1$. In distinction to the usual use of the recursive scheme which constructs a single sample with $L \gg M$, we shall have to use many different *cubic* samples with $L = M$.

We note that it is also possible to calculate the localisation length $\xi(E)$ by the Green's function method. The value of $\xi(E)$ is connected to the matrix $\mathbf{G}^{+}_{1N+1}$,

$$\frac{1}{\xi(E)} = -\lim_{\gamma \to 0} \lim_{N \to \infty} \frac{1}{2N} \ln \left| \mathrm{Tr}\, \mathbf{G}^{+}_{1N}(E) \right|^2 . \tag{21}$$

The recursion relation for $\xi^{(N+1)}(E)$ is

$$\frac{1}{\xi^{(N+1)}(E)} = -\frac{1}{N+1} s_\xi^{(N+1)} , \tag{22a}$$

$$s_\xi^{(N+1)} = s_\xi^{(N)} + \ln \left| \mathrm{Tr}\, \mathbf{G}^{+(N+1)}_{N+1,N+1} \right| . \tag{22b}$$

4 Finite-size scaling

For finite systems there can be no singularities induced by a phase transition and the divergences at the MIT are always rounded off [42]. Fortunately, the MIT can still be studied using a technique known as finite-size scaling [2]. Here we briefly review the main results taking the dimensionless four-terminal conductance g_4 of a large cubic sample of size $L \times L \times L$ as an example. We note that similar scaling ideas can also be applied to the reduced localisation length ξ/L. In order to obtain g_4 of the disordered region only, we have to subtract the contact resistance due to the leads. This gives

$$\frac{1}{g_4} = \frac{1}{g_2} - \frac{1}{\mathcal{N}} . \tag{23}$$

Here $\mathcal{N} = \mathcal{N}(E)$ is the number of propagating channels at the Fermi energy E which is determined by the quantization of wave numbers in transverse direction in the leads [43, 44].

Near the MIT one expects a one-parameter scaling law for the dimensionless conductance [7, 30, 42]

$$g_4(L, \varepsilon, b) = \mathcal{F}\left[\frac{L}{b}, \chi(\varepsilon)b^{1/\nu}\right] , \tag{24}$$

where b is the scale factor in the renormalisation group, χ is a relevant scaling variable and $\nu > 0$ is the critical exponent. The parameter ε measures the distance from the mobility edge E_c as in (2). However, recent advances in numerical precision have shown that in addition *corrections to scaling* due to the finite sizes of the sample need to be taken into account so that the general scaling form is

$$g_4(L, \varepsilon, b) = \mathcal{F}\left[\frac{L}{b}, \chi(\varepsilon)b^{1/\nu}, \phi(\varepsilon)b^{-y}\right] , \tag{25}$$

where ϕ is an irrelevant scaling variable and $y > 0$ is the corresponding irrelevant scaling exponent. The choice $b = L$ leads to the standard scaling form[5]

$$g_4(L, \varepsilon) = F\left[L^{1/\nu}\chi(\varepsilon), L^{-y}\phi(\varepsilon)\right] \tag{26}$$

with F being related to $\mathcal{F}$. For E close to E_c we may expand F up to order n_R in its first and up to order n_I in its second argument such that

$$g_4(L, \varepsilon) = \sum_{n'=0}^{n_I} \phi^{n'} L^{-n'y} F_{n'}\left(\chi L^{1/\nu}\right) \text{ with} \tag{27a}$$

$$F_{n'}(\chi L^{1/\nu}) = \sum_{n=0}^{n_R} a_{n'n}\chi^n L^{n/\nu} . \tag{27b}$$

Additionally χ and ϕ may be expanded in terms of the small parameter ε up to orders m_R and m_I, respectively. This procedure gives

$$\chi(\varepsilon) = \sum_{m=1}^{m_R} b_m \varepsilon^m, \quad \phi(\varepsilon) = \sum_{m'=0}^{m_I} c_{m'}\varepsilon^{m'} . \tag{28}$$

From (26) and (27) one can see that a finite system size results in a systematic shift of $g_4(L, \varepsilon = 0)$ with L, where the direction of the shift depends on the boundary conditions [42]. Consequently, the curves $g_4(L, \varepsilon)$ do not necessarily intersect at the critical point $\varepsilon = 0$ as one would expect from the scaling law (24). Neglecting this effect in high precision data will give rise to wrong values for the exponents.

Using a least-squares fit of the numerical data to (27) and (28) allows us to extract the critical parameters ν and E_c with high accuracy. One also obtains the finite-size corrections and can subtract these to show the anticipated scaling behaviour. This finite-size scaling analysis has been successfully applied to numerical calculations of the localisation length and the conductance within the Anderson model [3, 14].

[5] The choice of b is connected to the iteration of the renormalisation group [42].

5 MIT at $E = 0.5\,t$ for varying disorder

5.1 Scaling of the conductance

We first investigate the standard case of varying disorder at a fixed energy [14]. We choose $t_{ij} \neq 0$ for nearest neigbours i, j only, set $t_{ij} = t$ and $E = 0.5\,t$ which is close to the band centre. We impose hard wall boundary conditions in the transverse direction. For each combination of disorder strength W and system size L we generate an ensemble of 10000 samples. The systems under investigation are cubes of size $L \times L \times L$ for $L = 4, 6, 8, 10, 12$ and 14. For each sample we calculate the DOS $\rho(E, L)$ and the dimensionless two-terminal conductance g_2 using the recursive Green's function method explained in Section 3. Finally we compute the average DOS $\langle\rho(E, L)\rangle$, the *average* conductance $\langle g_4(E, L)\rangle$ and the *typical* conductance $\exp\langle\ln g_4(E, L)\rangle$.

The results for the different conductance averages are shown in Figs. 3 and 4 together with respective fits to the standard scaling form (26). Shown are the best fits that we obtained for various choices of the orders of the expansions (27, 28). The expansion orders and the results for the critical exponent and the critical disorder are given in Table 1. In Fig. 5 we show the same data as in Figs. 3 and 4 after the corrections to scaling have been subtracted indicating that the data points for different system sizes fall onto a common curve with two branches as it is expected from the one-parameter scaling theory. The

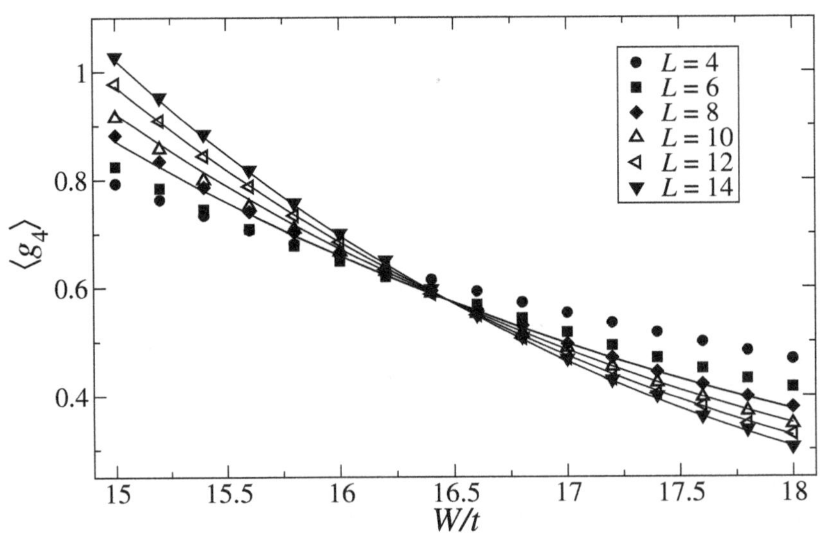

Fig. 3. Average dimensionless conductance vs disorder strength for $E = 0.5\,t$. System sizes are given in the legend. Errors of one standard deviation are obtained from the ensemble average and are smaller than the symbol sizes. Also shown (solid lines) are fits to (26) for $L = 8, 10, 12$ and 14

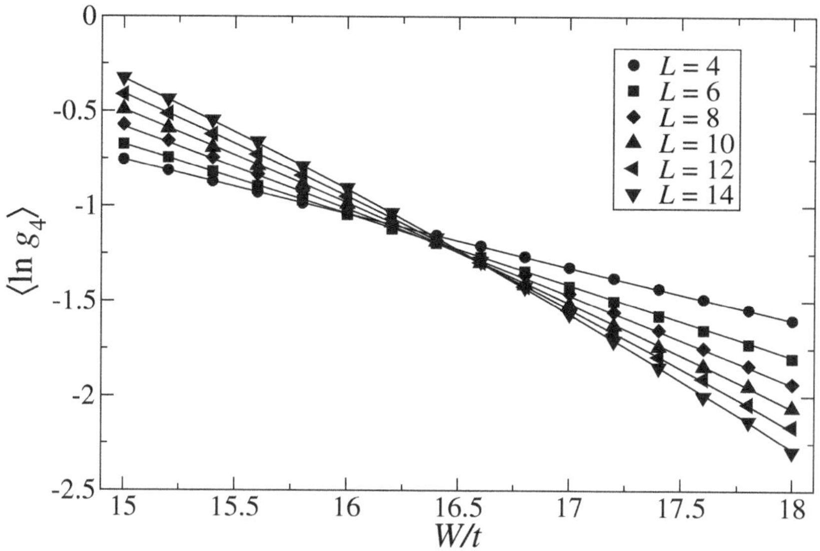

Fig. 4. Logarithm of the typical dimensionless conductance vs disorder strength for $E = 0.5\,t$. System sizes are given in the legend. Errors of one standard deviation are obtained from the ensemble average and are smaller than the symbol sizes. The solid lines are fits to (26)

results for the conductance averages and also the critical values are in good agreement with transfer-matrix calculations [9, 14].

Table 1. Best-fit estimates of the critical exponent and the critical disorder for both averages of g_4 using (26). The system sizes used were $L = 8, 10, 12, 14$ and $L = 4, 6, 8, 10, 12, 14$ for $\langle g \rangle$ and $\langle \ln g \rangle$, respectively. For each combination of disorder strength W and system size L we generate an ensemble of 10000 samples

average	$W_{\min}/t$	$W_{\max}/t$	n_R	n_I	m_R	m_I	ν	W_c/t	y
$\langle g_4 \rangle$	15.0	18.0	2	0	2	0	1.55 ± 0.11	16.47 ± 0.06	–
$\langle \ln g_4 \rangle$	15.0	18.0	3	1	1	0	1.55 ± 0.18	16.8 ± 0.3	0.8 ± 1.0

5.2 Disorder dependence of the density of states

The Green's function method enables us to compute the DOS of the disordered system. It should be independent of L. Figure 6 shows the average DOS at $E = 0.5\,t$ for different system sizes. There are still some fluctuations present. These can in principle be reduced by using larger system sizes and increasing

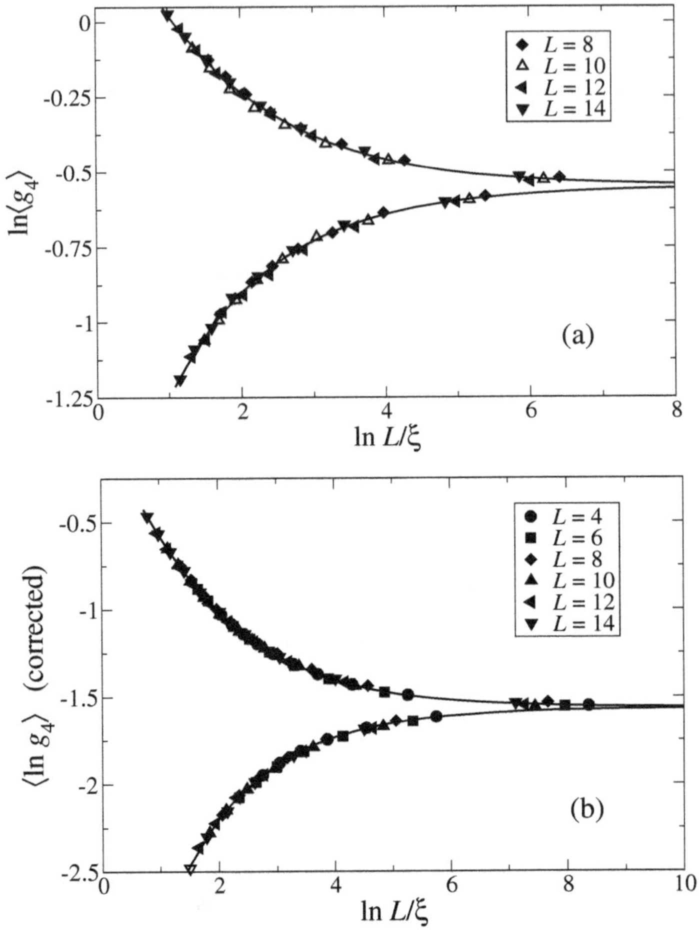

Fig. 5. Same data as in Figs. 3 and 4 after corrections to scaling are subtracted, plotted vs L/ξ to show single-parameter scaling. Different symbols indicate the system sizes given in the legend. The lines show the scaling function (26)

the number of samples. The fluctuations will be particularly inconvenient when trying to compute $\sigma(E)$.

The reduction of the DOS with increasing disorder strength can be understood from a simple argument. If the DOS were constant for all energies its value would be given by the inverse of the band width. In the Anderson model with box distribution for the on-site energies the band width increases linearly with the disorder strength W. The DOS in the Anderson model is not constant as a function of energy, nevertheless let us assume that for energies in the vicinity of the band centre the exact shape of the tails is not important. Therefore,

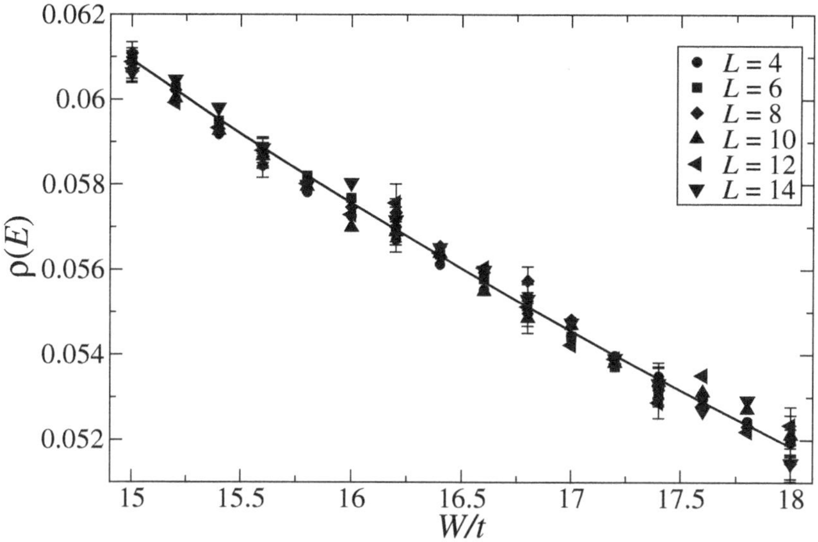

Fig. 6. Density of states vs disorder strength for $E = 0.5\,t$ and $L = 4, 6, 8, 10, 12, 14$. Errors of one standard deviation are obtained from the ensemble average and shown for every 4th data point only. The solid line shows a fit to (29) for $L = 6$ to illustrate the reduction of ρ with increasing disorder strength

$$\rho(W) \propto \frac{1}{B + \alpha W}\,, \tag{29}$$

shows a decrease of the DOS with W. Here B is an effective band width taking into account that the DOS is not a constant even for W=0. The parameter α allows for deviations due to the shape of the tails. In Fig. 6, we show that the data are indeed well described by (29).

6 Influence of the metallic leads

As mentioned in the introduction most numerical studies of the conductance have been focused on the disorder transition at or in the vicinity of the band centre. Let us now set $E = -5\,t$ and calculate the conductance averages as before. The results for the typical conductance are shown in Fig. 7a. Earlier studies of the localization length provided evidence of a phase transition around $W = 16.3\,t$ although the accuracy of the data was relatively poor [23]. Surprisingly, in Fig. 7a there seems to be no evidence of any transition nor of any systematic size dependence. The order of magnitude is also much smaller than in the case of $E = 0.5\,t$, although one expects the conductance at the MIT to be roughly similar.

The origin of this reduction can be understood from Fig. 8, which shows the DOS of a disordered sample and a clean system (i.e. without impurities and therefore without disorder) such as in the metallic leads. As already

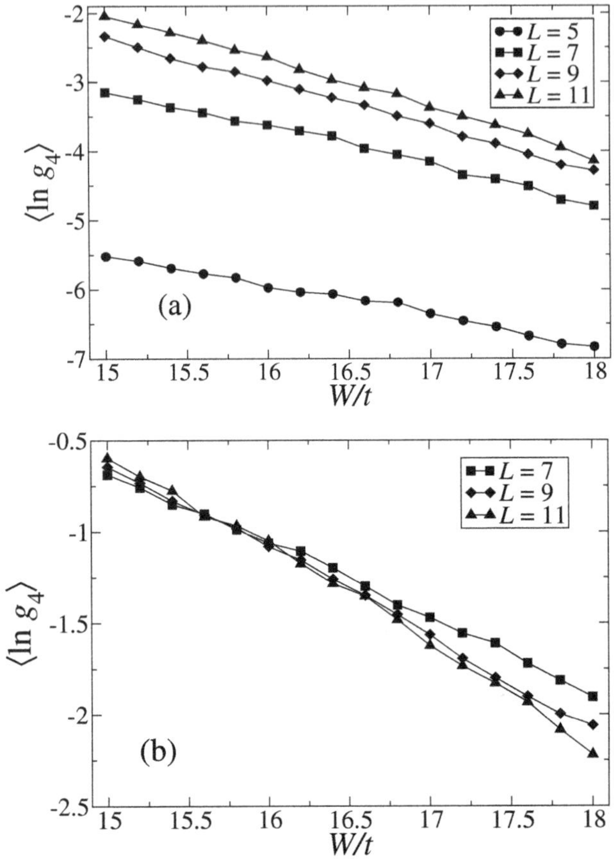

Fig. 7. System size dependence of the logarithm of the typical conductance for fixed energy $E = -5\,t$. Errors of one standard deviation are obtained from the ensemble average and are smaller than the symbol sizes. The lines are guides to the eye only. The upper plot was calculated using the metallic leads "as they are", i.e. the band centre of the leads coincides with the band centre in the disordered region. In the lower plot the band centre of the leads was "shifted" to the respective Fermi energy

pointed out in Ref. [19], the difference between the DOS in the leads and in the disordered region may lead to false results for the transport properties. Put to an extreme, if there are no states available at a certain energy in the leads, e.g. for $|E| \geq 6\,t$, there will be no transport regardless of the DOS and the conductance in the disordered system at that energy. The DOS of the latter system becomes always broadened by the disorder. Therefore, using the standard setup of system and leads, it appears problematic to investigate transport properties at energies outside the ordered band. Additionally, for energies $3\,t \lessapprox |E| < 6\,t$ the DOS of the clean system is smaller than the disorder broadened DOS. Thus the transport properties that crucially depend

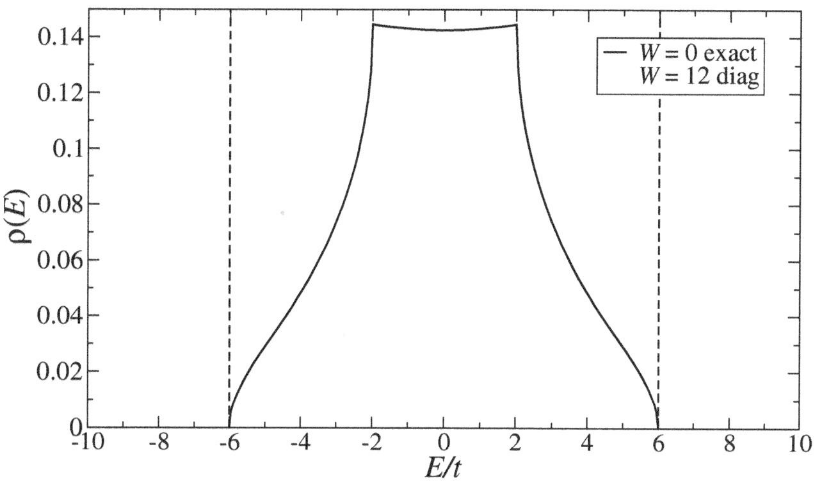

Fig. 8. DOS of a clean system (full black line) and a disordered system (grey) with $W = 12\,t$ and $L = 21$, obtained from diagonalising the Hamiltonian (1). The dashed lines indicate the band edges of the ordered system

on the DOS might be also changed in that energy range. The problems can be overcome by shifting the energy of the disordered region while keeping the Fermi energy in the leads in the lead-band centre (or vice versa). This is somewhat equivalent in spirit to applying a gate voltage to the disordered region and sweeping it — a technique similar to MOSFET experiments. The results for the typical conductance using this method are shown in Fig. 7(b). One can see some indication of scaling behaviour and also the order of magnitude is found to be comparable to the case of $E = 0.5\,t$. Another possibility of avoiding the DOS mismatch is choosing a larger hopping parameter in the leads [45], which results in a larger bandwidth, but also a lower DOS.

7 The MIT outside the band centre

Knowing the difficulties involving the metallic leads and using the "shifting technique" explained in the last section, we now turn our attention to the less-studied problem of the MIT at fixed disorder. We set the disorder strength to $W = 12\,t$ and again impose hard wall boundary conditions in the transverse direction [14]. We expect $E_c \approx 8\,t$ from the earlier studies of the localization length [23]. Analogous to the transition for varying W we generate for each combination of Fermi energy and system size an ensemble of 10000 samples (except for $L = 19$ and $L = 21$, where 4000 and 2000 samples, respectively, were generated) and examine the energy and size dependence of the average and the typical conductance, $\langle g_4 \rangle$ and $\exp\langle \ln g_4 \rangle$, respectively.

7.1 Energy dependence of the DOS

Before looking at the scaling behaviour of the conductance we have to make sure that the "shifting technique" indeed gives the right DOS outside the ordered band. Additionally, we have to check the average DOS for being independent of the system size. In Fig. 9 we show the DOS obtained from diagonalisation of the Anderson Hamiltonian with 30 configurations (using standard LAPACK subroutines) and the Green's function calculations. The Green's function data agree very well with the diagonalisation results, although there are still bumps around $E = -8.2\,t$ for small DOS values. The average DOS

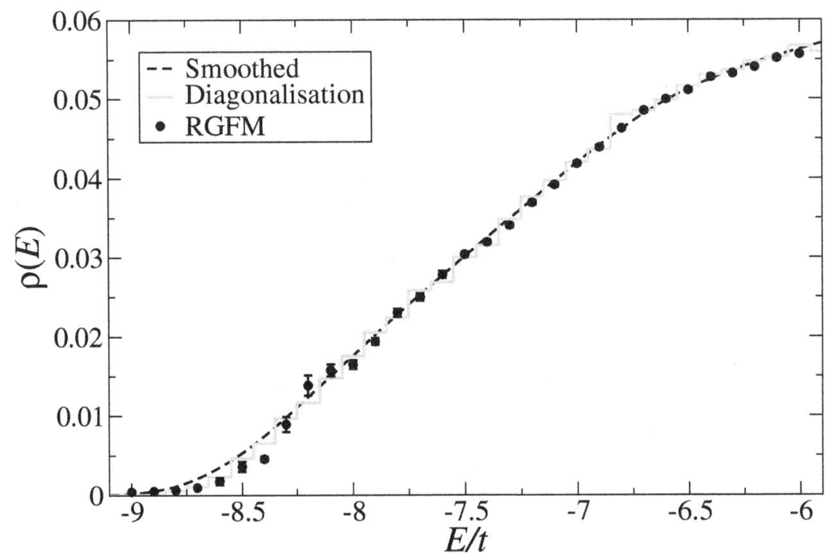

Fig. 9. DOS vs energy for $W = 12\,t$ obtained from the recursive Green's function method (circles with error bars obtained from the sample average) and from diagonalising the Anderson Hamiltonian (histogram), for $L^3 = 21^3 = 9261$. Also shown is a smoothed DOS (dashed line) obtained from the diagonalisation data using a Bezier spline

for different system sizes is shown in Fig. 10. For large energies the DOS is nearly independent of the system size. However, close to the band edge one can see fluctuations because in the tails there are only few states and thus many more samples are necessary to obtain a smooth DOS.

7.2 Scaling behaviour of the conductance

The size dependence of the average and the typical conductance is shown in Fig. 11. We find that for $E/t \leq -8.2$ the typical conductance is proportional

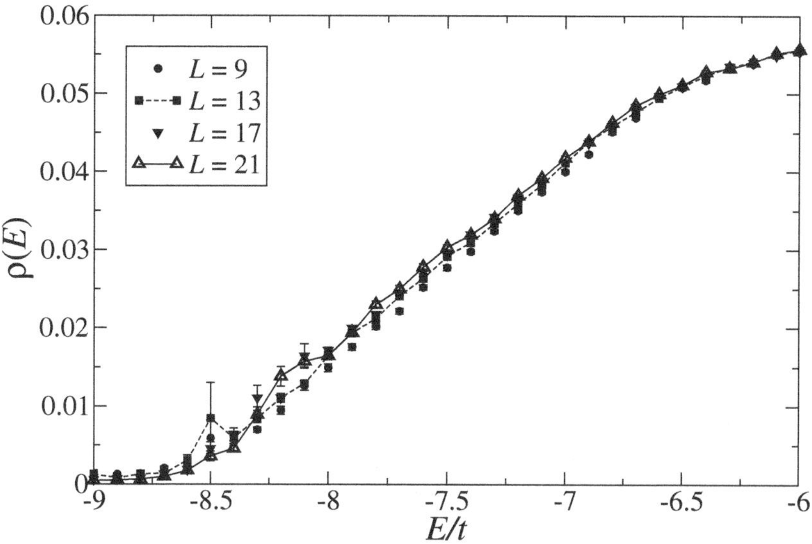

Fig. 10. DOS vs energy for different system sizes and $W = 12t$ calculated with the recursive Green's function method. The data are averaged over 10000 disorder configurations except for $L = 21$ when 2000 samples have been used. The lines are guides to the eye only. Error bars are obtained from the sample average

to the system size L and the constant of proportionality is negative. This corresponds to an exponential decay of the conductance with increasing L and is characteristic for *insulating* behaviour. Moreover, the constant of proportionality is the localisation length ξ. We find that $\xi(E)$ diverges at some energy, which indicates a phase transition. This energy dependence of ξ is shown in Fig. 12.

For $E/t \geq -8.05$, $\langle g_4 \rangle$ is proportional to L. This indicates the *metallic* regime and the slope of $\langle g_4 \rangle$ vs L is related to the d.c. conductivity. We fit the data in the respective regimes to the standard scaling form (26). The results for the critical exponent and the mobility edge are given in Table 2. The obtained values from both averages, $\langle g_4 \rangle$ and $\langle \ln g_4 \rangle$, are consistent. The

Table 2. Best-fit estimates of the critical exponent and the mobility edge for both averages of g_4 using (26) with $n_R = m_I = 0$. The system sizes used are $L = 11, 13, 15, 17, 19, 21$

average	E_{min}/t	E_{max}/t	n_R	m_R	ν	E_c/t
$\langle g_4 \rangle$	-8.2	-7.4	3	2	1.60 ± 0.18	-8.14 ± 0.02
$\langle \ln g_4 \rangle$	-8.8	-7.85	3	2	1.58 ± 0.06	-8.185 ± 0.012

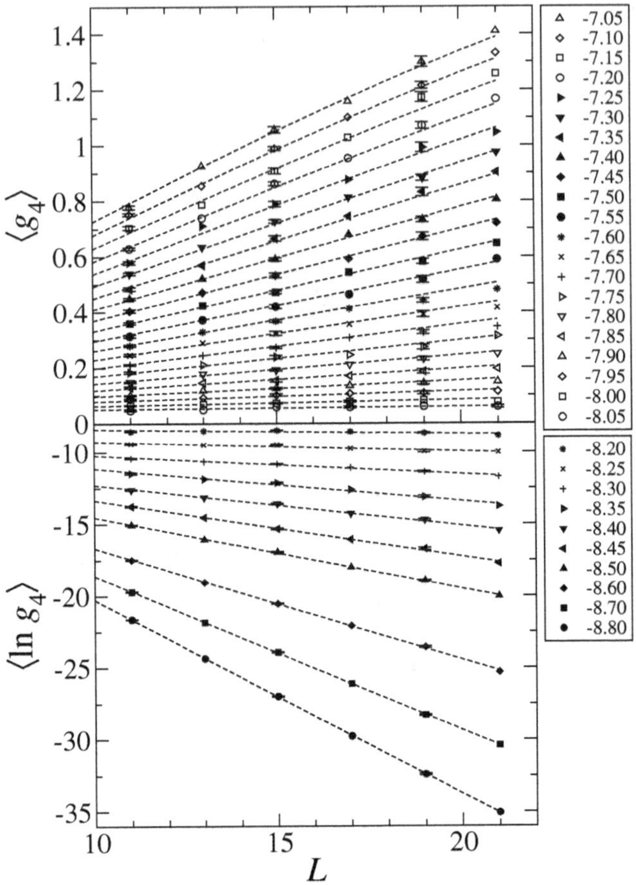

Fig. 11. System size dependence of the 4-point conductance averages $\langle g_4 \rangle$ and $\langle \ln g_4 \rangle$ for $W = 12\,t$ and Fermi energies as given in the legend. Error bars are obtained from the ensemble average and shown for every second L. The dashed lines in the metallic regime indicate the fit result to (30) using the parameters of Table 2. In the insulating regime, linear functions for $\langle \ln g_4 \rangle = -L/\xi + c$ have been used for fitting

average value of $\nu = 1.59 \pm 0.18$ is in agreement with results for conductance scaling at $E/t = 0.5$ and transfer-matrix calculations [9, 14].

7.3 Calculation of the d.c. conductivity

Let us now compute the d.c. conductivity from the conductance $\langle g_4(E, L) \rangle$. From Ohm's law, one naively expects the macroscopic conductivity to be the ratio of $\langle g_4(E, L) \rangle$ and L. There are, however, several complications. First, the mechanism of weak localisation gives rise to corrections to the classical behaviour for $g_4 \gg 1$. Second, it is a priori not known if this relation still

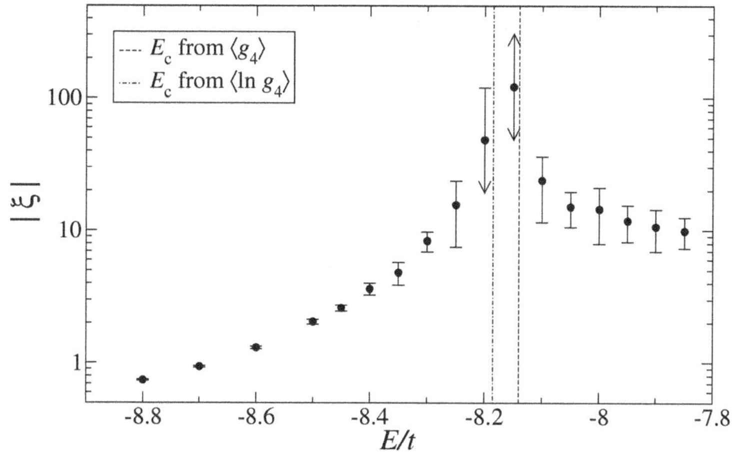

Fig. 12. Localisation length (for $E < E_c$) and correlation length (for $E > E_c$) vs energy obtained from a linear fit to $\langle \ln g_4 \rangle = \mp L/\xi + \text{const.}$, respectively. The error bars close to the transition have been truncated (arrows), because they extend beyond the plot boundaries

holds in the critical regime. And third, the expansion (27) does not yield a behaviour of the form $g_4 \propto \varepsilon^\nu$.

In order to check our data for consistence with the anticipated power law (2) for the conductivity $\sigma(E)$ in the critical regime, we assume the following scaling law for the conductance,

$$\langle g_4 \rangle = f(\chi^\nu L) \,, \tag{30}$$

which results from setting $b = \chi^{-\nu}$ in (24). Due to the relatively large error bars of $\langle g_4 \rangle$ at the MIT as shown in Fig. 11, we might as well neglect the irrelevant scaling variable. Then we expand f as a Taylor series up to order n_R and χ in terms of ε up to order m_R in analogy to (27b) and (28). The best fit to our data is determined by minimising the χ^2 statistic. Using $n_R = 3$ and $m_R = 2$ we obtain for the critical values, $\nu = 1.58 \pm 0.18$ and $E_c/t = -8.12 \pm 0.03$. These values are consistent with our previous fits. The linear term of the expansion of f corresponds to the conductivity close to the MIT. To estimate the quality of this procedure we also calculate the conductivity from the slope of a linear fit to $\langle g_4 \rangle$ throughout the metallic regime, and from the ratio $\langle g_4 \rangle / L$ as well.

The resulting estimates of the conductivity are shown in Fig. 13. We find that the power law is in good agreement with the conductivity obtained from the linear fit to $\langle g_4 \rangle = \sigma L + \text{const.}$ for $E \leq -7t$. In this range it is also consistent with the ratio of $\langle g_4 \rangle$ and L for the largest system computed ($L = 21$). Deviations occur for energies close to the MIT and for $E > -7t$. In the critical regime one can argue that for finite systems the conductance will always

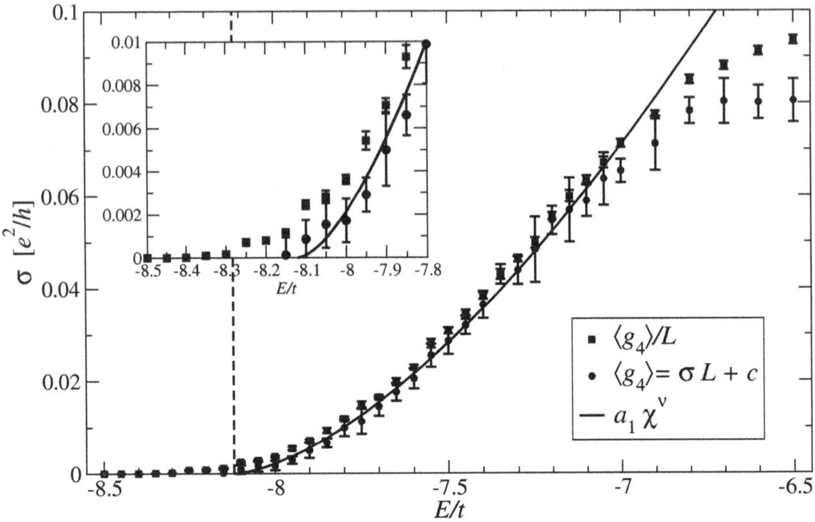

Fig. 13. Conductivity σ vs energy computed from $\langle g_4 \rangle / L$ for $L = 21$ (□), a linear fit with $\langle g_4 \rangle = \sigma L + \text{const.}$ (•) and a fit to the scaling law (30) (solid line). The dashed line indicates $E_c/t = -8.12$. Error bars of $\langle g_4 \rangle / L$ represent the error-of-mean obtained from an ensemble average and are shown for every third E value. The dashed line indicates the position of E_c and the inset shows the region close to E_c in more detail

be larger than zero in the insulating regime because the localisation length becomes eventually larger than the system size.

8 Conclusions

We computed the conductance at $T = 0$ and the DOS of the 3D Anderson model of localisation. These properties were obtained from the recursive Green's function method in which semi-infinite metallic leads at both ends of the system were taken into account.

We demonstrated how the difference in the DOS between the disordered region and the metallic leads has a significant influence on the results for the electronic properties at energies outside the band centre. This poses a big problem for the investigation of the MIT outside the band centre. We showed that by shifting the energy levels in the disordered region the mismatch can be reduced. In this case the average conductance and the typical conductance were found to be consistent with the one-parameter scaling theory at the transition at $E_c \neq 0$. Using a finite-size-scaling analysis of the energy dependence of both conductance averages we obtained an average critical exponent $\nu = 1.59 \pm 0.18$, which is in accordance with results for conductance scaling at $E/t = 0.5$, transfer-matrix calculations [9, 11, 14, 26] and diagonalisation studies [10,12]. However, a thorough investigation of the influence of the leads is still lacking. It would also be interesting to see if these effects can be related to studies of 1D multichannel systems with impurities [46].

We calculated the d.c. conductivity from the system-size dependence of the average conductance and found it consistent with a power-law form at the MIT [47]. This strongly supports previous analytical and numerical calculations of thermoelectric properties reviewed in Ref. [3]. Similar results for topologically disordered Anderson models [48–50], random-hopping models [37,51–53] and the interplay of disorder and many-body interaction [54–57] have been reported elsewhere.

Acknowledgments

We thankfully acknowledge fruitful and localized discussions with P. Cain, F. Milde, M.L. Ndawana and C. de los Reyes Villagonzalo.

References

1. P. W. Anderson. Absence of diffusion in certain random lattices. *Phys. Rev.*, 109:1492–1505, 1958.
2. B. Kramer and A. MacKinnon. Localization: theory and experiment. *Rep. Prog. Phys.*, 56:1469–1564, 1993.
3. R. A. Römer and M. Schreiber. Numerical investigations of scaling at the Anderson transition. In T. Brandes and S. Kettemann, editors, *The Anderson Transition and its Ramifications — Localisation, Quantum Interference, and Interactions*, volume 630 of *Lecture Notes in Physics*, pages 3–19. Springer, Berlin, 2003.
4. I. Plyushchay, R. A. Römer, and M. Schreiber. The three-dimensional Anderson model of localization with binary random potential. *Phys. Rev. B*, 68:064201, 2003.
5. J. E. Enderby and A. C. Barnes. Electron transport at the Anderson transition. *Phys. Rev. B*, 49:5062, 1994.
6. C. Villagonzalo, R. A. Römer, and M. Schreiber. Thermoelectric transport properties in disordered systems near the Anderson transition. *Eur. Phys. J. B*, 12:179–189, 1999. ArXiv: cond-mat/9904362.
7. E. Abrahams, P. W. Anderson, D. C. Licciardello, and T. V. Ramakrishnan. Scaling theory of localization: absence of quantum diffusion in two dimensions. *Phys. Rev. Lett.*, 42:673–676, 1979.
8. P. A. Lee and T. V. Ramakrishnan. Disordered electronic systems. *Rev. Mod. Phys.*, 57:287–337, 1985.
9. K. Slevin and T. Ohtsuki. Corrections to scaling at the Anderson transition. *Phys. Rev. Lett.*, 82:382–385, 1999. ArXiv: cond-mat/9812065.
10. F. Milde, R. A. Römer, and M. Schreiber. Energy-level statistics at the metal-insulator transition in anisotropic systems. *Phys. Rev. B*, 61:6028–6035, 2000.
11. F. Milde, R. A. Römer, M. Schreiber, and V. Uski. Critical properties of the metal-insulator transition in anisotropic systems. *Eur. Phys. J. B*, 15:685–690, 2000. ArXiv: cond-mat/9911029.

12. M. L. Ndawana, R. A. Römer, and M. Schreiber. Finite-size scaling of the level compressibility at the Anderson transition. *Eur. Phys. J. B*, 27:399–407, 2002.
13. M. L. Ndawana, R. A. Römer, and M. Schreiber. Effects of scale-free disorder on the Anderson metal-insulator transition. *Europhys. Lett.*, 68:678–684, 2004.
14. K. Slevin, P. Markoš, and T. Ohtsuki. Reconciling conductance fluctuations and the scaling theory of localization. *Phys. Rev. Lett.*, 86:3594–3597, 2001.
15. D. Braun, E. Hofstetter, G. Montambaux, and A. MacKinnon. Boundary conditions, the critical conductance distribution, and one-parameter scaling. *Phys. Rev. B*, 64:155107, 2001.
16. C. Villagonzalo, R. A. Römer, and M. Schreiber. Transport properties near the Anderson transition. *Ann. Phys. (Leipzig)*, 8:SI-269–SI-272, 1999. ArXiv: cond-mat/9908218.
17. C. Villagonzalo, R. A. Römer, M. Schreiber, and A. MacKinnon. Behavior of the thermopower in amorphous materials at the metal-insulator transition. *Phys. Rev. B*, 62:16446–16452, 2000.
18. A. MacKinnon. The conductivity of the one-dimensional disordered Anderson model: a new numerical method. *J. Phys.: Condens. Matter*, 13:L1031–L1034, 1980.
19. A. MacKinnon. The calculation of transport properties and density of states of disordered solids. *Z. Phys. B*, 59:385–390, 1985.
20. B. Mehlig and M. Schreiber. Energy-level and wave-function statistics in the Anderson model of localization. In K.H. Hoffmann and A. Meyer, editors, *Parallel Algorithms and Cluster Computing - Implementations, Algorithms, and Applications, Lecture Notes in Computational Science and Engineering*. Springer, Berlin, 2006.
21. P. Karmann, R. A. Römer, M. Schreiber, and P. Stollmann. Fine structure of the integrated density of states for Bernoulli-Anderson models. In K.H. Hoffmann, and A. Meyer, editors, *Parallel Algorithms and Cluster Computing - Implementations, Algorithms, and Applications, Lecture Notes in Computational Science and Engineering*. Springer, Berlin, 2006.
22. B. Bulka, B. Kramer, and A. MacKinnon. Mobility edge in the three dimensional Anderson model. *Z. Phys. B*, 60:13–17, 1985.
23. B. Bulka, M. Schreiber, and B. Kramer. Localization, quantum interference, and the metal-insulator transition. *Z. Phys. B*, 66:21, 1987.
24. T. Ohtsuki, K. Slevin, and T. Kawarabayashi. Review on recent progress on numerical studies of the Anderson transition. *Ann. Phys. (Leipzig)*, 8:655–664, 1999. ArXiv: cond-mat/9911213.
25. T. Ando. Numerical study of symmetry effects on localization in two dimensions. *Phys. Rev. B*, 40:5325, 1989.
26. P. Cain, R. A. Römer, and M. Schreiber. Phase diagram of the three-dimensional Anderson model of localization with random hopping. *Ann. Phys. (Leipzig)*, 8:SI-33–SI-38, 1999. ArXiv: cond-mat/9908255.
27. F. Milde, R. A. Römer, and M. Schreiber. Multifractal analysis of the metal-insulator transition in anisotropic systems. *Phys. Rev. B*, 55:9463–9469, 1997.
28. H. Stupp, M. Hornung, M. Lakner, O. Madel, and H. v. Löhneysen. Possible solution of the conductivity exponent puzzle for the metal-insulator transition in heavily doped uncompensated semiconductors. *Phys. Rev. Lett.*, 71:2634–2637, 1993.

29. S. Waffenschmidt, C. Pfleiderer, and H. v. Löhneysen. Critical behavior of the conductivity of Si:P at the metal-insulator transition under uniaxial stress. *Phys. Rev. Lett.*, 83:3005–3008, 1999. ArXiv: cond-mat/9905297.
30. F. Wegner. Electrons in disordered systems. Scaling near the mobility edge. *Z. Phys. B*, 25:327–337, 1976.
31. D. Belitz and T. R. Kirkpatrick. The Anderson-Mott transition. *Rev. Mod. Phys.*, 66:261–380, 1994.
32. R. A. Römer, C. Villagonzalo, and A. MacKinnon. Thermoelectric properties of disordered systems. *J. Phys. Soc. Japan*, 72:167–168, 2002. Suppl. A.
33. C. Villagonzalo. *Thermoelectric Transport at the Metal-Insulator Transition in Disordered Systems.* PhD thesis, Chemnitz University of Technology, 2001.
34. P. Cain, F. Milde, R.A. Römer, and M. Schreiber. Applications of cluster computing for the Anderson model of localization. In S.G. Pandalai, editor, *Recent Research Developments in Physics*, volume 2, pages 171–184. Transworld Research Network, Trivandrum, India, 2001.
35. P. Cain, F. Milde, R. A. Römer, and M. Schreiber. Use of cluster computing for the Anderson model of localization. *Comp. Phys. Comm.*, 147:246–250, 2002.
36. B. Kramer and M. Schreiber. Transfer-matrix methods and finite-size scaling for disordered systems. In K. H. Hoffmann and M. Schreiber, editors, *Computational Physics*, pages 166–188, Springer, Berlin, 1996.
37. A. Eilmes, R. A. Römer, and M. Schreiber. The two-dimensional Anderson model of localization with random hopping. *Eur. Phys. J. B*, 1:29–38, 1998.
38. U. Elsner, V. Mehrmann, F. Milde, R. A. Römer, and M. Schreiber. The Anderson model of localization: a challenge for modern eigenvalue methods. *SIAM J. Sci. Comp.*, 20:2089–2102, 1999. ArXiv: physics/9802009.
39. M. Schreiber, F. Milde, R. A. Römer, U. Elsner, and V. Mehrmann. Electronic states in the Anderson model of localization: benchmarking eigenvalue algorithms. *Comp. Phys. Comm.*, 121–122:517–523, 1999.
40. E. N. Economou. *Green's Functions in Quantum Physics.* Springer-Verlag, Berlin, 1990.
41. G. Czycholl, B. Kramer, and A. MacKinnon. Conductivity and localization of electron states in one dimensional disordered systems: further numerical results. *Z. Phys. B*, 43:5–11, 1981.
42. J. L. Cardy. *Scaling and Renormalization in Statistical Physics.* Cambridge University Press, Cambridge, 1996.
43. M. Büttiker. Absence of backscattering in the quantum Hall effect in multiprobe conductors. *Phys. Rev. B*, 38:9375, 1988.
44. A. Croy. Thermoelectric properties of disordered systems. M.Sc. thesis, University of Warwick, Coventry, United Kindgom, 2005.
45. B. K. Nikolić. Statistical properties of eigenstates in three-dimensional mesoscopic systems with off-diagonal or diagonal disorder. *Phys. Rev. B*, 64:14203, 2001.
46. D. Boese, M. Lischka, and L.E. Reichl. Scaling behaviour in a quantum wire with scatterers. *Phys. Rev. B*, 62:16933, 2000.
47. P. Cain, M. L. Ndawana, R. A. Römer, and M. Schreiber. The critical exponent of the localization length at the Anderson transition in $3D$ disordered systems is larger than 1. 2001. ArXiv: cond-mat/0106005.
48. J. X. Zhong, U. Grimm, R. A. Römer, and M. Schreiber. Level spacing distributions of planar quasiperiodic tight-binding models. *Phys. Rev. Lett.*, 80:3996–3999, 1998.

49. U. Grimm, R. A. Römer, and G. Schliecker. Electronic states in topologically disordered systems. *Ann. Phys. (Leipzig)*, 7:389–393, 1998.
50. U. Grimm, R. A. Römer, M. Schreiber, and J. X. Zhong. Universal level-spacing statistics in quasiperiodic tight-binding models. *Mat. Sci. Eng. A*, 294-296:564, 2000. ArXiv: cond-mat/9908063.
51. A. Eilmes, R. A. Römer, and M. Schreiber. Critical behavior in the two-dimensional Anderson model of localization with random hopping. *phys. stat. sol. (b)*, 205:229–232, 1998.
52. P. Biswas, P. Cain, R. A. Römer, and M. Schreiber. Off-diagonal disorder in the Anderson model of localization. *phys. stat. sol. (b)*, 218:205–209, 2000. ArXiv: cond-mat/0001315.
53. A. Eilmes, R. A. Römer, and M. Schreiber. Localization properties of two interacting particles in a quasi-periodic potential with a metal-insulator transition. *Eur. Phys. J. B*, 23:229–234, 2001. ArXiv: cond-mat/0106603.
54. R. A. Römer and A. Punnoose. Enhanced charge and spin currents in the one-dimensional disordered mesoscopic Hubbard ring. *Phys. Rev. B*, 52:14809–14817, 1995.
55. M. Leadbeater, R. A. Römer, and M. Schreiber. Interaction-dependent enhancement of the localisation length for two interacting particles in a one-dimensional random potential. *Eur. Phys. J. B*, 8:643–652, 1999.
56. R. A. Römer, M. Schreiber, and T. Vojta. Disorder and two-particle interaction in low-dimensional quantum systems. *Physica E*, 9:397–404, 2001.
57. C. Schuster, R. A. Römer, and M. Schreiber. Interacting particles at a metal-insulator transition. *Phys. Rev. B*, 65:115114–7, 2002.

Optimizing Simulated Annealing Schedules for Amorphous Carbons

Peter Blaudeck and Karl Heinz Hoffmann

Technische Universität Chemnitz, Institut für Physik
09107 Chemnitz, Germany
blaudeck@physik.tu-chemnitz.de

1 Introduction

Annealing, carried out in simulation, has taken on an existence of its own as a tool to solve optimization problems of many kinds [1–3]. One of many important applications is to find local minima for the potential energy of atomic structures, as in this paper, in particular structures of amorphous carbon at room temperature. Carbon is one of the most promising chemical elements for molecular structure design in nature. An infinite richness of different structures with an incredibly wide variety of physical properties can be produced. Apart from the huge variety of organic substances, even the two crystalline inorganic modifications, graphite and diamond, show diametrically opposite physical properties. Amorphous carbon continues to attract researchers for both the fundamental understanding of the microstructure and stability of the material and the increasing interest in various applications as a high performance coating material as well as in electronic devices.

Simulated Annealing (SA) of such systems requires to apply methods of molecular dynamics (MD) to the ensemble of atoms [4]. The first fundamental step is to find a suitable approach to the interatomic potentials and forces. In the spectrum of useful methods there are on the one hand the empirical potential models, for example the Tersoff potential [5]. These are simple enough to allow computations with many thousands of atoms but are limited in accuracy and chemical transferability to situations, that have not been included in the parametrization. On the other hand, fully self-consistent quantum mechanical approaches for general structures of any kind of atoms [6,7] are, due to their complexity, limited to short time simulations of only about one hundred atoms. The method used in this paper [8] takes an intermediate position between the empirical and the *ab initio* approaches and is briefly outlined in Sect. 2.

In Sect. 3 we give a short description of how to use parallel computation to overcome the general problem of the limited computer power by optimizing and performing SA schedules for any given computational resource and, in principle, for any finite system of atoms or molecules. Finally, an example of the effect of the optimization is shown in Sect. 4, using the new optimized schedule and comparing its results with old data.

2 Density-functional molecular dynamics

We apply the well established *ab initio* based non-selfconsistent tight-binding scheme [8], which, compared with fully self–consistent calculations, reduces the computational effort by at least two orders of magnitude. The approximate calculation of the MD interatomic forces is based on the density functional theory (DFT) within the local density approximation (LDA) using a localized atomic orbital (LCAO) basis [9]. The scheme includes first-principle concepts in relating the Kohn-Sham orbitals of the many-atom configuration to a minimal basis of the localized atomic-like valence orbitals of all atoms. We take into account only two-center Hamiltonian matrix elements h and overlap matrix elements S in the secular equation resulting in a general eigenvalue problem

$$\sum_{\mu} c_{\mu}^{i}(h_{\mu\nu} - \epsilon_i S_{\mu\nu}) = 0 \tag{1}$$

for the eigenvalues ϵ_i and eigenfunctions c_{μ}^{i} of the system [10]. The total potential energy of the system as a function of the atomic coordinates $\{\mathbf{R}_l\}$ can now be decomposed into two parts,

$$E_{\text{pot}}(\{\mathbf{R}_l\}) = E_{\text{bind}}(\{\mathbf{R}_l\}) + E_{\text{rep}}(\{\mathbf{R}_l - \mathbf{R}_k\}). \tag{2}$$

The first term E_{bind} is the sum of all occupied cluster electronic energies and represents the so-called band-structure energy. The second term, as a repulsive energy E_{rep}, comprises the core-core repulsion between the atoms and corrections due to the Hartree double-counting terms and the non-linearity in the superposition of exchange-correlation contributions. This term is fitted to the two-particle self-consistent LDA cohesive energy curves of corresponding diatomic molecules and crystalline modifications [11].

To apply the scheme to a part of an infinite system we carry out the simulation for a number of atoms in a fixed-volume cubic supercell using periodic boundary conditions. Furthermore, the temperature has to be controlled according to a proper function $T(t)$. To adjust the temperature, all atomic velocities are renormalized after a time step to achieve the correct mean kinetic energy per degree of freedom. The so defined temperature as a function of time has to follow the SA schedule.

A generally unsolved problem in any high performance computation for density-functional based or fully self-consistently SA of real material problems is, also in making use of the fastest computers, that it is only possible to simulate the annealing of 100 or 1000 atoms for timescales of just some picoseconds. However, in reality the typical time for cooling down atomic systems from gas to solid state is about two orders of magnitude larger.

Nevertheless, it is absolutely necessary to simulate this cooling down to realistic structures, to satisfy the demand of knowledge about the microscopic structure and properties. Therefore, the only remaining way is to accept the lack of computational power and at least to design the annealing schedule as well as possible, as shown in Sect. 3.

3 Optimizing simulated annealing

We give an approximate solution of the problem by performing four steps in chronological order in the following way.

1. Constructing a small system of atoms, just large enough to model the microscopic interactions of the atoms sufficiently.
2. Mapping its continuous configuration space to a discrete set of connected states.
3. Solving the optimization task for the discrete set, which is much simpler to handle.
4. Applying the so found schedule to any larger system with the same microscopic properties.

1. We know from practical experience that relatively small systems with a nearly correct atomic structure are sufficient to reflect the relevant properties of the configuration space of atomic clusters. Our example systems are arrangements of 64 carbon atoms in a cubic supercell with periodic boundary conditions. The mass densities are 2.7 g/cm^3, 3.0 g/cm^3, and the value of diamond 3.52 g/cm^3, which are of particular interest for practical purposes.

2. For the purpose of step two we have developed a computer algorithm that is described in detail in [12] and illustrated in Fig. 1. There is one master process performing a molecular dynamics run at high temperature (distinctly above the melting point of the system). After a fixed time interval (typically the time needed to move each atom for a distance larger than the mean bond length) a worker-process is created which has to
a) find the nearest significant energy minimum by rapidly cooling down the system without changing the atomic topology,
b) suddenly heat up the system to a fixed temperature given (escape-temperature) and measure the escape-time t^{esc}, i.e. the time needed by the system to escape from this minimum (criterion: first modification in the system's topology) to a "neighbour minimum" and investigate the properties of the latter. Subsequently, for the thermodynamic interpretation given below, this procedure has to be repeated with different escape-temperatures.

Now we divide the energy scale to intervals so that, for each pair of starting and final energy interval, the number of transitions N_{ij} is counted and a mean value $\overline{t_{ij}^{\text{esc}}}(T)$ can be calculated. The index j characterizes the starting interval

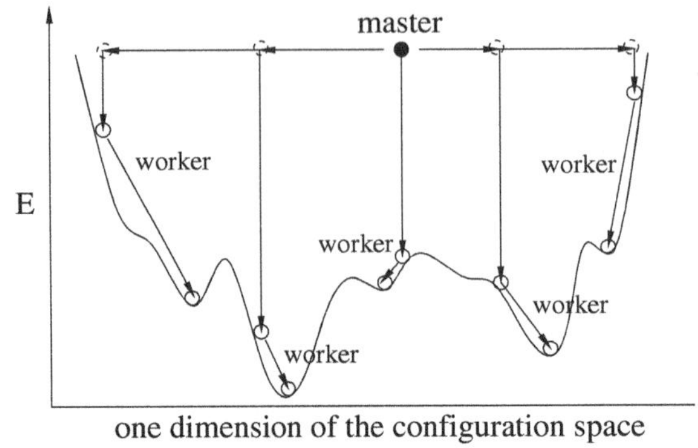

Fig. 1. A master process is walking through the configuration space at high temperature. After a certain period of time it initializes worker-processes. Each worker has to find the nearest essential energy minimum and the depth of this minimum

with the average energy E_j, the index i the destination interval respectively. We stress the suggestion, that for all transitions from the primary minimum E_j to the new minimum E_i a wall of height ΔE_{ij} has to be overcome. That wall influences the transition probability according to a Boltzmann factor. Therefore, we take an ansatz for the mean transition probability per time unit

$$\lambda_{ij}(T) = \frac{1}{t_{ij}^{\text{esc}}(T)} = C \exp(-\frac{\Delta E_{ij}}{kT}) \tag{3}$$

as a function of the temperature. The parameters to be fitted are the common (for all i, j) constant value C and the set ΔE_{ij}. It turns out that this fit is successful with a surprisingly small deviation from the original data. Finally, we derive the transition probabilities $G_{ij}(T)$ according to

$$\begin{aligned} G_{ij}(T) &= \frac{N_{ij}}{\sum_i N_{ij}} \lambda_{ij}(T) \quad j \neq i \\ G_{ii} &= 1 - \sum_{j \neq i} G_{ij} \ . \end{aligned} \tag{4}$$

The result is some kind of ladder with walls between its steps as shown in Fig. 2 For amorphous systems this ladder of states has some well understandable properties. Firstly, it is possible to skip one or more steps. Secondly, the lower the starting energy the higher the walls to be overcome are, because the system is already in a deeper minimum at the start. And thirdly, for the same starting energy, if the system will go downwards, this is more difficult with a larger number of steps to be skipped. Those properties reflect clearly the real thermal behaviour of an atomic or molecular system. The structures are able to descent very easily in the upper part of the energy scale. But, once arrived at any amorphous state with sufficiently low energy (for example in our picture at

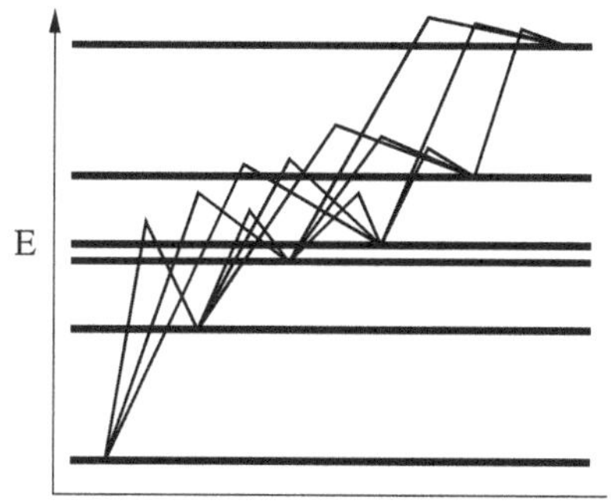

Fig. 2. Scheme of the resulting "ladder" system in energy space

the second lowest step of the ladder), it will become troublesome to go still lower in energy to the ground state or to another distinctly lower amorphous state.

3. For the third step recent numerical methods exist, which allow the determination of an optimal annealing schedule, provided that the properties, e.g. the energies E_j, of the states and all transition probabilities G_{ij} for a random walker from one state j to another state i within one time step ΔT are known. The G_{ij} depend on the temperature. With the transition probabilities G_{ij} the probabilities P_j for being at state j change like

$$P_i(t + \Delta t) = \sum_j G_{ij}(T(t))P_j(t) . \tag{5}$$

Now we can define the optimal annealing schedule $T(t)$ as the time dependent temperature for which

$$\overline{E} = \sum_j E_j P_j(t_{\mathrm{end}}) = \text{minimum} , \tag{6}$$

for fixed initial $P_j(0)$ and process time t_{end}. This optimal $T(t)$ exists and can be computed by using a variational principle for $\overline{E}$ [13].

4. With $T(t)$ now under control, this schedule can be used for any larger atomic system, provided the microscopic properties, first of all the atomic density and composition, remains unchanged.

4 Results

4.1 Optimized schedules

The calculations for the scheme discussed in Sect. 3, particularly the most expensive second step, have been performed using a parallel cluster with up to 36 Dual-Pentium-Boards and the send-receive functions of the common MPI-software.

In Fig. 3 the new optimized schedules $T(t)$ are displayed for different mass densities. Note that our LDA scheme generally overestimates forces and potential differences by a factor of about 1.5. Therefore, the mean values of both the potential and the kinetic energy, and consequently also the temperature differ from the real scale by this factor. The little time interval with constant temperature at the end of the schedule has been added artificially to allow the exact comparison of the final energies.

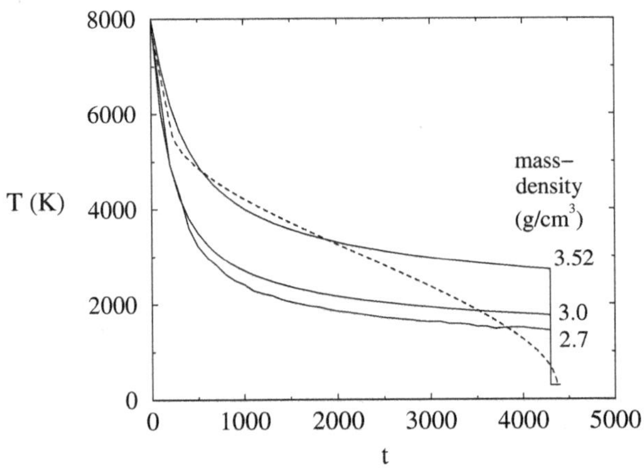

Fig. 3. Optimized annealing schedules (solid) for different mass densities, compared with a previously used schedule (dotted)

The dotted line in Fig. 3 is an artificially constructed schedule used in previous work. It was constructed based on the idea to find a sensible decreasing function $T(t)$ with a minimal descent at temperatures empirically known as most important for structure formation processes. The full lines in Fig. 3 show, that this physical purpose has to be fulfilled still more rigorously by the new optimized schedules. It seems quite clear that, compared with the old schedules, the decrease in temperature is large at the very beginning of the annealing process but much weaker in a temperature range where the most important freezing of the structure is expected to take place. The schedules depend on the mass density in a very plausible manner. The regions of weakest descent consistently follow the temperatures which correspond sequently to different "melting regions" for different binding energies per atom of these structures. The lower the mass density, the less the average number of bonds per atom, and, consequently, the lower the walls between the minima of the potential energy.

4.2 Binding energy

In performing the fourth step we have picked out the mass density 3.0 g/cm^3 as an example, because of the presence of previous results [14] found by using the heuristically constructed schedule already mentioned above. As in these previous investigations, we have now applied the new schedule to an ensemble of 32 different clusters, with 128 carbon atoms each. The most important measure for the quality of the SA schedule are the final potential energies per atom. These energies are collected in Fig. 4 for the members of the old and new ensemble. Within each ensemble the members are sorted by their energy values. For most of the newly created structures we can find final energies lower than for the structures from previous results [14]. The structures we have generated (see Fig. 5, for example) are clusters of amorphous carbon with sensible structural properties and suitable for further investigations in mechanical and electronic properties.

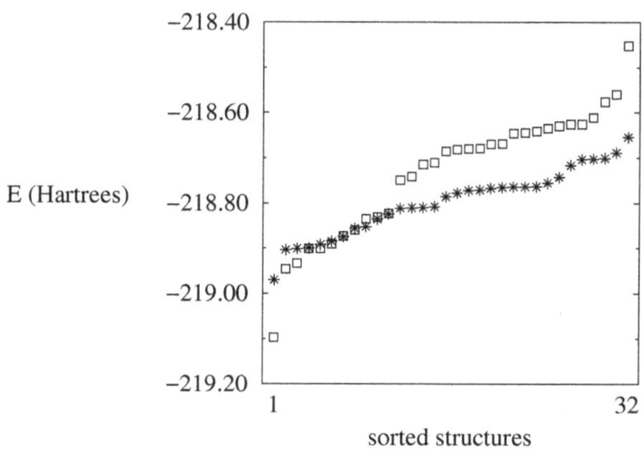

Fig. 4. Final potential energies, each set sorted, for the optimized annealing schedule (stars) compared with a set found by an old schedule (squares)

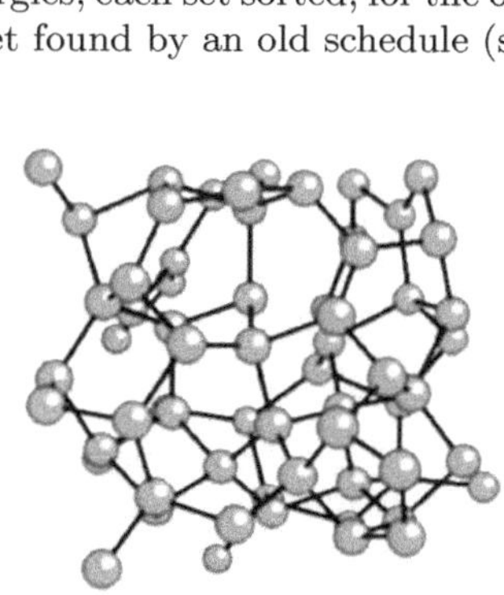

Fig. 5. Most common type of a resulting structure with mainly threefold or fourfold bound carbon atoms and formations of rings containing five, six, or seven atoms

5 Summary

We have found a way how to prepare an optimal simulated annealing schedule for modelling realistic amorphous carbon structures. Consequently, a scheme has been developed which can now be applied to any other atomic system of interest. One can take advantage of it, whenever systems have to be simulated to determine their structural, mechanical, or electronic properties.

References

1. S. Kirkpatrick, C. D. Gelatt, and M. P. Vecchi. Optimization by simulated annealing. *Science*, 220:671–680, 1983.
2. F. H. Stillinger and D. K. Stillinger. Cluster optimization simplified by interaction modification. *J. Chem. Phys.*, 93:6106, 1990.
3. X.-Y. Chang and R. S. Berry. Geometry, interaction range, and annealing. *J. Chem. Phys.*, 97:3573, 1992.
4. D. C. Rapaport. *The Art of Molecular Dynamics Simulation.* Cambridge University Press, 1995.
5. J. Tersoff. New empirical model for the structural properties of silicon. *Phys. Rev. Lett.*, 56:632, 1986.
6. R. Car and M. Parinello. Structural, dynamical, and electronic properties of amorphous silicon: An ab initio molecular-dynamics study. *Phys. Rev. Lett.*, 60:204–207, 1988.
7. G. Galli, R.M. Martin, R. Car, and M. Parrinello. Structural and electronic properties of amorphous carbon. *Phys. Rev. Lett.*, 62:555–558, 1989.
8. P. Blaudeck, Th. Frauenheim, D. Porezag, G. Seifert, and E. Fromm. A method and results for realistic molecular dynamic simulation of hydrogenated amorphous carbon structures using an lcao-lda scheme. *J. Phys.: Condens. Matter*, 4:6389–6400, 1992.
9. D. Porezag, Th. Frauenheim, Th. Köhler, G. Seifert, and R. Kaschner. Construction of tight-binding-like potentials on the basis of density-functional theory: Application to carbon. *Phys. Rev. B*, 51:12947–12957, 1995.
10. G. Seifert, H. Eschrig, and W.Biegert. An approximate variation of the lcao-x alpha method. *Z. Phys. Chem. (Leipzig)*, 267:529, 1986.
11. Th. Frauenheim, P. Blaudeck, U. Stephan, and G. Jungnickel. Atomic structure and physical properties of amorphous carbon and its hydrogenated analogs. *Phys. Rev. B*, 48:4823–4834, 1993.
12. P. Blaudeck and K. H. Hoffmann. Ground states for condensed amorphous systems: Optimizing annealing schemes. *Comp. Phys. Comm.*, 150(3):293–299, 2003.
13. Ralph E. Kunz, Peter Blaudeck, Karl Heinz Hoffmann, and Stephen Berry. Atomic clusters and nanoscale particles: from coarse-grained dynamics to optimized annealing schedules. *J. Chem. Phys.*, 108:2576–2582, 1998.
14. Th Köhler, Th. Frauenheim, P. Blaudeck, and K. H. Hoffmann. A density functional tight binding study of structure and property correlations in a-CH(:N) systems. DPG-Tagung, Regensburg, 23.-27.3.98, 1998.

Amorphisation at Heterophase Interfaces

Sibylle Gemming[1,2], Andrey Enyashin[1,3], and Michael Schreiber[1]

[1] Technische Universität Chemnitz, Institut für Physik, 09107 Chemnitz, Germany
sibylle.gemming@physik.tu-chemnitz.de
schreiber@physik.tu-chemnitz.de
[2] Forschungszentrun Rossendorf, P.O. 51 01 19, 01314 Dresden, Germany
[3] Technische Universität Dresden, Fachbereich Chemie, 01062 Dresden, Germany
enyashin@chm.theory.tu-dresden.de

1 Reactive heterophase interfaces

Heterophase interfaces are boundaries, which join two material types with different physical and chemical nature. Therefore, heterophase interfaces can exhibit a large variety of geometric morphologies ranging from atomically sharp boundaries to gradient materials, in which an interface-specific phase is formed, which provides a continuous change of the structural parameters and thus reduces elastic strains and deformations. In addition, also the electronic properties of the two materials may be different, e.g. at boundaries between an electronically conducting metal and a semiconductor or an insulating material. Due to the deviations in the electronic structure, various bonding mechanisms are observed, which span the range from weakly interacting systems to boundaries with strong, directed bonding and further to reactively bonding systems which exhibit a new phase at the interface. Thus, both elastic and electronic factors may contribute to the formation of a new, often amorphous phase at the interface. Numerical simulations based on electronic structure theory are an efficient tool to distinguish and quantify these different influence factors, and massively parallel computers nowadays provide the required numerical power to tackle structurally more demanding systems. Here, this power has been exploited by the parallelisation over an optimised set of integration points, which split the solution of the Kohn-Sham equations into a set of matrix equations with equal matrix sizes. In this way, the analysis and prediction of material properties at the nanoscale has become feasible.

1.1 Structure and stability of heterophase interfaces

Junctions between two different metals or semiconductors or insulators also belong to the class of heterophase interfaces and have been studied to elucidate the influence of more subtle differences of the electronic structure.

The observation of the giant magnetoresistance, for instance, has excited several theoretical investigations on metallic multilayers [1], such as Ni|Ru [2], Pt|Ta [3], Cu|Ta [4], or Ag|Pt [5]. Another active field of research are the phenomena related to the quantised conductance in mixed-metal nanostructures from materials such as AuPd or AuAg [6–8]. Hetero- and superstructures of III-V and II-VI semiconductors provide the possibility to generate specific optical properties by adjusting the electronic and elastic properties via the layer thickness and composition, e.g. the III-V combinations GaAs|AlAs [9, 10], GaSb|InAs [10], or InAs|GaSb [11], and mixed multilayers of III-V and II-VI semiconductors like GaAs|ZnSe [12], or of III-V on IV materials, such as SiC|AlN and SiC|BP [13]. In these and other more covalently bonded systems, like SiC|Si [14], the main focus of the investigations is on the correlation of lattice strain and electronic properties. Other insulator-insulator boundaries come from the heteroepitaxy of ferroic oxides on an oxide template, such as $SrTiO_3$|MgO [15], or multilayers such as $BaTiO_3$|$SrTiO_3$ [16], or from the doping of homophase boundaries with electronically active elements, such as Fe at the $\Sigma3(111)[1\text{-}10]$ boundary in $SrTiO_3$ [17, 18].

The present discussion will focus on contacts between a metal and a non-metallic material, where the electronic structures of the components differ most strongly, and an electronic structure theory can best exploit its modelling power. Nevertheless, it has demonstrated its applicability also for semiconductor-insulator boundaries such as C|Si [19], Si|SiO_2 [20], TiN|MgO [21], or ZrO_2|Si and $ZrSiO_4$|Si [22].

1.2 Interactions at heterophase interfaces

A strong interest in a theory-based microscopic understanding of the interactions at metal-insulator interfaces has developed over the last decades, motivated by the application of ceramic materials in various industrial applications, for instance as thermal barrier coatings [23], or even as medical implants [24]. The motivation to study metal-semiconductor interfaces stems mostly from the further miniaturisation of microelectronic devices and the concomitant need to control interface properties between the semiconductor substrate and metallic functional layers or the metallisation at the nanoscale.

This interest has lead to various attempts for theoretical modelling of the relevant contributions which influence the bonding behaviour at the interface. Theoretical studies span the whole range from finite-element modelling to understand macroscopic elastic properties [25], over atomistic simulations of dislocation networks, plasticity, and fracture [26], to the investigation of the electronic structure with ab-initio band-structure techniques [27–32].

From the theoretical modelling and the experimental observations on metal-to-non-metal bonding at flat, atomically abrupt interfaces three scenarios can be distinguished:

(A) Strong adhesion, which is for instance found for main-group or early transition metals on oxide-based insulators with a propensity of the metal to bind on top of the O atom. This behaviour is, for instance, observed for Ti on MgO(100) [33], for Al on $Al_2O_3(000.1)$ [34], for V on MgO(001) [35], and for Al and Ti on $MgAl_2O_4(001)$ [37–42]. At reactive interfaces, such as metal-silicon contacts, the interaction between the two constituents is even larger, leading to a thermodynamic driving force for the formation of the interface phase.

(B) Weak adhesion is obtained for late transition metals at non-polar insulator surfaces for instance for Ag|MgO(001) [33,43,44], for Cu|MgO(001) [26], for the VO_x|Pd(111) interface [45], or for Ag|$MgAl_2O_4(001)$ [37,39,42]. Due to the also experimentally apparent lack of strong bonding [46], the underlying adhesion mechanism has been ascribed to image charge interactions [47–51].

(C) A moderately strong bonding is obtained, when additional elastic interactions interfere with the metal-to-oxygen bonding. For the adhesion of Cu on the polar (111) surface of MgO, however, evidence for a direct metal-oxygen interaction was given and the occurrence of metal-induced gap states was postulated [52]. Presently, transition-metal oxides such as $BaTiO_3$, $SrTiO_3$, or ZnO are studied as substrates for the adhesion of transition metals like Pd, Pt, or Mo [53, 54]. In contrast, at reactive interfaces the elastic interaction does not lead to a weakening of the interface, but rather assists the formation of the interface-specific phase, e.g. by the introduction of misfit dislocations at the boundary.

Similar adhesive interactions are also monitored for metal-insulator contacts, where the insulator is not an oxide, but a carbide [55,56] or nitride [57].

2 Modelling reactive interfaces

2.1 Macroscopic modelling

From a phenomenological point of view, the three adhesion regimes can be characterised by the growth mode during metal deposition and by the so-called wetting angle. If a metal droplet is deposited on an unreactive insulator surface, the balance of three major energy contributions determine its shape: the interface energy $\gamma_{\mathrm{met-ins}}$ at the contact area of the two materials, and the two surface energies of insulator γ_{ins} and metal γ_{met} in contact with the environment (e.g. air). The relation between these quantities is given by Young's equation [58]:

$$\gamma_{\mathrm{ins}} - \gamma_{\mathrm{met-ins}} = \gamma_{\mathrm{met}} \cos(\theta). \tag{1}$$

In this equation, θ is the "contact" or "wetting" angle, which is the inclination angle between the metal surface and the metal-insulator contact area at the triple line between metal, insulator and air. If the interface is more stable than the sum of the two relaxed free surfaces, complete wetting of the ceramic by a (thin) metal film is achieved, and the wetting angle tends to zero. In this case, the metal can be deposited in a layer-by-layer or Franck-van-der-Merwe growth mode. On the other hand, if the free surfaces are more stable than the interface, the metal will form (nearly spherical) droplets and thus minimise the area of the unfavourable interface with the insulator surface. The wetting angle will approach 180°, and a Volmer-Weber island growth is obtained.

In the case of reactive wetting, additional factors come into play as outlined in detail in [58]. The most important factors are the diffusion of elements and the formation of additional phases at the metal-insulator interface and on the metal and insulator surfaces. This phenomenon can even lead to the formation of a rim at the triple line, where the metal partially dissolves the insulator underneath the droplet and diffusion above the insulator surface accumulates material on the other side of the triple line. Also, grain-boundary grooving has been observed below the metal droplet, where the metal preferentially dissolves material at boundaries (defects) of the substrate. Close to the wetting-dewetting transition point with $\theta \approx 90°$, the growth mode depends very sensitively on the deposition conditions. Experimentally, it is difficult to distinguish this phenomenon from the stress-driven mixed-mode growth of islands on a thin wetting layer (Stranski-Krastanov growth) [59], but electronic structure theory is particularly well suited to elucidate and quantify the different contributions. Thus, the basic features of electronic structure theory will shortly be summarised in the following. More details on mesoscopic modelling is provided e.g. by Stoneham and Harding [60] or in a recent textbook by Finnis [61].

2.2 Density-functional-based modelling

Density-functional theory (DFT) has proven an extremely efficient approach to obtain the properties of a given model system in its electronic ground state. A more detailed introduction to DFT and its extensions to non-equilibrium systems is given in review articles such as [62–64] or textbooks such as [65,66].

Generically, the parameter-free or "ab-initio" quantum-mechanical treatment of the electronic properties of a system employs a wave function $|\Psi\rangle$ to represent all relevant electronic degrees of freedom. This wave function depends on the spatial variables of all interacting particles, thus computations involving the wave function become rather demanding with increasing system size. As shown by Hohenberg and Kohn [67], in DFT the relevant interaction terms can be expressed as functionals of the electron density, which is a function of only three spatial variables. Based on the two Hohenberg-Kohn theorems, the ground state electron density can be obtained from the minimisation of the total energy functional, and all properties derived from this density can accordingly be calculated. In the original work these theorems were proven only under specific constraints, which were later alleviated by Levy [68]. Other generalisations were introduced for spin-polarised systems [69], for systems at finite temperature [70], and for relativistic systems [71,72]. Furthermore, most electronic structure calculations make use of the Kohn-Sham formalism, in which the original density-based functional is re-expressed in a basis of non-interacting single-particle states. The resulting matrix equation is a generalised eigenvalue problem, the solution of which yields the ground-state single-particle energies. Excitation spectra have also become accessible by time-dependent DFT and Bethe-Salpeter formalisms; detailed descriptions of those developments are available in [64, 65].

2.3 Approximative modelling methods

Routine applications of full DFT schemes comprise model systems with up to a few hundreds of atoms. Larger systems of up to several thousands of atoms are still tractable with approximations to the full DFT methods, such as the tight-binding (TB) description. The standard TB method also relies on a valence-only treatment, in which the electronic valence states are represented as a symmetry-adapted superposition of atomic orbitals. The Hamilton operator of the full Kohn-Sham formalism is approximated by a parametrised Hamiltonian, whose matrix elements are fitted to the properties of reference systems. A short-range repulsive potential includes the ionic repulsion and corrections due to approximations made in the band-structure term. It can be determined as a parametrised function of the interatomic distance, which reproduces the cohesive energy and elastic constants like the bulk modulus for crystalline systems.

Although the results of a TB calculation depend on the parametrisation, successful applications include high accuracy band structure evaluations of bulk semiconductors [73], band calculations in semiconductor heterostructures [74], device simulations for optical properties [75], simulations of amorphous solids [76], and predictions of low-energy silicon clusters [77, 78] (for a review, see [79]). Due to the simple parametrisation, the TB methods allow routine calculations with up to 1000 to 2000 atoms. Thus they allow an extension to more irregular interface structures, which have a larger repeat unit, and a more detailed assessment of the energy landscape, which includes a larger number of structure models. Another advantage is the simple derivation of additional quantities from TB data, because the TB approximations also simplify the mathematical effort for the calculation of material properties. Even the transport in a nanodevice of conducting and semiconducting segments along a nanotube could be modelled within a Landauer-Büttiker approach [80].

2.4 Efficient parallel computing for material properties

In the ground-state electronic-structure treatment the translational symmetry of a regular crystalline solid can be exploited. Only a small unit cell of the whole crystal is calculated explicitly, and the properties of the extended crystal are retrieved by applying periodic boundary conditions in three dimensions. The lattice periodicity suggests plane waves as an optimal basis for the representation of the electronic states. However, the strongly localised inner core electrons, such as 1s states of Mg or Al, require a very high number of plane waves (or kinetic energy cutoff) for an adequate description of the short-range oscillations in the vicinity of the nucleus. In the investigations described here this problem has been overcome by the pseudopotential technique. Only the valence and semicore states are treated explicitly, while an effective core potential accounts for the core-valence interaction, as implemented e.g. also in the Car-Parrinello approach [81]. A very successful compromise between efficiency and accuracy is provided by the use of norm-conserving pseudopotentials, which by construction exhibit the same electron scattering properties as the all-electron potentials [82, 83].

With a plane-wave basis the terms of the total-energy functional are most easily evaluated numerically on a grid of k-vectors in reciprocal (or momentum) space, where each plane wave is associated with its wave vector. In standard DFT band-structure calculations the choice of special integration points allows one to exploit the crystal symmetry. In this fashion the dimensions of the matrices, which enter the resulting secular equations at each integration point are kept small to minimise the computational effort. For the calculations presented here, however, a different strategy was followed. Since the matrix equations can be solved independently from each other at each of the integration points, a parallelisation over the integration grid is the most obvious choice. Thus, in a (massively) parallel application the integration point with the largest eigenvalue problem determines the parallelisation time step, and the gain at the high-symmetry points is not easily retrieved by load-balancing algorithms. The simplest solution is to redesign the integration grid in such a manner, that the matrix dimensions at each integration point are roughly of the same size. The sampling method derived by Moreno and Soler [84] fulfills this requirement, thus it was chosen for the investigation of the metal-semiconductor boundary described here.

3 Influence factors for interface reactivity

Reactive metal-to-non-metal interfaces often involve strongly electropositive metals with a high propensity to release electrons, but also easily accessible d-type conduction bands for metallic bonding. Thus reactive bonding has been observed for material combinations in which an early main-group or transition metal or a rare-earth element is employed as the contact metal or as a metallic adhesive layer. As mentioned above a mismatch between the lattice constants a_0 of the metal and the non-metallic bonding partner increases the elastic energy stored at the interface upon epitaxial deposition. This energy may then be lowered by the formation of periodically repeated misfit dislocations in the vicinity of the interface [43, 46]. In order to distinguish between the electronic and the elastic driving force for interface reactivity, two model systems were investigated, in which the same metal reactively bonds to a non-metallic surface, once in the absence and once in the presence of additional elastic strains. The early, electropositive transition element titanium is

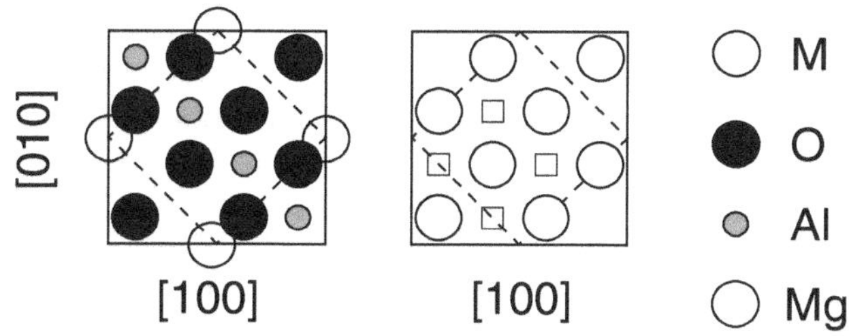

Fig. 1. Interface repeat unit of the AlO_2-terminated $MgAl_2O_4$ spinel in contact with the metals M = Ti, Al, Ag. On the left, the top (001) layer of spinel is shown. The Mg atoms depicted schematically are located by $1/4\ a_0$ below the AlO_2 termination plane. The right panel shows the optimum atom arrangement in the first metal (001) plane at the interface. As indicated the most stable bonding is obtained for Ti and Al on top of the spinel O ions. Additional O atoms can be inserted into the metal film at the sites denoted by the squares

chosen as reactive metal component. A suitable low-misfit non-metallic substrate is the spinel surface $MgAl_2O_4(001)$, whereas the a high lattice mismatch is encountered at the interface between Ti and the unreconstructed Si(111) surface.

3.1 Low lattice mismatch

The interaction of metals with the spinel (001) surface has been extensively studied both by first-principles DF calculations and by high-resolution transmission electron microscopy experiments [36–39, 85]. Because of the low lattice mismatch of 1% at the utmost, the interface $M(001)|MgAl_2O_4(001)$ with M = Al, Ti, Ag can be modelled with a rather small repeat unit parallel to the interface. The supercells for the study of the M|spinel boundaries are depicted schematically in Fig. 1.

In accordance with high-resolution transmission electron microscopy experiments [85] DFT band-structure calculations indicate that the termination of the spinel by a layer composed of Al and O ions is favored over a termination by a layer sparsely occupied by Mg ions. As outlined in Subsect. 1.2, Al and Ag, respectively, represent the extreme cases of strong bonding (A) by directed Al-O electron transfer and weak bonding (B) due to Pauli repulsion between the Ag(4d) and O(2p) shells and a slightly larger misfit of 1%. Additionally, a comparison with M=Ti was performed, because it allows for a better discrimination of the interaction-determining factors: charge transfer, Pauli repulsion, and elastic contributions. The relative ordering of the three metals Ti, Al, and Ag is as follows:

- electronegativity (charge transfer): Ti $<$ Al $<$ Ag
- electron gas parameter (Pauli repulsion): Al $\approx$ Ti $>$ Ag
- lattice mismatch (elastic contribution): Al $<$ Ag $<$ Ti.

The electronegativity of an atom is a measure for the binding strength of its valence electrons. At metal-oxide interfaces, metals with a low electronegativity can act as an electron donor for the undercoordinated oxygen ions of the contact plane, which enhances the interfacial bonding. The electron gas parameter is a measure of the local electron density and it is defined as the radius of a sphere which would contain one valence electron if the electron density were homogeneous; thus, a low electron gas parameters is characteristic of atoms with a high spatial density of electrons, which exhibit the higher propensity for Pauli repulsion.

The calculated interface distances of $d(\text{Al-O}) = 1.9$ Å $< d(\text{Ti-O}) = 1.99$ Å $< d(\text{Ag-O}) = 2.3$ Å indicate, that the Ti-containing adhesion system exhibits an average position between the two pristine boundaries. The work of separation W_{sep} was calculated as difference of the total energies of the interface system and the two free component slabs within the same supercell geometry. The values amount to $W_{\text{sep}}(\text{Al|Sp}) = 2.25\ \text{J/m}^2 > W_{\text{sep}}(\text{Ti|Sp}) = 1.81\ \text{J/m}^2 > W_{\text{sep}}(\text{Ag|Sp}) = 1.10\ \text{J/m}^2$ and confirm the intermediate nature of the Ti|spinel boundary. Therefore, the reduction of the Pauli repulsion by the low valence electron density of the Ti film is the main driving force for the adhesion enhancement (*titanium effect*), whereas the other two contributions balance each other. A stepwise exchange of Ag atoms by Ti atoms within the first metal layer indicated that the major stabilisation is already reached, when every second Ag atom is replaced by Ti.

Thus, the application of Ti as adhesive buffer layer in non-strained systems is based on its ability to form polar bonds with electronegative partners such as the O ions in oxide ceramics, but also metallic bonds with electron-rich elements such as Ag. This high reactivity has also drawbacks as described in [41, 42]: The high oxygen affinity of Ti can also induce an autocatalytic uptake of oxygen into a Ti adhesive layer. DFT bandstructure calculations have shown that the most stable position of the additional O atoms is the interstitial site indicated by the squares in Fig. 1. In this way, the metallic adhesive Ti layer is transformed into a brittle titanium suboxide, which exhibits both a higher Pauli repulsion and a stronger elastic strain contribution than the free metal. In this manner the reactivity of the interlayer degrades the long-term stability of the interface in an oxidative environment.

3.2 High lattice mismatch

Many examples for reactive interfaces including amorphous phases are found at heterophase boundaries between metallic and semiconducting materials, especially at the contacts between a microelectronic device and the bonding metal. Almost all early transition and rare earth metals bind to the Si(111) or Si(110) surfaces under the formation of binary and ternary silicides. The most extensively studied material combination is the interface Co|Si, where DFT-based modelling has helped to quantify the thermodynamic driving force for the formation of silicides like $CoSi_2$ [27–29,86,87]. The first steps of the silicide formation are the migration of Co atoms to interstitial sites in the Si lattice, the diffusion of those atoms along grain boundaries in Si, and the growth of larger $CoSi_2$ crystals from $CoSi_2$ seeds [27]. The interface reconstruction has also been explained within a DFT framework [28]. Slab calculations of $CoSi_2(100)|Si(100)$ [87], $CoSi_2(111)|Si(111)$ [86], and of up to two monolayers of Co on Si(001) [29] confirm that the formation of the reactive oxide does not depend on a particular orientation of the substrate. Also the interaction of Y and Gd [30], Sr [31], Ni [88] and Fe [32] with Si has been investigated theoretically. A first-principles study of the boundary between Si and an early transition metal such as Ti is, however, rendered difficult by the large mismatch of the lattice constants of Ti and Si. With a lattice mismatch of 24% large interface unit cells have to be employed in the interface modelling. Only with the application of parallel computing power, such demanding tasks have lately become feasible.

At elevated temperature a deposited layer of Ti reacts with Si(111) or Si(110) surfaces to form several stable silicides at the boundary, which lie within a composition range of Ti_3Si to $TiSi_2$ [89–92]. Depending on deposition and preparation conditions, the silicide interlayer is either crystalline or amorphous, and the structural phase transition between those two states occurs at a temperature well below the melting point of either of the two constituents or of the resultant silicide phase [89, 91, 93–97]. The structural features, thermodynamic stability, and electronic properties of several bulk $TiSi_x$ compounds have also been investigated theoretically [98]. Yet, the quantum-mechanical treatment of the Ti|Si interfaces is difficult, as the lattice parameters of hexagonally close-packed Ti ($a_0 = 2.95$ Å) and Si in the diamond structure ($a_0 = 3.85$ Å) are quite different. Thus, only every third Si atom roughly coincides with every fourth Ti atom along the close-packed [1-10.0] direction of Ti. An interface model with low remanent elastic stress can be constructed, which, however, contains a high number of atoms in the supercell and is computationally demanding. Yet, with the emergence of carbon nanotube integration in Si-based semiconductor devices, the interface between Ti and Si has become an important topic for investigation, because thin Ti interlayers act as adhesion enhancers both for the nanotube growth catalyst and for Au contacts to the individual tube [99]. Thus, it is of great technological relevance to study the corresponding interfaces quantitatively, as outlined in the following section.

4 The Ti(000.1)|Si(111) interface

4.1 Model structures

Due to the considerable lattice mismatch of 26% between the Si(111) and the Ti(000.1) crystals the appropriate interface supercell spans several lattice spacings. For this purpose, the so-called coincidence site lattice is constructed

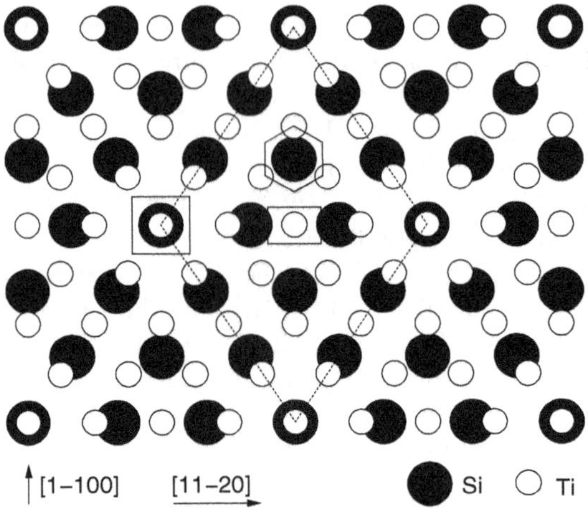

Fig. 2. Top view on the atom arrangement in the interface region of the supercell employed for the study of the Ti(000.1)|Si(111) boundary. The interface repeat unit is shaded grey, and only the Si and Ti planes adjacent to the interface are shown for clarity. The square denotes the most favourable position of Ti on top of Si. The less favourable bridging and three-fold hollow sites are indicated by the rectangle and the hexagon

from a superposition of the atom positions in the layers adjacent to the heterophase boundary, a (000.1) plane of Ti and a (111) plane of Si.

The unreconstructed Si crystal is terminated by a buckled honeycomb layer. This structure element has also been observed in layered silicides, such as the recently studied $CaSi_2$ [100]. As the rumpling of the Si layer amounts to 0.79 Å only the upper half of the atoms with a lateral spacing of d(Si-Si) = 3.85 Å are included in the construction of the coincidence site lattice, as depicted in Fig. 2. The Ti-Ti nearest-neighbour distance d(Ti-Ti) is only 2.95 Å. Figure 2 shows that coincident points occur at every third Si atom and every fourth Ti atom along the Si $\langle 110 \rangle \approx$ Ti $\langle$1-100$\rangle$ directions. In-between, both bridging and hollow-site arrangements occur in the same model structure, thus, a lateral shift of the two constituents did not lead to any more favourable local atom arrangements at the interface. The corresponding supercell, indicated by the area shaded in grey, contains 16 Ti atoms and 9 Si atoms per layer (18 Si per buckled double layer) parallel to the boundary. It is spanned by the vectors $a_1 = const\cdot$ [1-12.0] and $a_2 = const\cdot$ [1-21.0], with $const = 4\ d$(Ti-Ti) $= 3\,d$(Si-Si). Along the interface normal, the supercell consists of five (000.1) layers of Ti and four buckled (111) layers of Si with stacking sequences of A-B-A-B-A and B-C-A-B, respectively. A smaller model with the approximation that 2 d(Si-Si) $\approx$ 3 d(Ti-Ti) did not yield any interface binding. The considerable misfit of 33% induces tensile strain for Ti(000.1) and compressive strain for Si(111) and prevents a bonding interaction.

4.2 Stability at low temperature

Due to the difference between the real lattice mismatch of 26% and the mismatch in the model of only 25% there are remanent elastic deformations, tensile for the Si slab and compressive for the Ti slab. As the bulk elastic properties of both elements are almost equal, there exists no natural choice whether the lengths of the in-plane vectors $|a_1| = |a_2|$ spanning the supercell ought to amount to 3 d(Si-Si) = 11.55 Å or to 4 d(Ti-Ti) = 11.8 Å. Thus, the structure optimisation of the interface was carried out for both cases and also for the arithmetic mean. Within this range of values the total energy of the system does not significantly depend on the choice of a_1 and a_2. This effect is presumably related to the similarity of the bulk moduli (both about 110 GPa), which means that in this range of values the lattice expansion of Si and the lattice compression of Ti lead to a comparable energy change. Thus, the results obtained for $|a_1| = |a_2| = 3\ d$(Si-Si) will be discussed in the following, because this choice reflects best the experimental setup, where a Ti contact layer is evaporated onto a comparatively massive Si substrate.

Structure optimisation yields a supercell height of c = 24.06 Å, which corresponds to an interface contraction of 0.16 Å. On the Si side the buckling of the double layer next to the interface is strongly diminished by 0.21 Å, whereas the spacing to the next lower double layer is reduced by only 0.06 Å. This indicates that all Si atoms of the buckled layer interact with the Ti atoms. On the Ti side, the last layer is slightly curved by 0.09 Å, and the average distance between this Ti layer and the next one is reduced by 0.06 Å. The Ti-Si distances vary from 2.55 Å at the on-top position to 2.71 Å at the hollow sites. There, an additional Si atom from the lower part of the Si double layer relaxes towards the interface such that the Ti-Si distance amounts to 2.76 Å, and the effective coordination number of Ti with respect to Si partners is enhanced to four.

The work of separation W_{sep} with respect to the free Si(111) and Ti(000.1) slabs is calculated as the difference of the total energies of the interface, E_{tot}(Ti(000.1)|Si(111)), and the free slabs, E_{tot}(Ti(000.1)) and E_{tot}(Si(111)). The obtained low value of $W_{sep} = 0.27$ J/m^2 reflects the only weakly attractive interaction at the unreacted interface. The low work of separation is, of course, related to the low density of only 11% of favourable, on-top interaction sites at the boundary. When normalised to this low density of binding sites, an upper bound for the separation energy is obtained, which is of the order of 2 J/m^2. This value compares well with the work of separation calculated for the strain-free M|spinel systems. This argument is corroborated by an analysis of the binding electron density, obtained as difference between the electron density of the total interface and the densities of the free, unrelaxed slabs. An electron transfer from Ti to Si takes place, which is limited to the first Si layer and exhibits the highest accumulation at the Si atom bonded on top of a Ti atom. Within the Ti slab the density difference decays rapidly as a result of the very effective metallic screening. In the Si slab, further, but smaller electron density oscillations can be monitored also in the second layer below the interface. This observation reflects the expectation that the screening of a charge accumulation is less effective in semiconductors. At the interface, some electron density accumulation occurs also at the bridge and three-fold hollow-site positions. Thus, of all adhesion sites only the ones with the shortest Ti-Si distances along the [1-10.0] direction contribute to the binding, while for the other sites no favourable interaction is detected. These observations also rationalise the low, rather anisotropic binding energy.

4.3 High-temperature behaviour

Experimental investigations indicate the formation of the modifications Ti_5Si_3, TiSi, and $TiSi_2$ at elevated temperatures, therefore full DFT molecular-dynamics calculations were carried out for the Ti(000.1)|Si(111) interface, employing the supercell described above. In order to study the amorphisation process at the interface, the optimised, unreacted structure was employed as starting geometry. The temperature was raised in one step and controlled by a velocity rescaling algorithm at constant volume. These simulation conditions resemble best the setup of the laser heating experiments, which yielded detailed experimental data on the phase transformations at the Ti|Si interface [95].

For $T = 300$ K initially the stress tensor σ is anisotropic, with components $\sigma_{||} = 13$ GPa parallel and $\sigma_{\perp} = 8$ GPa normal to the interface. Even after several picoseconds of relaxation time the two stress tensor components do not equilibrate. Therefore, the simulation was also performed at an elevated temperature of $T = 600$ K, at which the formation of films with the stoichiometry TiSi has been reported [89]. For this temperature, the initial stress tensor components exhibit a higher value of about 23 GPa due to the boundary condition which is given by the lateral lattice periodicity of Si as substrate material. However, the internal stresses are equilibrated and lowered to 20 GPa after a relaxation time of only 1 ps. The stress reduction is accompanied by the reorientation of atoms in the boundary region. At the higher temperature the increased kinetic energy of the atoms induces a roughening of the interface, which now comprises the Ti(000.1) layer and the full Si(111) double layer adjacent to the interface. In this way, a non-crystalline film is obtained at the boundary. The area density of atoms within this boundary film yields indeed a stoichiometry of about Ti : Si = 1 : 1.

Since the supercell is repeated periodically, the model film obtained here is only an approximant of the structure of the real, amorphous boundary phase. However, one can evaluate the short-range part of the radial distribution function within the limits set by the minimum image convention, i.e. one half of the supercell dimensions. For the cell employed here this amounts to about 5 Å, thus the first coordination sphere can safely be analysed. In order to cover only the bond lengths at the interface, the bond lengths were calculated in real space and weigthed according to their frequency of occurence. For the

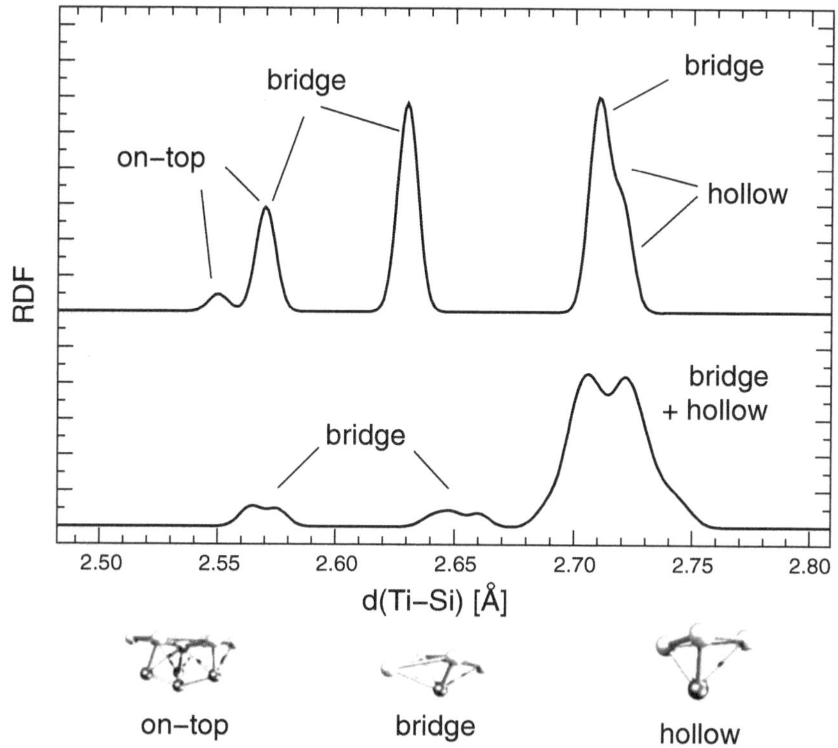

Fig. 3. Radial distribution function (RDF) of the Ti-Si bond lengths at the interface. The upper curve corresponds to the optimised structure, the lower curve gives the RDF of the interface tempered at 600 K

MD simulation at 600 K, an additional time average over the last picosecond of the simulation time was performed. Fig. 3 gives a comparison of the radial distribution function of the Ti-Si bond length between the optimised interface structure (upper curve) and the structure obtained after the molecular dynamics equilibration at 600 K (lower curve). The curves have been broadened by convolution with a Gaussian function of 0.01 Å full width at half maximum, and the upper curve is shifted for clarity. For the unreacted interface structure several maxima of the radial Ti-Si distribution function are obtained between 2.55 Å and 2.71 Å. These maxima correspond to the bond lengths at the different adsorption sites as indicated in Fig. 3. The shortest bonds occur at the on-top and bridge sites, but they are outnumbered by the longer Ti-Si contacts at the hollow-site position. Additional longer bonds range between 2.76 and 2.9 Å, but they are not related to the first bonding shell, thus they are omitted in Fig. 3. For the interface reacted at 600 K, no sharp local maxima are obtained, but the bond-length distribution is more uniform with the most pronounced maximum at 2.7 Å. In addition, the average number of bonds is by 10% lower than the corresponding number of bonds at the optimised interface. These findings reflect the equilibration of the bond lengths and the local coordination spheres upon reaction at the interface.

Concomitant with the geometry change, the total energy of the supercell relaxed at 600 K is lower than the total energy of the starting structure obtained from the geometry optimisation (formally at 0 K). This leads to a doubling of the work of separation to 0.52 J/m^2. Thus, the release of the elastic energy, which is stored in the atomically flat structure model yields a significant stabilisation of the boundary. The results also confirm experimental findings that TiSi occurs as an intermediate reaction product towards the final high-temperature phase, $TiSi_2$. Since the further reaction to a full $TiSi_2$ layer at the boundary would require the use of even larger supercells, further investigations are currently carried out on two separate model systems, Si(111)|$TiSi_2$(000.1) and Ti(000.1)|$TiSi_2$(000.1). Preliminary results from the structure optimisation indicate that the values of the work of separation will exceed 1 J/m^2. Thus, a further stabilisation of the interface by the formation of the disilicide is predicted.

5 Conclusions

The detailed description of the stability of and the structure formation at reactive interfaces requires an electronic-structure based method, which can properly account for local electron redistributions and subtle changes of the coordination number. Band-structure calculations based on the DFT provide the suitable theoretical framework for the investigations of such systems, and the application of parallel computing supplies the required numerical power. The most efficient way to exploit this power is provided by parallelisation over an optimised set of integration points, which split the solution of the Kohn-Sham equations into a set of matrix equations with equal matrix sizes. With this approach two reactive systems, both employing the element Ti as metallic component, have been studied to elucidate and quantify the role, which electronic and elastic contributions play in interface reactivity.

For an interface with low lattice misfit such as M|$MgAl_2O_4$(001) (M = Ti, Al, Ag) the balance of electronic factors dominates the structure and stability. If the electronegativity difference between metal and substrate favours electron transfer, a directed bonding of the metal on top of the O ions is obtained. This situation occurs for M = Ti and Al. Otherwise, the Pauli repulsion between the Ag(4d) shell and the O(2p) shell leads to a weaker adhesion on the hollow sites of the spinel surface. Furthermore, the reactive metal Ti can undergo an autocatalytic oxygen uptake at the octahedral interstitial site, which then leads to an oxidative corrosion of Ti interlayers.

Ti(000.1)|Si(111) is a system with a high lattice mismatch and a rich interface chemistry. The ternary interface phase ranges from Ti-rich Ti_5Si_3 to

Si-rich $TiSi_2$, which is the thermodynamically most stable phase. DFT investigations show, that the electron transfer at the unreacted interface leads to a weakly bonding interaction, because only a small number of favourable bonding sites can be saturated. DFT molecular-dynamics simulations at an elevated temperature of 600 K facilitate the release of elastic stresses still stored in the unreacted interface. The concomitant interface roughening leads to an equilibration of the Ti-Si bond lengths and a doubling of the interface stability. Thus, at this boundary, electron transfer processes and elastic contributions influence the binding energy equally strongly.

The comparison of the two model cases in which the same metal exhibits quite different reactivity shows that a realistic material modelling has to account for both electron transfer across the interface and elastic factors parallel to the interface equally well. Simplified approaches, neglecting the details of the electronic structure, may predict the properties of non-reactive boundaries. However, the fine balance between influence factors at reactive boundaries indeed requires the more accurate treatment provided by the density-functional first-principles modelling.

References

1. R.C. Longo, V.S. Stepanyuk, W. Hegert, A. Vega, L.J. Gallego, J. Kirschner. Interface intermixing in metal heteroepitaxy on the atomic scale. *Phys. Rev. B*, 69:073406, 2004.
2. J.E. Houston, J.M. White, P.J. Feibelman, D.R. Hamann. Interface-state properties for strained-layer Ni adsorbed on Ru(0001). *Phys. Rev. B*, 38:12164, 1988.
3. R.E. Watson, M. Weinert, J.W. Davenport. Structural stabilities of layered materials: Pt–Ta. *Phys. Rev. B*, 35:9284, 1987.
4. H.R. Gong, B.X. Liu. Interface stability and solid-state amorphization in an immiscible Cu–Ta system. *Appl. Phys. Lett.*, 83:4515, 2003.
5. S. Narasimham. Stress, strain, and charge transfer in Ag/Pt(111): A test of continuum elasticity theory. *Phys. Rev. B*, 69:045425, 2004.
6. S. Gemming, M. Schreiber. Nanoalloying in mixed Ag_mAu_n nanowires. *Z. Metallkd.*, 94:238, 2003.
7. S. Gemming, G. Seifert, M. Schreiber. Density functional investigation of gold-coated metallic nanowires. *Phys. Rev. B*, 69:245410, 2004.
8. S. Gemming, M. Schreiber. Density-functional investigation of alloyed metallic nanowires. *Comp. Phys. Commun.*, 169:57, 2005.
9. P.J. Lin-Chung, T.L. Reinecke. Antisite defect in GaAs and at the GaAs–AlAs interface. *J. Vac. Sci. Technol.*, 19:443, 1981.
10. S. Das Sarma, A. Madhukar. Ideal vacancy induced band gap levels in lattice matched thin superlattices: The GaAs–AlAs(100) and GaSb–InAs(100) systems. *J. Vac. Sci. Technol.*, 19:447, 1981.
11. Y. Wei, M. Razeghi. Modeling of type-II InAs/GaSb superlattices using an empirical tight-binding method and interface engineering. *Phys. Rev. B*, 69:085316, 2004.
12. A. Kley, J. Neugebauer. Atomic and electronic structure of the GaAs/ZnSe(001) interface. *Phys. Rev. B*, 50:8616, 1994.

13. W.R.L. Lambrecht, B. Segall. Electronic structure and bonding at SiC/AlN and SiC/BP interfaces. *Phys. Rev. B*, 43:7070, 1991.
14. L. Pizzagalli, G. Cicero, A. Catellani. Theoretical investigations of a highly mismatched interface: SiC/Si(001). *Phys. Rev. B*, 68:195302, 2003.
15. P. Cásek, S. Bouette-Russo, F. Finocchi, C. Noguera. $SrTiO_3(001)$ thin films on MgO(001): A theoretical study. *Phys. Rev. B*, 69:085411, 2004.
16. R.R. Das, Y.I. Yuzyuk, P. Bhattacharya, V. Gupta, R.S. Katiyar. Folded acoustic phonons and soft mode dynamics in $BaTiO_3/SrTiO_3$ superlattices. *Phys. Rev. B*, 69:132301, 2004.
17. S. Hutt, S. Köstlmeier, C. Elsässer. Density functional study of the $\Sigma3/(111)$ grain boundary in strontium titanate. *J. Phys.: Condens. Matter*, 13:3949, 2001.
18. S. Gemming, M. Schreiber. Impurity and vacancy clustering at the $\Sigma3(111)[1\text{-}10]$ grain boundary in strontium titanate. *Chem. Phys.*, 309:3, 2005.
19. M. Sternberg, W.R.L. Lambrecht, T. Frauenheim. Molecular-dynamics study of diamond/silicon (001) interfaces with and without graphitic interface layers. *Phys. Rev. B*, 56:1568, 1997.
20. T. Sakurai, T. Sugano. Theory of continuously distributed trap states at $Si–SiO_2$ interfaces. *J. Appl. Phys.*, 52:2889, 1981.
21. D. Chen, X.L. Ma, Y.M. Wang, L. Chen. Electronic properties and bonding configuration at theTiN/MgO(001) interface. *Phys. Rev. B*, 69:155401, 2004.
22. R. Puthenkovilakam, E.A. Carter, J.P. Chang. First-principles exploration of alternative gate dielectrics: Electronic structure of ZrO_2/Si and $ZrSiO_4/Si$ interfaces. *Phys. Rev. B*, 69:155329, 2004.
23. M. Rühle, A.G. Evans. High toughness ceramics and ceramic composites. *Progr. Mat. Sci.*, 33:85, 1989.
24. G. Willmann, N. Schikora, R.P. Pitto. Retrieval of ceramic wear couples in total hip arthroplasty. *Bioceram.*, 15:813, 1994.
25. A.M. Freborg, B.L. Ferguson, W.J. Brindley, G.J. Petrus. Modeling oxidation induced stresses in thermal barrier coatings. *Mater. Sci. Eng. A*, 245:182, 1998.
26. R. Benedek, M. Minkoff, L.H. Yang. Adhesive energy and charge transfer for MgO/Cu heterophase interfaces. *Phys. Rev. B*, 54:7697, 1996.
27. A. Horsfield, H. Fujitani. Density-functional study of the initial stage of the anneal of a thin Co film on Si. *Phys. Rev. B*, 63:235303, 2001.
28. B.D. Yu, Y. Miyamoto, O. Sugino, A. Sakai, T. Sasaki, T. Ohno. Structural and electronic properties of metal-silicide/silicon interfaces: A first-principles study. *J. Vac. Sci. Technol. B*, 19:1180, 2001.
29. B.S. Kang, S.K. Oh, H.J. Kang, K.S. Sohn. Energetics of ultrathin $CoSi_2$ film on a Si(001) surface. *J. Phys.: Condens. Matter*, 15:67, 2003.
30. C. Rogero, C. Koitzsch, M.E. Gonzalez, P. Aebi, J. Cerda, J.A. Martin-Glago. Electronic structure and Fermi surface of two-dimensional rare-earth silicides epitaxially grown on Si(111). *Phys. Rev. B*, 69:045312, 2004.
31. C.R. Ashman, C.J. Först, K. Schwarz, P.E. Blöchl. First-principles calculations of strontium on Si(001). *Phys. Rev. B*, 69:075309, 2004.
32. S. Walter, F. Blobner, M. Krause, S. Muller, K. Heinz, U. Starke. Interface structure and stabilization of metastable B2-FeSi/Si(111) studied with low-energy electron diffraction and density functional theory. *J. Phys.: Condens. Matter*, 15:5207, 2003.
33. U. Schoenberger, O.K. Andersen, M. Methfessel. Bonding at metal ceramic interfaces – Ab-initio density-functional calculations for Ti and Ag on MgO. *Acta Metall. Mater.*, 40:S1, 1992.

34. K. Kruse, M.W. Finnis, J.S. Lin, M.C. Payne, V.Y. Milman, A. DeVita, M.J. Gillan. First-principles study of the atomistic and electronic structure of the niobium–α-alumina(0001) interface. *Phil. Mag. Lett.*, 73:377, 1996.
35. Y. Ikuhara, Y. Sugawara, I. Tanaka, P. Pirouz. Atomic and electronic structure of V/MgO interface. *Interface Science*, 5:5, 1997.
36. R. Schweinfest, S. Köstlmeier, F. Ernst, C. Elsässer, T. Wagner, M.W. Finnis. Atomistic and electronic structure of $Al/MgAl_2O_4$ and $Ag/MgAl_2O_4$ interfaces. *Phil. Mag. A*, 81:927, 2000.
37. S. Köstlmeier, C. Elsässer. Ab-initio investigation of metal-ceramic bonding. $M(001)/MgAl_2O_4$, M=Al, Ag. *Interface Science*, 8:41, 2000.
38. S. Köstlmeier, C. Elsässer, B. Meyer, M.W. Finnis. Ab initio study of electronic and geometric structures of metal/ceramic heterophase boundaries. *Mat. Res. Soc. Symp. Proc.*, 492:97, 1998.
39. S. Köstlmeier, C. Elsässer, B. Meyer, M.W. Finnis. A density-functional study of interactions at the metal-ceramic interfaces $Al/MgAl_2O_4$ and $Ag/MgAl_2O_4$. *phys. stat. sol. (a)*, 166:417, 1998.
40. S. Köstlmeier, C. Elsässer. Density functional study of the "titanium effect" at metal/ceramic interfaces. *J. Phys.: Condens. Matter*, 12:1209, 2000.
41. C. Elsässer, S. Köstlmeier-Gemming. Oxidative corrosion of adhesive interlayers. *Phys. Chem. Chem. Phys.*, 3:5140, 2001.
42. S. Köstlmeier, C. Elsässer. Oxidative corrosion of adhesive interlayers. *Mat. Res. Soc. Symp. Proc.*, 586:M3.1, 1999.
43. V. Vitek, G. Gutekunst, J. Mayer, M. Rühle. Atomic structure of misfit dislocations in metal-ceramic interfaces. *Phil. Mag. A*, 71:1219, 1996.
44. J.-H. Cho, K.S. Kim, C.T. Chan, Z. Zhang. Oscillatory energetics of flat Ag films on MgO(001). *Phys. Rev. B*, 63:113408, 2001.
45. C. Klein, G. Kresse, S. Surnev, F.P. Netzer, M. Schmidt, P. Varga. Vanadium surface oxides on Pd(111): A structural analysis. *Phys. Rev. B*, 68:235416, 2003.
46. A. Trampert, F. Ernst, C.P. Flynn, H.F. Fischmeister and M. Rühle. High-resolution transmission electron microscopy studies of the Ag/MgO interface. *Acta Metall. Mater.*, 40:S227, 1992.
47. A.M. Stoneham, P.W. Tasker. Metal non-metal and other interfaces – The role of image interactions. *J. Phys. C*, 18:L543, 1985.
48. D.M. Duffy, J.H. Harding, A.M. Stoneham. Atomistic modeling of the metal-oxide interface with image interactions. *Acta Metall. Mater.*, 40:S11, 1992.
49. D.M. Duffy, J.H. Harding, A.M. Stoneham. Atomistic modeling of metal-oxide interfaces with image interactions. *Phil. Mag. A*, 67:865, 1993.
50. M.W. Finnis. Metal ceramic cohesion and the image interaction. *Acta Metall. Mater.*, 40:S25, 1992.
51. A.M. Stoneham, P.W. Tasker. Image charges and their influence on the growth and the nature of thin oxide-films. *Phil. Mag. B*, 55:237, 1987.
52. D.A. Muller, D.A. Shashkov, R. Benedek, L.H. Yang, J. Silcox, D.N. Seidman. Adhesive energy and charge transfer for MgO/Cu heterophase interfaces. *Phys. Rev. Lett.*, 80:4741, 1998.
53. T. Ochs, S. Köstlmeier, C. Elsässer. Microscopic structure and bonding at the $Pd/SrTiO_3(001)$ interface. *Integr. Ferroelectr.*, 30:251, 2001.
54. A. Zaoui. Energetic stabilities and the bonding mechanism of ZnO(0001)/Pd(111) interfaces. *Phys. Rev. B*, 69:115403, 2004.

55. M. Christensen, S. Dudiy, G. Wahnström. First-principles simulation of metal-ceramic interface adhesion: Cu/WC versus Cu/TiC. *Phys. Rev. B*, 65:045408, 2002.
56. M. Christensen, G. Wahnström. Co-phase penetration of WC(10-10)/WC(10-10) grain boundaries from first principles. *Phys. Rev. B*, 67:115415, 2003.
57. J. Hartford. Interface energy and electron structure for Fe/VN. *Phys. Rev. B*, 61:2221, 2000.
58. E. Saiz, A.P. Tomsia, R.M. Cannon. Ridging effects on wetting and spreading of liquids on solids. *Acta Mater.*, 46:2349, 1998.
59. J.A. Venables, G.D.T. Spiller, M. Hanbucken. Nucleation and growth of thin films. *Rep. Prog. Phys.*, 47:399, 1984.
60. A.M. Stoneham, J.H. Harding. Not too big, not too small: The appropriate scale. *Nature Materials*, 2:65, 2003.
61. M.W. Finnis. *Interatomic Forces in Condensed Matter.* Oxford University Press, Oxford, 2003.
62. R.O. Jones, O. Gunnarsson. Density-functional theory. *Rev. Mod. Phys.*, 61:689, 1989.
63. H. Eschrig. *The Fundamentals of Density Functional Theory.* Edition am Gutenbergplatz, Leipzig, 2003.
64. G. Onida, L. Reining, A. Rubio. Electronic excitations: density-functional versus many-body Green's-function approaches. *Rev. Mod. Phys.*, 74:601, 2002.
65. R.M. Dreizsler, E.K.U. Gross. *Density Functional Theory.* Springer, Berlin, 1990.
66. R.G. Parr, W. Yang. *Density-Functional Theory of Atoms and Molecules.* Oxford University Press, New York, 1989.
67. P. Hohenberg, W. Kohn. Inhomogeneous electron gas. *Phys. Rev.*, 136:B864, 1964.
68. M. Levy. Electron densities in search of Hamiltonians. *Phys. Rev. A*, 26:1200, 1982.
69. U. von Barth, L. Hedin. The energy density functional formalism for excited states. *J. Phys. C*, 5:1629, 1972.
70. N.D. Mermin. Thermal properties of the inhomogeneous electron gas. *Phys. Rev.*, 137:A1441, 1965.
71. S.H. Vosko, L. Wilk, M. Nusair. Accurate spin-dependent electron liquid correlation energies for local spin density calculations: a critical analysis. *Can. J. Phys.*, 58:1200, 1980.
72. O. Gunnarsson, M. Jonson, B.I. Lundqvist. Descriptions of exchange and correlation effects in inhomogeneous electron systems. *Phys. Rev. B*, 20:3136, 1979.
73. J.-M. Jancu, R. Scholz, F. Beltram, F. Bassani. Empirical *spds** tight-binding calculation for cubic semiconductors: General method and material parameters. *Phys. Rev. B*, 57:6493, 1998.
74. R. Scholz, J.-M. Jancu, F. Bassani. Superlattice calculation in an empirical spds* tight-binding model. *Mat. Res. Soc. Symp. Proc.*, 491:383, 1998.
75. A. Di Carlo. Time-dependent density-functional-based tight-binding. *Mat. Res. Soc. Symp. Proc.*, 491:391, 1998.
76. C.Z. Wang, K.M. Ho, C.T. Chan. Tight-binding molecular-dynamics study of amorphous carbon. *Phys. Rev. Lett.*, 70:611, 1993.

77. P. Ordejón, D. Lebedenko, M. Menon. Improved nonorthogonal tight-binding Hamiltonian for molecular-dynamics simulations of silicon clusters. *Phys. Rev. B*, 50:5645, 1994.
78. M. Menon, K.R. Subbaswamy. Nonorthogonal tight-binding molecular-dynamics scheme for silicon with improved transferability. *Phys. Rev. B*, 55:9231, 1997.
79. C.M. Goringe, D.R. Bowler, E. Hernandez. Tight-binding modelling of materials. *Rep. Prog. Phys.*, 60:1447, 1997.
80. A. Di Carlo, M. Gheorghe, P. Lugli, M. Sternberg, G. Seifert, T. Frauenheim. Theoretical tools for transport in molecular nanostructures. *Physica B*, 314:86, 2002.
81. R. Car, M. Parrinello. Unified approach for molecular dynamics and density-functional theory. *Phys. Rev. Lett.*, 55:2471, 1985.
82. D. R. Hamann, M. Schlüter, C. Chiang. Norm-conserving pseudopotentials. *Phys. Rev. Lett.*, 43:1494, 1979.
83. G.B. Bachelet, D.R. Hamann, M. Schlüter. Pseudopotentials that work: From H to Pu. *Phys. Rev. B*, 26:4199, 1082.
84. J. Moreno, J.M. Soler. Optimal meshes für integrals in real- and reciprocal-space unit cells. *Phys. Rev. B*, 45:13891, 1992.
85. R. Schweinfest, Th. Wagner, F. Ernst. *Annual Report to the VW Foundation on the Project Progress.* Stuttgart, 1997.
86. R. Stadler, D. Vogtenhuber, R. Podloucky. Ab initio study of the $CoSi_2$(111)/Si(111) interface. *Phys. Rev. B*, 60:17112, 1999.
87. R. Stadler, R. Podloucky. *Ab initio* studies of the $CuSi_2$(100)/Si(100) interface. *Phys. Rev. B*, 62:2209, 2000.
88. H. Fujitani. First-principles study of the stability of the $NiSi_2$/Si(111) interface. *Phys. Rev. B*, 57:8801, 1998.
89. B. Chenevier, O. Chaix-Pluchery, P. Gergaud, O. Thomas, F. La Via. Thermal expansion and stress development in the first stages of silicidation in Ti/Si thin films. *J. Appl. Phys.*, 94:7083, 2003.
90. G. Kuri, Th. Schmidt, V. Hagen, G. Materlik, R. Wiesendanger, J. Falta. Subsurface interstitials as promoters of three-dimensional growth of Ti on Si(111): An x-ray standing wave, x-ray photoelectron spectroscopy, and atomic force microscopy investigation. *J. Vac. Sci. Technol. A*, 20:1997, 2002.
91. J.M. Yang, J.C. Park, D.G. Park, K.Y. Lim, S.Y. Lee, S.W. Park, Y.J. Kim. Epitaxial C49-$TiSi_2$ phase formation on the silicon (100). *J. Appl. Phys.*, 94:4198, 2003.
92. O.A. Fouad, M. Yamazato, H. Ichinose, M. Nagano. Titanium disilicide formation by rf plasma enhanced chemical vapor deposition and film properties. *Appl. Surf. Sci.*, 206:159, 2003.
93. R. Larciprete, M. Danailov, A. Barinov, L. Gregoratti, M. Kiskinova. Thermal and pulsed laser induced surface reactions in Ti/Si(001) interfaces studied by spectromicroscopy with synchrotron radiation. *J. Appl. Phys.*, 90:4361, 2001.
94. M.S. Alessandrino, M.G. Grimaldi, F. La Via. C49-C54 phase transition in anometric titanium disilicide nanograins. *Microelec. Eng.*, 64:189, 2003.
95. L. Lu, M.O. Lai. Laser induced transformation of $TiSi_2$. *J. Appl. Phys.*, 94:4291, 2003.
96. S.L. Cheng, H.M. Lo, L.W. Cheng, S.M. Chang, L.J. Chen. Effects of stress on the interfacial reactions of metal thin films on (001)Si. *Thin Solid Films*, 424:33, 2003.

97. C.C. Tan, L. Lu, A. See, L. Chan. Effect of degree of amorphization of Si on the formation of titanium silicide. *J. Appl. Phys.*, 91:2842, 2002.
98. M. Ekman, V. Ozolins. Electronic structure and bonding properties of titanium silicides. *Phys. Rev. B*, 57:4419, 1998.
99. F. Wakaya, Y. Ogi, M. Yoshida, S. Kimura, M. Takai, Y. Akasaka, K. Gamo. Cross-sectional transmission electron microscopy study of the influence of niobium on the formation of titanium silicide in small-feature contacts. *Micr. Eng.*, 73:559, 2004.
100. S. Gemming, G. Seifert. Nanotube bundles from calcium disilicide - a DFT study. *Phys. Rev. B*, 68:075416-1–7, 2003.

Energy-Level and Wave-Function Statistics in the Anderson Model of Localization

Bernhard Mehlig[1] and Michael Schreiber[2]

[1] Gothenburg University, Department of Physics
41296 Göteborg, Sweden
mehlig@fy.chalmers.se

[2] Technische Universität Chemnitz, Institut für Physik
09107 Chemnitz, Germany
schreiber@physik.tu-chemnitz.de

1 Introduction

Universal aspects of correlations in the spectra and wave functions of closed, complex quantum systems can be described by random-matrix theory (RMT) [1]. On small energy scales, for example, the eigenvalues, eigenfunctions and matrix elements of disordered quantum systems in the metallic regime [2] or those of classically chaotic quantum systems [3] exhibit universal statistical properties very well described by RMT. It is now also well established that deviations from RMT behaviour are often significant at larger energy scales.

In the case of classically chaotic quantum systems, this was first discussed by Berry using a semiclassical approach and the so-called diagonal approximation (for a review see [3]). In the case of classically diffusive, disordered quantum systems, non-universal deviations from universal spectral fluctuations (as described by RMT) were first discussed in [4], using diagrammatic perturbation theory.

Andreev and Altshuler [5] have used an approach based on the non-linear sigma model to calculate non-universal deviations in the spectral statistics of disordered quantum systems from RMT behaviour, on all energy scales. On energy scales much larger than the mean level spacing, the results of diagrammatic perturbation theory are reproduced in this way, and thus the non-universal deviations (from the RMT predictions) derived in [4]. As has been shown in [6], diagrammatic perturbation theory and semiclassical arguments combined with the diagonal approximation are essentially equivalent in this regime.

In [5] it was also argued that non-universal corrections affect the spectral two-point correlation function not only on large energy scales, but also on small energy scales, of the order of the mean level spacing. This may appear surprising, but it was explained in [7] that this is just a consequence of the fact that the two-point correlation function is well approximated by a sum of shifted Gaussians.

Turning to eigenfunction statistics, deviations from RMT behaviour in classically chaotic quantum systems due to so-called scars (that are wave functions localized in the vicinity of unstable classical periodic orbits [8]) were analyzed in [9]. In disordered quantum systems, deviations from universal wave-function statistics from RMT behaviour, due to increased localisation, have been studied using the non-linear sigma model. For a summary of results see [10, 11]. The deviations are most significant in the tails of distribution functions [12], of wave-function amplitudes [13–18], of the local density of states [13,18], of inverse participation ratios [18], and of NMR line shapes [13].

In the following, exact-diagonalization results for the so-called Anderson model of localization are reviewed. The exact diagonalizations were made possible by employing the Lanczos algorithm described in [19]. We concentrate on deviations from RMT in spectral and wave-function statistics in the quasi-one-dimensional case. The remainder of this brief review is organized as follows. In Sect. 2 the Hamiltonian is written down, and the quantities computed are defined: the distribution of wave-function intensities and the spectral form factor. Theoretical expectations are briefly summarized in Sect. 3. The material in this section is largely taken from [11] and, to some extent, also from [7]. In Sect. 4 the exact-diagonalisation results are described. The numerical results on wave-function statistics described in this brief review were published in [9, 20, 21]. The analytical and numerical results on spectral statistics were obtained in collaboration with M. Wilkinson [7,22]. The problem of analyzing statistical properties of quantum-mechanical matrix elements is not addressed in this brief review. Results based on exact diagonalisation and semiclassical methods can be found in [23–25].

2 Formulation of the problem

2.1 The Hamiltonian

The Anderson model [26] is defined by the tight-binding Hamiltonian on a d-dimensional hypercubic lattice

$$\widehat{H} = \sum_{\boldsymbol{r},\boldsymbol{r}'} t_{\boldsymbol{r}\boldsymbol{r}'} c_{\boldsymbol{r}}^{\dagger}\, c_{\boldsymbol{r}'} + \sum_{\boldsymbol{r}} v_{\boldsymbol{r}} c_{\boldsymbol{r}}^{\dagger} c_{\boldsymbol{r}}\,. \tag{1}$$

Here $c_{\boldsymbol{r}}^{\dagger}$ and $c_{\boldsymbol{r}}$ are the creation and annihilation operators of an electron on site $\boldsymbol{r}$, the hopping amplitudes are usually chosen as $t_{\boldsymbol{r}\boldsymbol{r}'} = 1$ for nearest-neighbour sites and zero otherwise. Below we summarise the results of exact diagonalizations for lattices with $64 \times 4 \times 4$, $128 \times 4 \times 4$ and $128 \times 8 \times 8$ sites, using open boundary conditions in the longitudinal direction and periodic boundary conditions in the transversal directions. The on-site potentials $v_{\boldsymbol{r}}$ are Gaussian distributed with zero mean and

$$\langle v_{\boldsymbol{r}} v_{\boldsymbol{r}'} \rangle = \frac{W^2}{12} \delta_{\boldsymbol{r}\boldsymbol{r}'} . \tag{2}$$

As usual, the parameter W characterises the strength of the disorder and $\langle \cdots \rangle$ denotes the disorder average.

As is well-known, the eigenvalues E_j and eigenfunctions ψ_j of this Hamiltonian, in the metallic regime, exhibit fluctuations described by RMT. Depending on the symmetry, Dyson's Gaussian orthogonal or unitary ensembles [27] are appropriate. The former applies in the absence of magnetic fields or more generally when the secular matrix is real symmetric. With magnetic field, complex phases for the hopping matrix elements yield a unitary secular matrix. We refer to these cases by assigning, as usual, the parameter $\beta = 1$ to the former and $\beta = 2$ to the latter. The metallic regime is characterized by $g \gg 1$ where $g = 2\pi\nu_0 D L^{d-2}$ is the dimensionless conductance. Here $\nu_0 = 1/(V\Delta)$ is the average density of states per unit volume, Δ is the mean level spacing (that is the average spacing between neighbouring energy levels), and $V = L^d$ is the volume. $D = v_{\mathrm{F}} \tau_\ell / d$ is the (dimensionless) diffusion constant, determined by the collision time $\tau_\ell = \ell / v_{\mathrm{F}}$ (here ℓ is the mean free path, that is the average distance travelled between two subsequent collisions with the impurity potential), and the Fermi velocity v_{F}. Four length scales are important: the lattice spacing a, the linear dimension L, the localization length ξ, and the mean free path ℓ. In the following the diffusive limit is considered, where $\ell \ll L$. We also require that $L, \xi, \ell \gg a$.

2.2 Wave-function statistics

By diagonalizing the Hamiltonian $\widehat{H}$ using the Lanczos algorithm [19], one obtains the distribution $f_\beta(E, t; \boldsymbol{r})$ of normalized wave-function probabilities $t = |\psi_j(\boldsymbol{r})|^2 V$ corresponding to the eigenvalues $E_j \approx E$

$$f_\beta(E, t; \boldsymbol{r}) = \Delta \Big\langle \sum_j \delta(t - |\psi_j(\boldsymbol{r})|^2 V) \Big\rangle_{E_j \simeq E} . \tag{3}$$

Here $\langle \cdots \rangle_E$ denotes a combined disorder and energy average (over a small interval of width η centered around E). The spatial structure of the wave functions is described by means of

$$g_\beta(E, t; \boldsymbol{r}) = \Delta \Big\langle \sum_j |\psi_j(\boldsymbol{r})|^2 V \delta(t - |\psi_j(\boldsymbol{r})|^2 V) \Big\rangle_{E_j \simeq E} . \tag{4}$$

The normalized shape of the wave functions is given by the ratio of g and f which we denote by $\langle V |\psi(\boldsymbol{r})|^2 \rangle_t$, where $\langle \cdots \rangle_t$ denotes the average over a small interval around t,

$$\langle V |\psi(\boldsymbol{r})|^2 \rangle_t = g_\beta(E, t; \boldsymbol{r}) / f_\beta(E, t; \boldsymbol{r}) . \tag{5}$$

2.3 Spectral statistics

The spectral two-point correlation function is defined as

$$R_\beta(E,\epsilon) = \Delta^2 \left\langle \nu(E+\epsilon/2)\,\nu(E-\epsilon/2)\right\rangle_E - 1 \tag{6}$$

where $\nu(E) = \sum_j \delta(E-E_j)$ is the density of states. It is assumed that a sufficiently small energy window is considered, so that the mean level spacing Δ is approximately constant within the window. In [7,22] the so-called spectral form factor was computed, defined as

$$K_\beta(E,\tau) = \int_{-\infty}^{\infty} \mathrm{d}n\, \mathrm{e}^{-2\pi i n\tau}\, R_\beta(E,n\Delta)\,. \tag{7}$$

Here τ is a scaled time, related to the physical time t by $\tau = 2\pi\hbar t/\Delta \equiv t/t_{\mathrm{H}}$. Using the definition of the spectral two-point correlation function, one obtains

$$K_\beta(E,\tau) = \Big\langle \int_{-\infty}^{\infty} \frac{\mathrm{d}\epsilon}{\Delta}\left[\Delta^2\,\nu(E+\epsilon/2)\nu(E-\epsilon/2) - 1\right]\,\exp(-2\pi\mathrm{i}\epsilon\tau/\Delta)\Big\rangle\,. \tag{8}$$

For a finite spectral window, the form factor can be expressed as

$$K_\beta(E,\tau) = \Big\langle\Big|\sum_j w(E-E_j)\exp(2\pi\mathrm{i}E_j\tau/\Delta) - \hat{w}(\tau)\Big|^2\Big\rangle \tag{9}$$

where $w(E)$ is a spectral window function centred around zero and normalized according to

$$\int \frac{\mathrm{d}E}{\Delta} w^2(E) = 1\,. \tag{10}$$

Furthermore $\hat{w}$ is the Fourier transform of w

$$\hat{w}(\tau) = \int \frac{\mathrm{d}E}{\Delta}\, w(E)\,\exp(2\pi\mathrm{i}E\tau/\Delta)\,. \tag{11}$$

Thus for large τ the form factor converges to unity:

$$K_\beta(E,\tau) \simeq \sum_j w^2(E-E_j) \simeq \int \frac{\mathrm{d}E}{\Delta}\, w^2(E-E_j) = 1\,. \tag{12}$$

In [22] a Hann window [28] was used

$$w(E) = \begin{cases} \sqrt{2/(3\eta)}\,[1+\cos(2\pi(E/\eta))] & \text{for } |E| \le \eta/2\,,\\ 0 & \text{otherwise}\,. \end{cases} \tag{13}$$

3 Summary of theoretical expectations

In this section we briefly summarise the theoretical expectations for wave-function statistics, $f_\beta(E,t)$ and $g_\beta(E,t)$, and for the spectral form factor $K(\tau)$ in quasi-one-dimensional disordered quantum systems. The results in 3.1 are taken from the review article by Mirlin [11]; they were derived by means of a non-linear sigma model for a white-noise Hamiltonian. The discussion of spectral statistics in Sect. 3.2 follows [7].

3.1 Wave-function statistics

The wave-function statistics depends on the index β. For $\beta = 2$ one obtains for a quasi-one-dimensional conductor

$$f_2(E,t;x) = \frac{\mathrm{d}^2}{\mathrm{d}t^2}\left[\mathcal{W}^{(1)}(t/X,\theta_+)\mathcal{W}^{(1)}(t/X,\theta_-)\right] \tag{14}$$

and (assuming $|\boldsymbol{r}| \equiv r \gg \ell$)

$$g_2(E,t;x) = -X\frac{\mathrm{d}}{\mathrm{d}t}\left[\frac{\mathcal{W}^{(2)}(t/X,\theta_1,\theta_2)\mathcal{W}^{(1)}(t/X,\theta_-)}{t}\right] \tag{15}$$

where $\theta_+ = (L-x)/\xi$, $\theta_- = x/\xi$, $\theta_1 = r/\xi$ and $\theta_2 = (L-x-r)/\xi$. Here, $X = (\beta/2)L/\xi$ (see [20]), L is the length of the sample, and ξ is the localisation length. Moreover, x is the distance of the observation point from the nearest boundary of the sample in the perpendicular direction. The function $\mathcal{W}^{(1)}(z,\theta)$ obeys the differential equation

$$\frac{\partial}{\partial\theta}\mathcal{W}^{(1)}(z,\theta) = \left(z^2\frac{\partial^2}{\partial z^2} - z\right)\mathcal{W}^{(1)}(z,\theta) \tag{16}$$

with initial condition $\mathcal{W}^{(1)}(z,0) = 1$. The function $\mathcal{W}^{(2)}(z,\theta,\theta')$ obeys the same differential equation, but with the initial condition $\mathcal{W}^{(2)}(z,0,\theta) = z\mathcal{W}^{(1)}(z,\theta)$.

In the case of $\beta = 1$, one obtains

$$f_1(E,t;x) = \frac{2\sqrt{2}}{\pi\sqrt{t}}\frac{\mathrm{d}^2}{\mathrm{d}t^2}\int_0^\infty \frac{\mathrm{d}z}{\sqrt{z}}\,\mathcal{W}^{(1)}\Big((z+t/2)/X,\theta_-\Big)\,\mathcal{W}^{(1)}\Big((z+t/2)/X,\theta_+\Big)\,. \tag{17}$$

In the metallic regime (where $X \to 0$) the usual RMT results $f_1^{(0)}(t) = \exp(-t/2)/\sqrt{2\pi t}$ and $f_2^{(0)}(t) = \exp(-t)$ are obtained. The former distribution ($\beta = 1$) is often referred to as the Porter-Thomas distribution [29]. For increasing localization (finite but still small X), one has approximately $f_\beta(E,t;x) = f_\beta^{(0)}(t)[1 + \delta f_\beta(E,t;x)]$ with

$$\delta f_\beta(E,t;x) \simeq P(x;0)\begin{cases} 3/4-3t/2+t^2/4 & \text{for } \beta=1\,,\\ 1-2t+t^2/2 & \text{for } \beta=2\,,\end{cases} \tag{18}$$

valid for $t \ll X^{-1/2}$. Here $P(x;t)$ is the one-dimensional diffusion propagator [30]. In the tails ($t \gg X^{-1} > 1$) of $f_\beta(E,t;x)$, Eqs. (14,17) simplify to [15]

$$f_\beta(E,t;x) \simeq A_\beta(x,X)\exp(-2\beta\sqrt{t/X})\,. \tag{19}$$

This result may also be obtained within a saddle-point approximation to the non-linear sigma model [16]. The prefactors $A_\beta(x,X)$ for $\beta = 1,2$ are given in [15,16].

We finally turn to the shape of the wave functions. In close vicinity of the localisation center, the anomalously localized wave functions exhibit a very narrow peak (of width less than ℓ). The above expressions apply for $r \gg \ell$ and thus describe the smooth background intensity, but not the sharp peak itself [18]. For large values of t, for instance, it was suggested by Mirlin [18] that the background intensity should be given by

$$\langle V|\psi(\boldsymbol{r})|^2\rangle_t \approx \frac{1}{2}\sqrt{tX}\left(1 + r\sqrt{t/(L\xi)}\right)^{-2} \tag{20}$$

where in accordance with the above, $\ell \ll r$ is assumed, and also $r \ll \xi$.

It is necessary to emphasize that the tails of the wave-function amplitude distribution [see for example (19)] may depend on the details of the model considered. In the case of two-dimensional Anderson models at the centre of the spectrum, i.e. at $E = 0$, for instance, the distribution is of the form predicted by the non-linear sigma model, albeit with modified coefficients [31]. This could explain anomalies observed in simulations of two- and also in three-dimensional Anderson models [20,32–34]. In [35] the wave-function amplitude distributions of random banded matrices near the band centre were analyzed. It was shown that they agree well with the predictions of the non-linear σ model for the quasi-one-dimensional case [11,15].

In 4.1, results of exact diagonalisations of the quasi-one-dimensional Anderson tight-binding Hamiltonian are compared to (14,17-20).

3.2 Spectral statistics

The spectral two-point correlation function $R_\beta(E,\epsilon)$ may be written as a sum of two contributions,

$$R_\beta(E,\epsilon) = R_\beta^{\mathrm{av}}(E,\epsilon) + R_\beta^{\mathrm{osc}}(E,\epsilon)\,. \tag{21}$$

$R_\beta^{\mathrm{av}}(E,\epsilon)$ is defined as

$$R_\beta^{\mathrm{av}}(E,\epsilon) = \int \mathrm{d}\epsilon'\, v(\epsilon-\epsilon')\, R_\beta(E,\epsilon)\,. \tag{22}$$

Here $v(\epsilon)$ is a Gaussian window centred around zero with variance much larger than the oscillations of $R_\beta(E,\epsilon)$ in ϵ, and normalized to Δ^{-1}, see [7]. $R_\beta^{\mathrm{osc}}(E,\epsilon)$ is the remaining oscillatory contribution.

The smooth contribution $R_\beta^{\rm av}(E,\epsilon)$ describes correlations on energy scales much larger than the mean level spacing and may be calculated within a semiclassical approach (using the diagonal approximation), diagrammatic perturbation theory, or Dyson's Brownian motion model. $R_\beta^{\rm osc}(E,\epsilon)$ cannot be calculated in this way.

It was first demonstrated by Andreev and Altshuler [5] that there is a very simple relation between smooth and oscillatory contributions to the spectral form factor. This relation describes non-universal deviations from RMT in disordered metals in terms of the spectral determinant $D_\beta(\epsilon)$ of the diffusion propagator $P(x,t)$

$$R_\beta^{\rm av}(E,\epsilon) = -\frac{1}{4\pi^2}\frac{\partial^2}{\partial\epsilon^2}\log\,D_\beta(\epsilon) \tag{23}$$

$$R_\beta^{\rm osc}(E,\epsilon) = \frac{1}{\pi^2}\cos(2\pi\epsilon/\Delta)\,D_\beta^2(\epsilon)\,. \tag{24}$$

This relation implies in particular that non-universal corrections affect the two-point correlation function not only at large energy scales, but also at small scales (of the order of the mean level spacing). In [7] it was shown that (23,24) are a consequence of the fact that the two-point correlation function is well approximated by a sum of shifted Gaussians.

For a quasi-one-dimensional diffusive conductor, the quantity $D_\beta(\epsilon)$ in (23,24) is approximately given by $D_\beta(\epsilon) = 4\pi^2\exp[-2\pi^2\sigma_\beta^2(\epsilon)/\Delta^2]$ with

$$\sigma_\beta^2(\epsilon) \simeq \frac{2\Delta^2}{\beta\pi^2}\left[\log(2\pi\epsilon) + s_\beta - \frac{1}{2}\sum_{\nu=1}^{\infty}\log\frac{(g\nu^2/2)^2}{(g\nu^2/2)^2+\epsilon^2}\right]\,. \tag{25}$$

Here s_β is a β-dependent constant [7]. For the spectral form factor one obtains (see [11] and references quoted therein)

$$K_\beta^{\rm av}(E,\tau) = \frac{2}{\beta}|\tau|\sum_{\nu=0}^{\infty}{\rm e}^{\,-\,\pi g\nu^2|\tau|} = \frac{2}{\beta}|\tau|\Big\{\frac{1}{2}+\frac{1}{2}\sum_{\mu=-\infty}^{\infty}\frac{{\rm e}^{\,-\,\pi\mu^2/(g|\tau|)}}{\sqrt{g|\tau|}}\Big\} \tag{26}$$

The asymptotic behaviour (for $\tau \ll g^{-1} \ll 1$) is given by [4–6]

$$K_\beta^{\rm av}(E,\tau) = \frac{2}{\beta}\sqrt{\frac{|\tau|}{4g}} \tag{27}$$

and [5]

$$K_\beta^{\rm osc}(E,\tau) = \sum_{\mu=1}^{\infty}\left({\rm e}^{-\pi g\mu^2|\tau+1|} + {\rm e}^{-\pi g\mu^2|\tau-1|}\right)(-1)^\mu\frac{2}{g\mu\sinh(\pi\mu)}\,. \tag{28}$$

An alternative derivation of (28) was given in [7].

In [22] these behaviours were compared to results of exact diagonalisations of the quasi-one-dimensional Anderson tight-binding Hamiltonian. The results of [22] are summarized in Sect. 4.2.

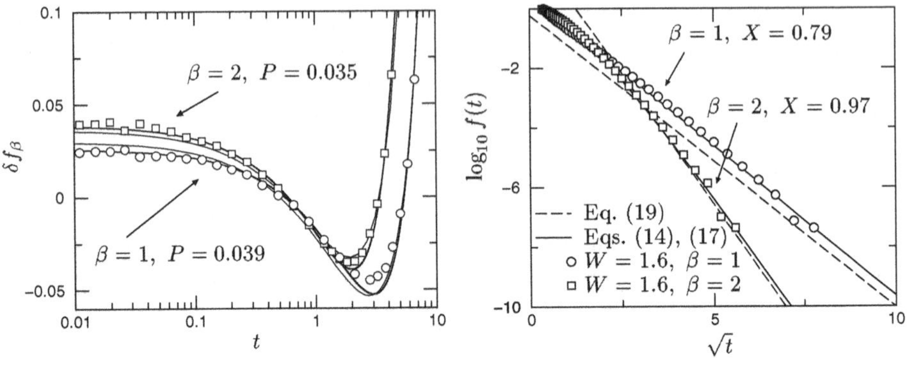

Fig. 1. (Left) $\langle \delta f_\beta(E,t;x)\rangle_x$ (see text) for a lattice with $128 \times 8 \times 8$ sites with $E = -1.7$, $\eta = 0.01$, 3000 samples, and $W = 1.0$. For $\beta = 1$ (○) and $\beta = 2$ (□), these results are compared to (14,17) (———) and (18) (———). (Right) $f_\beta(E,t;x)$ for a lattice with $128 \times 4 \times 4$ sites; for $X \lesssim 1$, $x \simeq L/2$, $W = 1.6$, $E = -1.7$, $\eta = 0.01$, 5000 samples, and $\beta = 1, 2$ compared to (14,17) and (19). Taken from [20]

4 Results of exact diagonalisations

4.1 Wave-function statistics

Figure 1 shows the exact diagonalization results for $\langle \delta f_\beta(E,t;x)\rangle_x$ (that is δf_β averaged over a small region around x) in comparison with (14,17) and (18). We observe very good agreement. The classical quantity $P \equiv \langle P(x;0)\rangle_x$ ought to be independent of β. In Fig. 1 it is determined by a best fit of (18) to the numerical data and is found to change somewhat with β, albeit weakly (Fig. 1). For narrower wires ($128 \times 4 \times 4$) we have observed that the ratio P_1/P_2 (determined by fitting $P \equiv P_\beta$ independently for $\beta = 1, 2$) becomes very small for small values of W (corresponding to $X \lesssim 0.1$) while it approaches unity for large values of W. A possible explanation for this deviation is given in [20]. Surprisingly, the form of the deviations is still very well described by (18) (not shown). The parameter dependence of the ratio P_1/P_2 in the case of smooth (correlated) disorder was studied in [36].

Figure 1 also shows the tails of $f_\beta(E,t;x)$ for weak disorder ($X \lesssim 1$) in comparison with (14,17) and (19). Since for very small values of X the tails decay so fast that we cannot reliably compute them, we decreased the wire cross section and increased the value of W in Fig. 1, thus increasing X. The quoted values of X were obtained by fitting (14,17). The values thus determined differ somewhat between $\beta = 1$ and 2 (see Fig. 1, right). As mentioned in [20], this difference was found to depend on the choice of E, W, and η.

The numerical results for $\langle V|\psi(\boldsymbol{r})|^2\rangle_t$ in the case $\beta = 2$ are summarized in Fig. 2. Apart from a sharp peak at the localisation center, numerical results

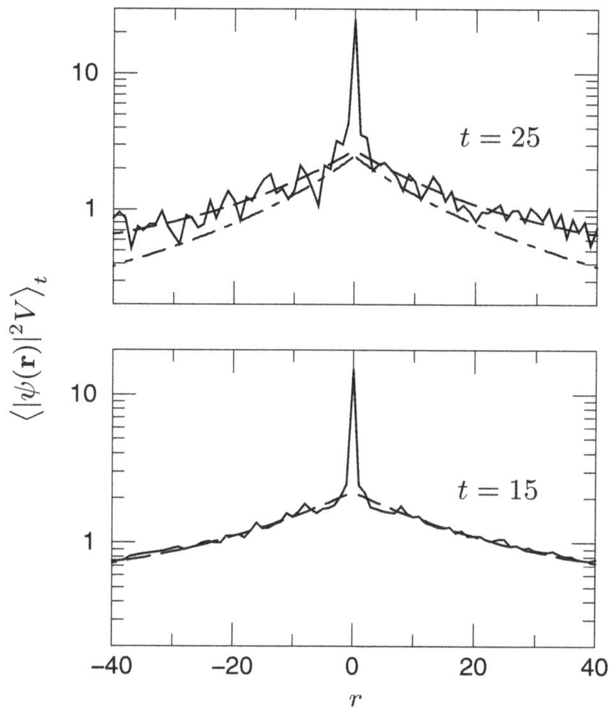

Fig. 2. Structure of anomalously localised wave functions with $t = |\psi(0)|^2V$. Solid lines: Numerical results for $V = 128 \times 4 \times 4$ sites, disorder $W = 1.6$ and energy $E \simeq -1.7$, averaged over 40 000 wave functions. Dashed lines: Analytical predictions with $X = 0.97$. The dash-dotted line shows the asymptotic formula (20) for $t = 25$. Taken from [21]

are very well described by (14-16). The asymptotic formula (20) considerably underestimates $\langle V|\psi(\boldsymbol{r})|^2\rangle_t$ for the values of t shown in Fig. 2.

4.2 The spectral form factor

Figure 3 shows the spectral form factor $K(\tau)$ for the Anderson model on a lattice with $64 \times 4 \times 4$ sites for $W = 1.4$, in the metallic regime. As expected, the data are very well described by RMT for both values of β. The same figure also shows $K(\tau)$ on a logarithmic scale for the same model, but for $W = 2.6$. At small times a crossover to (26) is observed. As expected, $K(\tau)$ differs by a factor of two between the models with $\beta = 1$ and $\beta = 2$. Figure 4 finally shows $\delta K(\tau) = K(\tau) - K_{\rm RMT}(\tau)$ on a linear scale. One observes that the singularity at the Heisenberg time $t_{\rm H}$ is well described by the theory (28).

It was thus demonstrated in [22] that spectral correlations in the quasi-one-dimensional Anderson model exhibit deviations from RMT behaviour on the scale of the Heisenberg time ($t \simeq t_{\rm H}$ corresponding to $\tau \simeq 1$).

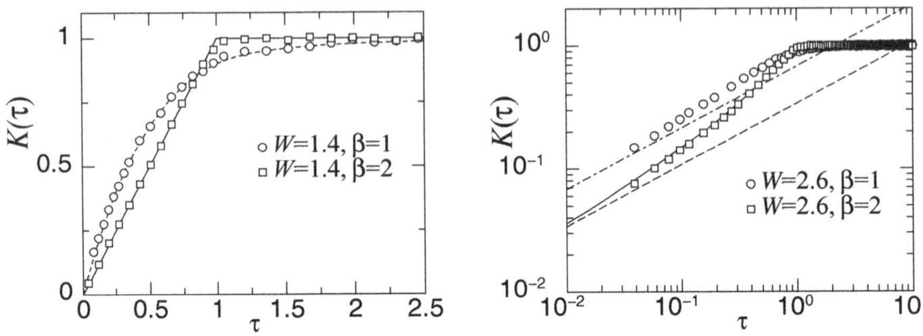

Fig. 3. (Left) Ensemble-averaged spectral form factor for a $64 \times 4 \times 4$ Anderson tight-binding model with $W = 1.4$ and $\beta = 1$ (○) and $\beta = 2$ (□). The form factor was averaged over 3×10^4 samples and an energy window of width $\eta = 0.31$ around $E = -2.6$. Also shown are the RMT expressions [27] appropriate in the metallic regime, for $\beta = 1$ (- - - -) and $\beta = 2$ (———). (Right) Same as the left plot, but for $W = 2.6$, and on a double logarithmic scale in order to emphasize the small-τ behaviour. Also shown are the small-τ asymptotes according to (27), for $\beta = 1$ (— · — · —) and $\beta = 2$ (— — —), as well as the result according to (26)(———). Taken from [22]

5 Summary

In summary, numerical studies of the quasi-one-dimensional Anderson tight-binding model with Gaussian disorder show that diffusion-driven deviations from RMT in spectral and wave-function amplitude statistics are of the

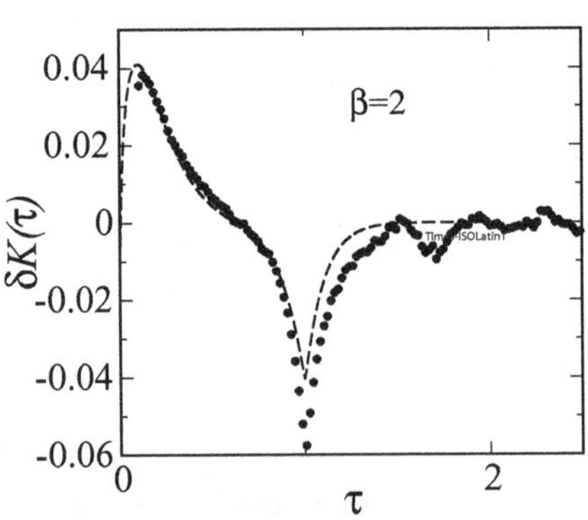

Fig. 4. Deviations of the computed spectral form factor from the RMT result $\delta K(\tau) \equiv K(\tau) - K_{\mathrm{RMT}}(\tau)$, for the data in Fig. 3 (right) for $\beta = 2$ and $W = 2.6$ (•). Also shown is the analytical result obtained from (28) (— — —). Taken from [22]

form predicted by the non-linear sigma model (see [11] for a review) and the Brownian motion model [7]. In higher spatial dimensions the situation is likely to be more complicated when the tails of the wave-function amplitude distribution are considered. In two spatial dimensions for instance, the non-linear sigma model would predict predict log-normal tails, that is $f_\beta(t) \simeq A_\beta \exp[-C_\beta(\ln t)^2]$. As mentioned above, this form is consistent with exact-diagonalisation results of the two-dimensional Anderson tight-binding model [25]. However, the dependence of A_β and C_β on the microscopic parameters of the model is likely to be non-universal, that is specific to the model considered [25, 31].

References

1. O. Bohigas. Random matrix theories and chaotic dynamics. In M. J. Gianoni, A. Voros, and J. Zinn-Justin, editors, *Chaos and quantum physics*, page 87, North–Holland, Amsterdam, 1991.
2. K. B. Efetov. Supersymmetry and theory of disordered metals. *Adv. Phys.*, 32:53, 1983.
3. M. Berry. Some quantum-to-classical asymptotics. In M. J. Giannoni, A. Voros, and J. Zinn-Justin, editors, *Chaos and quantum physics*, page 251, North–Holland, Amsterdam, 1991.
4. B. L. Altshuler and B. I. Shklovskii. Repulsion of energy-levels and the conductance of small metallic samples. *Sov. Phys. JETP*, 64:1, 1986.
5. A. V. Andreev and B. L. Altshuler. Spectral statistics beyond random-matrix theory. *Phys. Rev. Lett.*, 75:902, 1995.
6. N. Argaman, Y. Imry, and U. Smilansky. Semiclassical analysis of spectral correlations in mesoscopic systems. *Phys. Rev. B*, 47:4440, 1993.
7. B. Mehlig and M. Wilkinson. Spectral correlations: understanding oscillatory contributions. *Phys. Rev. E*, 63:045203(R), 2001.
8. E. Heller. Bound-state eigenfunctions of classically chaotic Hamiltonian-systems - scars of periodic-orbits. *Phys. Rev. Lett.*, 53:1515, 1984.
9. K. Müller, B. Mehlig, F. Milde, and M. Schreiber. Statistics of wave functions in disordered and in classically chaotic systems. *Phys. Rev. Lett.*, 78:215, 1997.
10. A. D. Mirlin. *Habilitation thesis.* University of Karlsruhe, 1999.
11. A. D. Mirlin. Statistics of energy levels and eigenfunctions in disordered systems. *Phys. Rep.*, 326:259, 2000.
12. B.L. Altshuler, V. E. Kravtsov, and I. V. Lerner. Distribution of mesoscopic fluctuations and relaxation processes in disordered conductors. In B.L. Altshuler, P. A. Lee, and R. A. Webb, editors, *Mesoscopic Phenomena in Solids*, page 449, North–Holland, Amsterdam, 1991.
13. B. L. Altshuler and V. N. Prigodin. Distribution of local density of states and shape of NMR line in a one-dimensional disordered conductor. *Sov. Phys. JETP*, 68:198, 1989.
14. A. D. Mirlin and Y. V. Fyodorov. The statistics of eigenvector components of random band matrices - analytical results. *J. Phys. A: Math. Gen.*, 26:L551, 1993.

15. Y. V. Fyodorov and A. Mirlin. Statistical properties of eigenfunctions of random quasi 1d one-particle Hamiltonians. *Int. J. Mod. Phys. B*, 8:3795, 1994.
16. V. I. Fal'ko and K. B. Efetov. Multifractality: generic property of eigenstates of 2d disordered metals. *Europhys. Lett.*, 32:627, 1995.
17. Y. V. Fyodorov and A. Mirlin. Mesoscopic fluctuations of eigenfunctions and level-velocity distribution in disordered metals. *Phys. Rev. B*, 51:13403, 1995.
18. A. D. Mirlin. Spatial structure of anomalously localized states in disordered conductors. *J. Math. Phys.*, 38:1888, 1997.
19. J. Cullum and R. A. Willoughby. *Lanczos Algorithms for Large Symmetric Eigenvalue Computations.* Birkhäuser, Boston, 1985.
20. V. Uski, B. Mehlig, R. Römer, and M. Schreiber. Exact diagonalization study of rare events in disordered conductors. *Phys. Rev. B*, 62:R7699, 2000.
21. V. Uski, B. Mehlig, and M. Schreiber. Spatial structure of anomalously localized states in disordered conductors. *Phys. Rev. B*, 66:233104, 2002.
22. V. Uski, B. Mehlig, and M. Wilkinson. Unpublished.
23. M. Wilkinson. Random matrix theory in semiclassical quantum mechanics of chaotic systems. *J. Phys. A: Math. Gen.*, 21:1173, 1988.
24. M. Wilkinson and P. N. Walker. A Brownian motion model for the parameter dependence of matrix elements. *J. Phys. A: Math. Gen.*, 28:6143, 1996.
25. V. Uski, B. Mehlig, R. Römer, and M. Schreiber. Smoothed universal correlations in the two-dimensional Anderson model. *Phys. Rev. B*, 59:4080, 1999.
26. P. W. Anderson. Absence of diffusion in certain random lattices. *Phys. Rev.*, 109:1492, 1958.
27. M. L. Mehta. *Random Matrices and the Statistical Theory of Energy Levels.* Academic Press, New York, 1991.
28. W. T. Vetterling W. H. Press, S. A. Teukolsky and B. P. Flannery. *Numerical recipes.* Cambridge University Press, Cambridge, 1992.
29. C. E. Porter. Fluctuations of quantal spectra. In C. E. Porter, editor, *Statistical Theories of Spectra*, page 2. Academic Press, New York, 1965.
30. N. G. van Kampen. *Stochastic processes in physics and chemistry.* North-Holland, Amsterdam, 1983.
31. V. M. Apalkov, M. E. Raikh, and B. Shapiro. Anomalously localized states in the Anderson model. *Phys. Rev. Lett.*, 92:066601, 2004.
32. V. Uski, B. Mehlig, R.A. Römer, and M. Schreiber. Incipient localization in the Anderson model. *Physica B*, 284 - 288:1934, 2000.
33. B. K. Nikolić. Statistical properties of eigenstates in three-dimensional mesoscopic systems with off-diagonal or diagonal disorder. *Phys. Rev. B*, 64:014203, 2001.
34. B. K. Nikolić. Quest for rare events in mesoscopic disordered metals. *Phys. Rev. B*, 65:012201, 2002.
35. V. Uski, R. A. Römer, and M. Schreiber. Numerical study of eigenvector statistics for random banded matrices. *Phys. Rev. E*, 65:056204, 2002.
36. V. Uski, B. Mehlig, and M. Schreiber. Signature of ballistic effects in disordered conductors. *Phys. Rev. B*, 63:241101, 2001.

Fine Structure of the Integrated Density of States for Bernoulli–Anderson Models

Peter Karmann[1], Rudolf A. Römer[2], Michael Schreiber[1], and Peter Stollmann[3]

[1] Technische Universität Chemnitz, Institut für Physik
09107 Chemnitz, Germany
[2] Department of Physics and Centre for Scientific Computing,
University of Warwick, Coventry, CV4 7AL, United Kingdom
[3] Technische Universität Chemnitz, Fakultät für Mathematik
09107 Chemnitz, Germany

1 Introduction

Disorder is one of the fundamental topics in science today. A very prominent example is Anderson's model [1] for the transition from metal to insulator under the presence of disorder.

Apart from its intrinsic physical value this model has triggered an enormous amount of research in the fields of random operators and numerical analysis, resp. numerical physics. As we will explain below, the Anderson model poses extremely hard problems in these fields and so mathematical rigorous proofs of many well substantiated findings of theoretical physics are still missing. The very nature of the problem also causes highly nontrivial challenges for numerical studies.

The transition mentioned above can be reformulated in mathematical terms in the following way: for a certain random Hamiltonian one has to prove that its spectral properties change drastically as the energy varies. For low energies there is pure point spectrum, with eigenfunctions that decay exponentially. This energy regime is called localization.

For energies away from the spectral edges, the spectrum is expected to be absolutely continuous, providing for extended states that can lead to transport. Sadly enough, more or less nothing has been proven concerning the second kind of spectral regime, called delocalization. An exception are results on trees [2, 3], and for magnetic models [4]. Anyway, there are convincing theoretical arguments and numerical results that support the picture of the metal–insulator transition (which is, in fact, a dimension-dependent effect and should take place for dimensions $d > 2$).

In our research we are dealing with a different circle of questions. The Anderson model is in fact a whole class of models: an important input is the measure μ that underlies the random onsite couplings. In all proofs of localization (valid for $d > 1$) one needs regularity of this measure μ. This excludes a prominent and attractive model: the Anderson model of a binary alloy, i.e. two kinds of atoms randomly placed on the sites of a (hyper)cubic lattice, known as Bernoulli–Anderson model in the mathematical community. Basically there are two methods of proof for localization: multiscale analysis [5], and the fractional moment method [6]. In both cases one needs an a-priori bound on the probability that eigenvalues of certain Hamiltonians cluster around a fixed energy, i.e., one has to exclude resonances of finite box Hamiltonians. Equivalently, one needs a weak kind of continuity of the integrated density of states (IDS). In multiscale analysis this a-priori estimate comes in a form that is known as Wegner's estimate [7]. Here we will present both rigorous analytical results and numerical studies of these resonances. We will take some time and effort to describe the underlying concepts and ideas in the next Section. Then we report recent progress concerning analytical results. Here one has to mention a major breakthrough obtained in a recent paper [8] of J. Bourgain and C. Kenig who prove localization for the continuum Bernoulli–Anderson model in dimensions $d \geq 2$. Finally, we display our numerical studies and comment on future directions of research.

We conclude this section with an overview over recent contributions in the physics literature concerning the binary-alloy model. These may be classified into mainly simulations or mainly theoretical analyses. The former are discussed in [9], albeit in the restricted setting of a Bethe lattice, providing a detailed analysis of the electronic structure of the binary-alloy and the quantum-percolation model, which can be derived from the binary alloy replacing one of the alloy constituents by vacancies. The study is based on a selfconsistent scheme for the distribution of local Green's functions. Detailed results for the local density of states (DOS) are obtained, from which the phase diagram of the binary alloy is constructed. The existence of a quantum-percolation threshold is discussed. Another study [10] of the quantum site-percolation model on simple cubic lattices focuses on the statistics of the local DOS and the spatial structure of the single particle wave functions. By using the kernel polynomial previous studies of the metal–insulator transition are refined and the nonmonotonic energy dependence of the quantum-percolation threshold is demonstrated. A study of the three-dimensional binary-alloy model with additional disorder for the energy levels of the alloy constituents is presented in [11]. The results are compared with experimental results for amorphous metallic alloys. By means of the transfer-matrix method, the metal–insulator transitions are identified and characterized as functions of Fermi-level position, band broadening due to disorder and alloy composition. The latter is also investigated in [12], which discusses the conditions to be put on mean-field-like theories to be able to describe fundamental physical phenomena in disordered electron systems. In particular, options for a consistent mean-field theory of electron localization and for a reliable description of transport properties are investigated. In [13] the single-site coherent potential approximation is extended to include the effects of non-local disorder correlations (i.e. alloy short-range order) on the electronic structure of random alloy systems. This is achieved by mapping the original Anderson disorder problem to that of a selfconsistently embedded cluster. The DOS of the binary-alloy model has been studied in [14], where also the mobility edge, i.e. the phase boundary between metallic and insulating behaviour was investigated. The critical behaviour, in particular the critical exponent with which the localization length of the electronic states diverges at the phase transition was analyzed in [15] in comparison with the standard Anderson model.

2 Resonances and the integrated density of states

2.1 Wegner estimates, IDS, and localization

In this Section we sketch the basic problem and introduce the model we want to consider. A major point of the rather expository style is to make clear, why the problem is as difficult as it appears to be. This also sheds some light on why it is intrinsically hard to study numerically. Let us first write down the Hamiltonian in an analyst's notation:

On the Hilbert space $\ell^2(\mathbb{Z}^d)$ we consider the random operator

$$H(\omega) = -\Delta + V_\omega,$$

where the discrete Laplacian incorporates the constant (nonrandom) off-diagonal or hopping terms. It is defined, for $\psi \in \ell^2(\mathbb{Z}^d)$ by

$$\Delta\psi(i) = \sum_{\langle i,j \rangle} \psi(j)$$

for $i \in \mathbb{Z}^d$. The notation reflects $\langle \cdot, \cdot \rangle$ that we are dealing with a nearest neighbor interaction, where the value of the wave function at site i is only influenced by those at the $2\,d$ neighbors on the integer lattice. Here we neglect the diagonal part $\psi(i)$ which would only contribute a shift of the energy scale. The random potential V_ω is, in its simplest form, given by independent identically distributed (short: i.i.d.) random variables at the different sites. A convenient representation is given in the following way:

$$\Omega = \prod_{i \in \mathbb{Z}^d} \mathbb{R}, \quad \mathbb{P} = \prod_{i \in \mathbb{Z}^d} \mu, \quad V_\omega(i) = \omega_i$$

where μ is a probability measure on the real line. This function $V_\omega(i)$ gives the random diagonal multiplication operator acting as

$$V_\omega\psi(i) = \omega_i\psi(i).$$

For simplicity we assume that the support of the so-called single-site measure μ is a compact set $K \subset \mathbb{R}$. Put differently, for every site i we perform a random experiment that gives the value ω_i distributed according to μ. In physicist's notation we get

$$H(\omega) = \sum_{\langle i,j \rangle} |i\rangle\langle j| + \sum_i \omega_i |i\rangle\langle i|,$$

where $|i\rangle$ denotes the basis functions in site representation.

In principle, one expects that the spectral properties should not depend too much on the specific distribution μ (apart from the very special case that μ reduces to a point mass, in which case there is no disorder present). Let us take a look at two very different cases. In the Bernoulli–Anderson model we have the single-site measure $\mu = \frac{1}{2}\delta_0 + \frac{1}{2}\delta_1$. In that case the value $\omega_i \in \{0, 1\}$ is determined by a fair coin. We will also consider a coupling parameter W in the random part, in which case we have either $\omega_i = 0$ or $\omega_i = W$ each with probability $\frac{1}{2}$. The resulting random potential is denoted by V_ω^{B}. In the second case the potential value is determined with respect to the uniform distribution so that we get $\mu(\mathrm{d}x) = \chi_{[0,1]}(x)\mathrm{d}x$. We write V_ω^{U} for this case.

Let us point out one source of the complexity of the problem: The two operators that sum up to $H(\omega)$ are of very different nature:

- The discrete Laplacian is a difference operator. It is diagonal in Fourier space $L^2([0, 2\pi]^d)$, where it is given by multiplication with the function

$$\sum_{k=1}^{d} 2 \cos x_k .$$

  Therefore, its spectrum is given by the range of this function, so that

$$\sigma(-\Delta) = [-2d, 2d],$$

  the spectrum being purely absolutely continuous.
- The random multiplication operator V_ω is diagonal in the basis $\{\delta_i | i \in \mathbb{Z}^d\}$. The spectrum is hence the closure of the range of V_ω, which is just the support K of the measure for $\mathbb{P}$–a.e.$\omega \in \Omega$ (a.e. stands for almost every, i.e., for all but a set of measure zero). In the aforementioned special cases we get

$$\sigma(V_\omega^{\mathrm{B}}) = \{0, 1\}$$

  for the Bernoulli–Anderson model and

$$\sigma(V_\omega^{\mathrm{U}}) = [0, 1],$$

  both for a.e. $\omega \in \Omega$. Clearly, the spectral type is pure point with perfectly localized eigenfunctions for every ω, the set of eigenvalues being given by $\{\omega_i | i \in \mathbb{Z}^d\}$.

One major problem of the analysis as well as the numerics is now obvious: We add two operators of the same size with completely different spectral type and there is no natural basis to diagonalize the sum $H(\omega) = -\Delta + V_\omega$, since one of the two terms is diagonal in position space while the other is diagonal in momentum space (the Fourier picture).

Another cause of difficulties is the expected spectral type of $H(\omega)$. In the localized regime it has a dense set of eigenvalues. These eigenvalues are extremely unstable. Rank-one perturbation theory gives the following fact which illustrates this instability: If we fix all values ω_j except one, say ω_i and vary the latter continuously in an interval, the resulting spectral measures will be mutually singular and for a dense set of values of ω_i the spectrum will contain a singular continuous component, cf. [16].

Moreover, the qualitative difference between the Bernoulli–Anderson model and the model with uniform distribution is evident: V_ω^{U} displays the spectral type we want to prove for $H(\omega)$: it has a dense set of eigenvalues for a.e. ω. If one can treat $-\Delta$ in some sense as a small perturbation we arrive at the desired conclusion. In view of the preceding paragraph, this cannot be achieved by standard perturbation arguments. In the Bernoulli–Anderson model V_ω^{B} has only eigenvalues 0 and 1, each infinitely degenerate.

In typical proofs of localization an important tool is the study of box Hamiltonians. To explain this, we consider the cube $\Lambda_L(i)$ of side length L centered at i. We restrict $H(\omega)$ to the sites in $\Lambda_L(i)$ which constitutes a subspace of dimension $|\Lambda_L(i)| = L^d$. We denote the restriction by $H_L(\omega)$ and suppress the boundary condition, since it does not play a role in asymptotic properties as $L \to \infty$.

These box Hamiltonians enter in resolvent expansions and it is important to estimate the probability that their resolvents have a large norm, i.e., we are dealing here with small-denominator problems. Since the resolvent has large norm for energies near the spectrum one needs upper bounds for

$$\mathbb{P}\{\sigma(H_L(\omega)) \cap [E-\epsilon, E+\epsilon] \neq \emptyset\} = p(\epsilon, L).$$

In fact, one wants to show that $p(\epsilon, L)$ is small for large L and small ϵ. At the same time, there is a clear limitation to such estimates: In the limit $L \to \infty$ the spectra of $H_L(\omega)$ converge to the spectrum of $H(\omega)$. This means that for fixed $\epsilon > 0$ and $E \in \sigma(H(\omega))$ (and only those energies E are of interest),

$$p(\epsilon, L) \to 1 \text{ for } L \to \infty.$$

The famous Wegner estimate [7] states that for absolutely continuous μ we get

$$p(\epsilon, L) \leq C\epsilon L^d, \tag{1}$$

where the last factor is the volume of the cube $|\Lambda_L(i)| = L^d$. This estimate is sufficient for a proof of localization for energies near the spectral edges. The proof is not too complicated for the discrete model. To see why it might be true, let us include a very simple argument in the case that there is no hopping term.

Then

$$\begin{aligned}\mathbb{P}\{\sigma(V_\omega) \cap [E-\epsilon, E+\epsilon] \neq \emptyset\} &= \mathbb{P}\{\exists j \in \Lambda_L \text{ s.t. } V_\omega(j) \in [E-\epsilon, E+\epsilon]\} \\ &\leq |\Lambda_L| \cdot \mu[E-\epsilon, E+\epsilon] \\ &\leq C \cdot |\Lambda_L| \cdot \epsilon,\end{aligned}$$

if μ is absolutely continuous. Here we see that the situation is completely different for the Bernoulli–Anderson model for which $\mu[E-\epsilon, E+\epsilon] \geq \frac{1}{2}$ whenever $E \in \{0,1\}$. Also, it is clear that a Wegner estimate of the type (1) above cannot hold. Since in

$$\mathbb{P}\{\sigma(H_L(\omega)) \cap [E-\epsilon, E+\epsilon] \neq \emptyset\}$$

only $2^{|\Lambda_L|}$ Bernoulli variables are comprised, this probability must at least be $2^{-|\Lambda_L|}$, unless it vanishes.

The Wegner estimate is intimately related to continuity properties of the integrated density of states (IDS), a function $N : \mathbb{R} \to [0, \infty)$ that measures the number of energy levels per unit volume:

$$N(E) = \lim_{L\to\infty} \frac{1}{|\Lambda_L|} \mathbb{E}(\mathrm{Tr}\chi_{(-\infty,E]}(H_L(\omega))).$$

Here $\mathbb{E}$ denotes the expectation value and $\chi_{(-\infty,E]}(H_L(\omega))$ is the projection onto the eigenspace spanned by the eigenvectors with eigenvalue below E and the trace determines the dimension of this space, i.e., the number of eigenvalues below E counted with their multiplicity. Since we are dealing with operators of rank at most $|\Lambda_L|$, we get

$$\begin{aligned}N(E+\epsilon) - N(E-\epsilon) &\approx \frac{1}{|\Lambda_L|} \mathbb{E}(\mathrm{Tr}\chi_{(E-\epsilon,E+\epsilon]}(H_L(\omega))) \\ &\leq \mathbb{P}\{\sigma(H_L(\omega)) \cap [E-\epsilon, E+\epsilon] \neq \emptyset\}.\end{aligned}$$

This means that Wegner estimates lead to continuity of the IDS. Although that is not clear from the above rather crude reasoning, the Wegner estimate for absolutely continuous μ yields differentiability of the IDS.

2.2 Recent rigorous analytical results

In this section we will mainly be dealing with continuum models,

$$H(\omega) = -\Delta + V_\omega,$$

where $-\Delta$ is now the unbounded Laplacian with domain $W^{2,2}(\mathbb{R}^d)$, the Sobolev space of square integrable functions with square integrable second partial derivatives. The random multiplication operator V_ω is defined by

$$V_\omega(x) = \sum_{i\in\mathbb{Z}^d} \omega_i u(x-i),$$

with a single-site potential $u(x) \geq 0$ bounded and of compact support (for simplicity reasons), and random coupling like above. Some results we mention are valid under more general assumptions, as can be seen in the original papers. For results concerning localization of these models we refer to [5] for a survey of the literature up to the year 2000. More recent results concerning the IDS and its continuity properties can be found in [17]. Here we will report on more recent developments. The first result is partly due to one of us [18].

Theorem 1. *Let $H(\omega)$ be an alloy-type model and $u \geq \kappa\chi_{[-1/2,1/2]^d}$ for some positive κ. Then for each $E_0 \in \mathbb{R}$ there exists a constant C_W such that, for all $E \leq E_0$ and $\varepsilon \leq 1/2$*

$$\mathbb{E}\{\mathrm{Tr}[\chi_{[E-\varepsilon,E+\varepsilon]}(H_L(\omega))]\} \leq C_W \; s(\mu,\varepsilon) \, (\log \tfrac{1}{\varepsilon})^d \; |\Lambda_L|, \tag{2}$$

where

$$s(\mu,\varepsilon) = \sup\{\mu([E-\varepsilon, E+\varepsilon]) \mid E \in \mathbb{R}\}. \tag{3}$$

The mentioned alloy-type models include the models we introduced as well as additional periodic exterior potentials and magnetic vector potentials. The idea of the proof is to combine methods from [19] with a technique to control the influence of the kinetic term: the estimate in [19] is quadratic in the volume of the cube and so it cannot be used to derive continuity of the IDS. On the other hand, there had been recent progress for models with absolutely continuous μ [20–22] using the spectral shift function. In [18] we present an improved estimate of the spectral shift function and apply it to arrive at the estimate (2). Of course, the latter is not really helpful, unless the measure μ shares a certain continuity. Still it is interesting in so far that it yields that N is nearly as continuous as μ with a logarithmically small correction. For more details we refer to [18], where the reader can also find a detailed account of how our result compares with recent results in this direction.

We now mention a major breakthrough obtained in the recent work [8] where the continuum Bernoulli–Anderson model is treated. For this model, the authors set up a multi-scale induction to prove a Wegner estimate of the following type:

Theorem 2. *For the Bernoulli–Anderson model and $\alpha, \beta > 0$ there exist $C, \gamma > 0$ such that*

$$\mathbb{P}\{\sigma(H_L(\omega)) \cap [E-\epsilon, E+\epsilon] \neq \emptyset\} \leq CL^{-\frac{1}{2}d+\alpha}$$

for

$$\epsilon \leq \exp(-\gamma L^{\frac{4}{3}+\beta}).$$

This can be found as Lemma 5.1 in [8]. It is important to note that the proof does not so far extend to the discrete case. The reason is that a major step in the proof is a quantitative unique continuation result that does not extend to the discrete setting. Therefore, Wegner estimates for the discrete Bernoulli–Anderson model are still missing.

3 Numerical studies

Most numerical studies have been performed for the standard Anderson model of localization with uniform distribution of the potential values. In this model the DOS changes smoothly with increasing disorder from the DOS of the Laplacian with its characteristic van Hove singularities to one featureless band [23]. If the distribution is chosen as usual with mean 0 and width W, then the theoretical band edges are given by $\pm(2d + W/2)$. Numerically these values are of course reached with vanishing probability. If the box distribution is replaced by an unbounded distribution like the Gaussian or the Lorentzian, then the band tails in principle extend to infinity, although numerically no significant change can be observed from the box-distribution case [24]. A dramatically different situation occurs in a binary alloy, where with increasing disorder W the DOS separates into two bands of width $4\,d$ each. Choosing the measure $\mu = \frac{1}{2}\delta_0 + \frac{1}{2}\delta_W$ as discussed in the previous section, the splitting of the band into subbands occurs theoretically for $W = 4\,d$, although the again numerically very small DOS in the tails of the subbands leads to the appearance of separated subbands already for smaller disorder values [14]. This, however, is not the topic of the current investigation. We rather concentrate on unexpected structures that we have found near the centre of the subbands.

The DOS is defined as usual

$$\rho(E) = \left\langle \sum_{i=1}^{L^d} \delta(E - E_i) \right\rangle$$

where E_i are the eigenvalues of the box Hamiltonian discussed in the previous section, and $\langle \cdots \rangle$ indicates the average over an ensemble of different configurations of the random potential, i.e. the disorder average. For the numerical diagonalization we use the Lanczos algorithm [25] which is very effective for sparse matrices. In the present case the matrices are extremely sparse, because except for the diagonal matrix element with the potential energy there are only $2\,d$ elements in each row and column of the secular matrix due to the Laplacian. In fact, for the standard model of Anderson localization which is of course as sparse as the Bernoulli–Anderson Hamiltonian matrix the Lanczos algorithm has been shown to be most effective also in comparison with more modern eigenvalue algorithms [26, 27]. One of the reasons for the difficulties which all eigenvalue algorithms encounter is our use of periodic boundary conditions in all directions, making a transformation of the secular matrix to a band matrix impossible. However, a severe problem arises for the Lanczos algorithm, because numerical inaccuracies due to finite precision arithmetics yield spurious eigenvalues which show up as incorrectly multiple eigenenergies. In principle these can be detected and eliminated in a straightforward way. The respective procedure, however, becomes ineffective in those parts of the spectrum where the Hamiltonian itself has multiple eigenvalues or an unusually large DOS. This happens to be the case in our investigation and

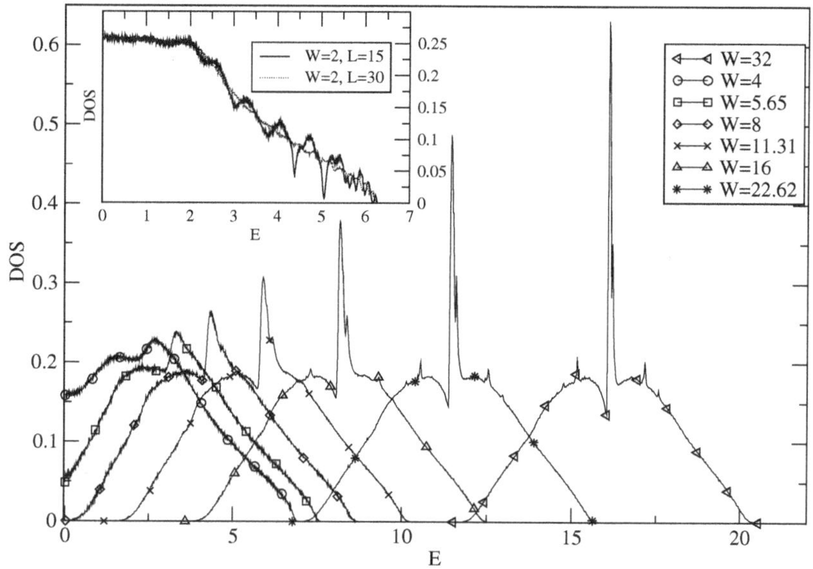

Fig. 1. DOS of the upper half of the spectrum for several disorder strengths W and system sizes $L = 30$ averaged over 250 configurations of disorder except for $W = 32$, where the system size is $L = 15$ and 2000 configurations have been used. The inset shows the DOS for $W = 2$, $L = 15$ and 30

turned out to be more significant for larger disorder and system sizes. As a consequence we have missed up to .09% of all eigenvalues in the data presented below. In general, the performance of the Lanczos algorithm is much better in the band tails, because the convergence is much faster. Therefore it turned out to be advantageous to calculate the DOS in the centre of the subbands separately with different settings of the parameters which control the convergence of the algorithm.

Our results are shown in Fig. 1 for various disorders. Here we have chosen a symmetric binary distribution, i.e. $\mu = \frac{1}{2}\delta_{-W/2} + \frac{1}{2}\delta_{W/2}$. The spectrum is thus symmetric with respect to $E = 0$ and only the upper subband is displayed in Fig. 1. With increasing disorder strength W, the subband moves of course to larger energies. Already for $W = 8$ the subbands appear to be separated, as the DOS is numerically zero around $E = 0$. We have also calculated the DOS for the system size $L = 15$ for disorders between $W = 4$ and $W = 22.6$. The data are not shown in Fig. 1, because they do not significantly differ from the data for the larger system size $L = 30$ shown in the plot. Only for $W = 2$ there are significant deviations due to finite-size effects: for vanishing disorder the finite size of the system with its periodic boundaries would yield only very few but highly multiple eigenvalues. Remnants of such structures can be seen in the inset of Fig. 1 for the smaller system size as somewhat smeared-out peaks.

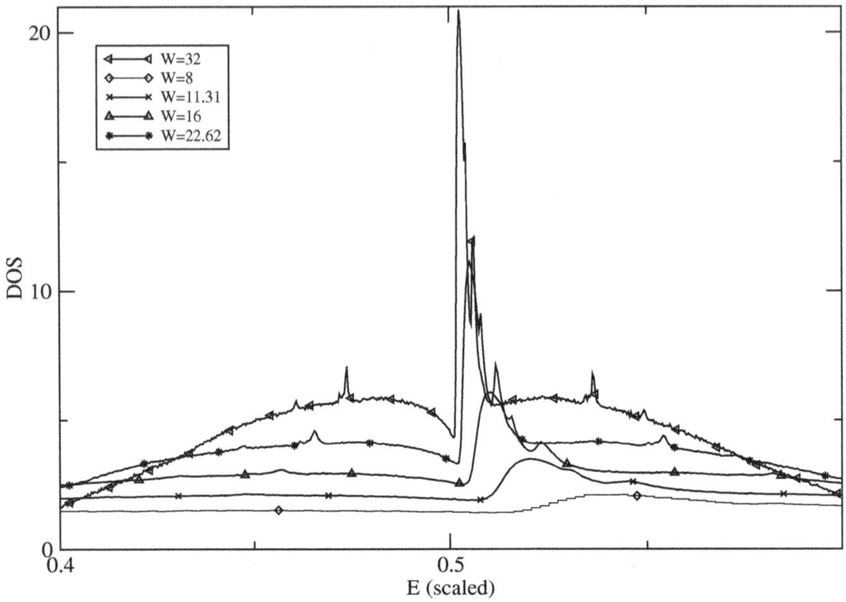

Fig. 2. DOS of the upper subband from Fig. 1 with all eigenvalues scaled by W. The DOS has been normalized after rescaling

For the larger system size $L = 30$ the DOS in the inset reflects the DOS of the pure Laplacian with only a weak smearing of the van Hove singularities.

The prominent feature of the spectra is a strong peak in the centre of the subband accompanied by a distinguished minimum on the low energy side and a side peak on the right hand shoulder. In order to study the emergence of these features we have plotted the central region of the subbands in Fig. 2 versus scaled energy thus eliminating the shift of subbands with increasing disorder. One can clearly see, that the peak and the minimum close to it approach the centre of the subband. The maximum and minimum values of the DOS can be described by power laws

$$\rho_{\text{max}} \propto W^{\beta}, \quad \beta = 1.79 \pm .03$$
$$\rho_{\text{min}} \propto W^{\beta}, \quad \beta = 0.75 \pm .03$$

as demonstrated in Fig. 3 where the data have been fitted by power laws for large W. The exponent $\beta > 1$ implies that in the limit of large disorder the DOS diverges. This is not surprising for the scaled DOS, because the scaled width of the subband shrinks. We note, however, that also in the unscaled plot the height of the peak increases with disorder W. It turns out that the approach of the peak and the minimum towards the exact centre of the subband at $E/W = \frac{1}{2}$ can also be described by power laws (see Fig. 4):

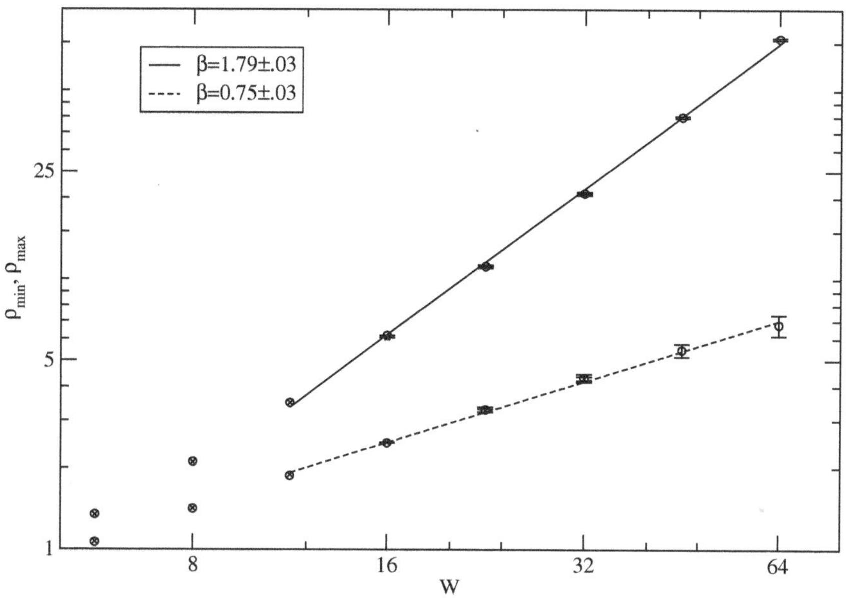

Fig. 3. Scaling of $\rho_{\min}$ and $\rho_{\max}$ with W for $L = 15$(o) and $L = 30$(x). The lines are least-squares linear fits for $L = 15$. Error bars are related to the number of missed eigenvalues

$$E(\rho_{\max})/W - 1/2 \propto W^{-\gamma}, \quad \gamma = 2.02 \pm .05$$
$$E(\rho_{\min})/W - 1/2 \propto W^{-\gamma}, \quad \gamma = 1.80 \pm .04$$

Both exponents are close to the value 2 and might be explained by perturbation theory [28].

In summary, we have seen that the DOS of the Bernoulli–Anderson model for sufficiently strong disorder shows two separate subbands with a strong sharp peak near the centre in a striking contrast to the standard Anderson model with box distribution or other continuous distributions for the potential energies. A more detailed analysis of these structures will have to be performed. It is reasonable to assume that they may be connected with certain local structures of the configuration like independent dimers and trimers or other clusters separated from the rest of the system by a neighbourhood of atoms of the other kind. In such a situation where all neighbouring atoms belong to the other subband, the wave function at those sites would approach zero for large disorder, i.e. the space related to those sites becomes inaccessible for the electrons from the other subband. This is exactly what happens also in the quantum percolation model.

We note that there are other smaller peaks to be seen in the DOS which might be related to larger separate clusters. A more detailed analysis of these structures is under investigation. If an additional disorder is applied randomizing the potential energy as in the standard Anderson model of localization,

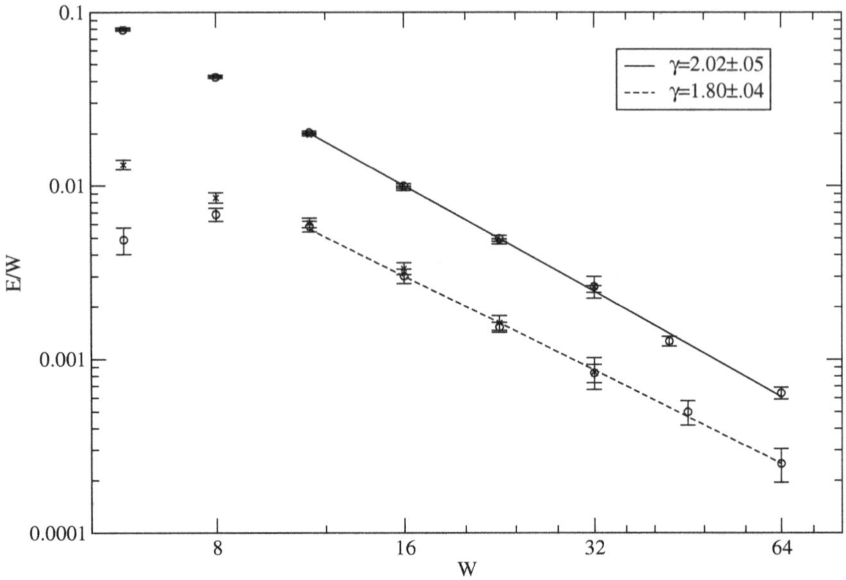

Fig. 4. Distance of the minimum (lower data) and the maximum (upper data) of the DOS from the centre of the subband. The lines are least-squares linear fits for $L = 15$. Error bars are related to the bin size of the histogram

then the peaks are quickly smeared out already for small values of this additional disorder [11].

Recently an efficient preconditioning algorithm has been proposed for the diagonalization of the Anderson Hamiltonian [29]. Previously respective shift-and-invert techniques had been shown to be significantly faster than the standard implementation of the Lanczos algorithm, but the memory requirements where prohibitively large for moderate system sizes already, even when only a very small number of eigenvalues and eigenvectors was calculated. The new implementation reduces this problem considerably, although the memory requirement is still larger than for the standard implementation [29]. It remains an open question whether that algorithm is also superior when the calculation of the complete spectrum is required.

References

1. P. W. Anderson. Absence of diffusion in certain random lattices. *Phys. Rev.*, 109:1492–1505, 1958.
2. M. Aizenman, R. Sims and S. Warzel. Fluctuation based proof of the stability of ac spectra of random operators on tree graphs. In N. Chernov, Y. Karpeshina, I.W. Knowles, R.T. Lewis, and R. Weikard, editors, *Recent Advances in Differential Equations and Mathematical Physics*, to appear in *Contemp. Math.*, Amer. Math. Soc., Providence, RI, 2006. ArXiv: math-ph/0510069.

3. A. Klein. Absolutely continuous spectrum in the Anderson model on the Bethe lattice. *Math. Res. Lett. 1*, 4:399–407, 1994.
4. F. Germinet, A. Klein, J. Schenker: announced.
5. P. Stollmann. Caught by disorder: bound states in random media. *Progress in Mathematical Physics, Vol. 20.* Birkhäuser, Boston, 2001.
6. M. Aizenman, A. Elgart, S. Naboko, J. Schenker, G. Stolz. Moment analysis for localization in random Schrödinger operators. ArXiv: math-ph/0308023. *Invent. Math.*, 163:343–413, 2006.
7. F. Wegner. Bounds on the DOS in disordered systems. *Z. Phys. B*, 44:9–15, 1981.
8. J. Bourgain and C. Kenig. Localization for the Bernoulli–Anderson model. *Invent. Math.*, 161:389–426, 2005.
9. A. Alvermann and H. Fehske. Local distribution approach to the electronic structure of binary alloys. *Eur. Phys. J. B*, 48:295–303, 2005. ArXiv: cond-mat/0411516.
10. G. Schubert, A. Weiße and H. Fehske. Localization effects in quantum percolation. *Phys. Rev. B*, 71:045126, 2005.
11. I. V. Plyushchay, R. A. Römer and M. Schreiber. Three-dimensional Anderson model of localization with binary random potential. *Phys. Rev. B*, 68:064201, 2003.
12. V. Janiš and J. Kolorenč. Mean-field theories for disordered electrons: Diffusion pole and Anderson localization. *Phys. Rev. B*, 71:245106, 2005. ArXiv: cond-mat/0501586.
13. M. S. Laad and L. Craco. Cluster coherent potential approximation for electronic structure of disordered alloys. *J. Phys.: Condens. Matter*, 17:4765–4777, 2005. ArXiv: cond-mat/0409031.
14. C. M. Soukoulis, Q. Li, and G. S. Grest. Quantum percolation in three-dimensional systems. *Phys. Rev. B*, 45:7724–7729, 1992.
15. E. Hofstetter and M. Schreiber. Finite-size scaling and critical exponents: A new approach and its application to Anderson localization. *Europhys. Lett.*, 27:933–939, 1993.
16. B. Simon. *Spectral analysis of rank one perturbations and applications. Mathematical quantum theory. II. Schrödinger operators (Vancouver, BC, 1993), CRM Proc. Lecture Notes Vol. 8*:109–149. Amer. Math. Soc., Providence, RI, 1995.
17. I. Veselić. Integrated density of states and Wegner estimates for random Schrödinger operators. In C. Villegas-Blas and R. del Rio, editors, *Schrödinger operators (Universidad Nacional Autonoma de Mexico, 2001)*, volume 340 of *Contemp. Math.*, pages 98–184. Amer. Math. Soc., Providence, RI, 2004. ArXiv: math-ph/0307062.
18. D. Hundertmark, R. Killip, S. Nakamura, P. Stollmann, I. Veselic: Bounds on the spectral shift function and the density of states. *Comm. Math. Phys.*, 262:489–503, 2006.
19. P. Stollmann. Wegner estimates and localization for continuum Anderson models with some singular distributions. *Arch. Math. (Basel)*, 75:307–311, 2000.
20. J.-M. Combes, P. D. Hislop, and F. Klopp. Hölder continuity of the integrated density of states for some random operators at all energies. *Int. Math. Res. Not.*, 4:179–209, 2003.

21. J.-M. Combes, P. D. Hislop, F. Klopp, and S. Nakamura. The Wegner estimate and the integrated density of states for some random operators. *Proc. Indian Acad. Sci. Math. Sci.*, 112:31–53, 2002. www.ias.ac.in/mathsci/.
22. J.-M. Combes, P. D. Hislop, and S. Nakamura. The L^p-theory of the spectral shift function, the Wegner estimate, and the integrated density of states for some random Schrödinger operators. *Commun. Math. Phys.*, 70:113–130, 2001.
23. A. Croy, R. A. Römer and M. Schreiber. Localization of electronic states in amorphous materials: recursive Green's function method and the metal–insulator transition at $E \neq 0$. In K. H. Hoffmann and A. Meyer, editors, *Parallel Algorithms and Cluster Computing - Implementions, Algorithm and Applications, Lecture Notes in Computational Science and Engineering.* Springer, Berlin, 2006.
24. B. Bulka, M. Schreiber and B. Kramer. Localization, quantum interference, and the metal–insulator transition. *Z. Phys. B*, 66:21–30, 1987.
25. J. K. Cullum and R. A. Willoughby. *Lanczos Algorithms for Large Symmetric Eigenvalues Computations, Vol. 1, Theory.* Birkhäuser, Basel, 1985.
26. U. Elsner, V. Mehrmann, F. Milde, R. A. Römer and M. Schreiber. The Anderson model of localization: A challenge for modern eigenvalue methods. *SIAM J. Sci. Comp.*, 20:2089–2102, 1999.
27. M. Schreiber, F. Milde, R. A. Römer, U. Elsner and V. Mehrmann. Electronic states in the Anderson model of localization: benchmarking eigenvalue algorithms. *Comp. Phys. Comm.*, 121-122:517–523, 1999.
28. V. Z. Cerovski. *(private communication).*
29. O. Schenk, M. Bollhöfer and R. A. Römer. On large scale diagonalization techniques for the Anderson model of localization. ArXiv: math.NA/0508111.

Modelling Aging Experiments in Spin Glasses

Karl Heinz Hoffmann[1], Andreas Fischer[1], Sven Schubert[1], and Thomas Streibert[1,2]

[1] Technische Universität Chemnitz, Institut für Physik
09107 Chemnitz, Germany
hoffmann@physik.tu-chemnitz.de
andreas.fischer@physik.tu-chemnitz.de
[2] T. Streibert has also published as T. Klotz

1 Introduction

Spin glasses are a paradigm for complex systems. They show a wealth of different phenomena including metastability and aging. Especially in the low temperature regime they reveal a very complex dynamical behaviour. For temperatures below the spin glass transition temperature one finds a variety of features connected to the inability of the systems to attain thermodynamic equilibrium with the ambient conditions on the observation time scale: aging and memory effects have been observed in many experiments [1–11]. Spin glasses are good model systems as their magnetism provides an easy and very accurate experimental probe into their dynamic behavior. In order to investigate such features different experimental techniques have been applied. Complicated setups including temperature and field changes with subsequent relaxation phases lead to more interesting effects such as age reinitialization and freezing [12, 13].

In order to understand the observed phenomena a number of different concepts have been advanced. They include real space models, such as the droplet model [14–17], mean field theory approaches [18–21], as well as state space approaches [22–28].

The latter are based on the picture of a mountainous energy landscape, with valleys separated by barriers of varying heights, containing other valleys of varying depth, which again contain other valleys and so forth. The concept of energy 'landscapes' is a powerful tool to describe phenomena in a number of different physical systems. All these systems are characterized by an energy function which possesses many local minima separated by barriers as a function of the state variables. If graphically depicted, the energy function thus looks very much like a mountainous landscape.

The thermal relaxation process on the energy landscape is modelled as a diffusion over the many different energy barriers, which often show a self

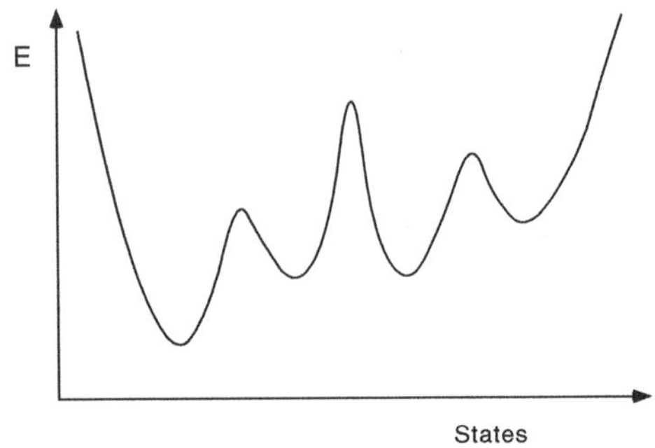

Fig. 1. A sketch of a complex state space. The energy is depicted as a function of a sequence of neighboring states. It resembles a cut through a mountainous landscape and thus the energy function is sometimes referred to as the energy landscape

similar structure of valleys inside valleys inside valleys etc. As a consequence one finds slow relaxation processes at low temperatures and a sequence of local equilibrations in the state space.

Here we present the steps which we took to map the path from basic spin glass models to modelling successfully the complex aging experiments performed over the last two decades. We start our presentation by studying the state space properties of microscopic Hamiltonians [29–34], then we investigate coarse graining procedures of such systems, as well as coarse grained dynamics on those state spaces. We present a dynamical study of aging experiments based on state spaces computed from an underlying microscopic Ising spin glass Hamiltonian. Finally we show that (even coarser) tree models provide not only an intuitive understanding of the observed dynamical features, but allow to model aging experiments astonishingly well. As becomes apparent in the following presentation only increased computational power and appropriate serial and especially parallel algorithms allowed us to execute this research effort.

2 Aging experiments

In the context discussed here 'aging' means that physical properties of the spin glass depend on the time elapsed since its preparation. The preparation time is the time at which the temperature of the spin glass was quenched below its glass temperature. The fact that aging occurs leads to the conclusion, that the spin glass is in thermal non-equilibrium during the observation time, which can last several days in experiments. Such aging effects have been observed in a large number of spin-glass experiments [1–4, 6, 7, 12, 13, 35–38], but are not confined to spin glasses. They have also been measured in high T_c superconductors [39] and CDW systems [40].

2.1 ZFC-experiments

A typical spin glass experiment is the so-called ZFC (Zero Field Cooled) experiment (see Fig.2). In this experimental setup the spin glass is quenched below its glass temperature. After a waiting time t_{w} an external magnetic field H is switched on and the magnetization M is measured as a function of measurement time t, which is counted from the application of the magnetic field. If the measurement is repeated for different waiting times, the results change: the magnetization depends on the waiting time t_{w} [5].

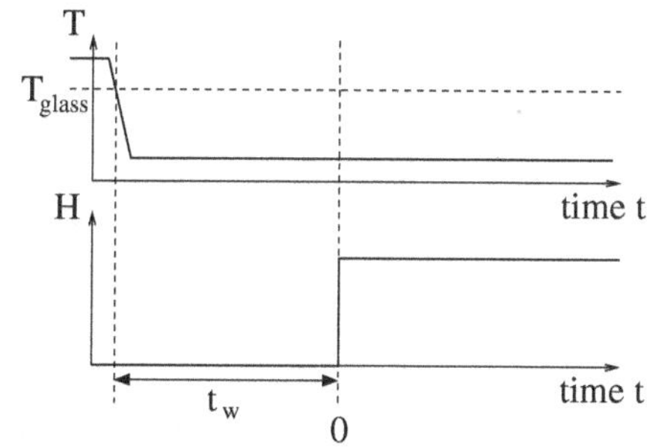

Fig. 2. Time schedule of a simple ZFC experiment. The probe is cooled below its glass temperature T_{glass} where no external magnetic field is applied. After a waiting time t_{w} a weak field is switched on. The magnetization is subsequently measured

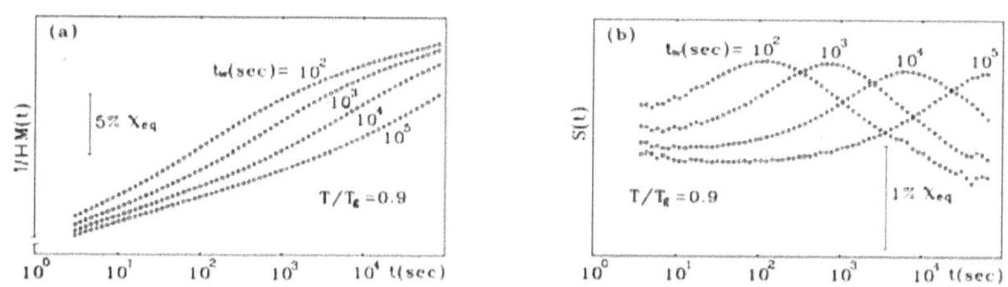

Fig. 3. Time dependent magnetization $M(t)$ and the corresponding relaxation rate $S(t) = \partial M/\partial \log t$ measured in a ZFC experiment [5]. The results of this measurement depend on the waiting time

In Fig. 3 the magnetization and the relaxation rate (the derivative of the magnetization with respect to the logarithm of time) are plotted versus the logarithm of the time after the application of the field. We see a kink in the magnetization $M(t, t_{\mathrm{w}})$ plotted as a function of logarithmic time at $t = t_{\mathrm{w}}$ or equivalently, a maximum in the derivative $S(t, t_{\mathrm{w}})$ of the magnetization with respect to the logarithm of the time at $t = t_{\mathrm{w}}$. This effect can be observed over a wide range of magnitudes of the waiting time.

2.2 ZFC-experiments with temperature step

In a second experimental setup, the temperature during the waiting time t_w is lowered to $T_M - \Delta T$ during the measurement time t [13]. At the end of the waiting time, when the external magnetic field is applied and the measurement of the magnetization is started, the temperature is increased in a steplike fashion to the measurement temperature T_M. The result of the temperature step is that the curves with a lower temperature during the waiting time seem to be 'younger', i.e. the maximum of the relaxation rate is shifted towards lower times.

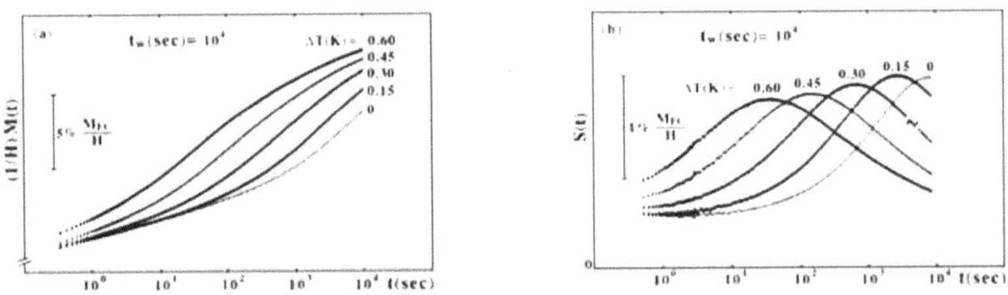

Fig. 4. Time dependent magnetization $M(t)$ and the corresponding relaxation rate $S(t) = \partial M/\partial \log t$ measured in a ZFC experiment [13] as a function of the temperature decrease during the waiting time. Note the shift of the maxima of the relaxation rate

2.3 TRM-experiments with temperature steps

There is a number of further experiments which measure the response as temperatures are changed during the waiting time. This leads to partial reinitialization effects as observed in the work of Vincent *et al.* [41], who studied temperature cycling in thermoremanent magnetization experiments. In these experiments the sample is quenched in a magnetic field, which is turned off after time t_w. Then the decay of the thermoremanent magnetization is measured. The temperature cycling consists of a temperature pulse added during the waiting time t_w.

The data of Vincent *et al.* [41] are shown in the left part of Fig. 5, for $t_w = 30$ and $t_w = 1000$ min. On top of this reference experiment, a very short temperature variation ΔT is imposed on the system 30 min before cutting the field. The important feature is that increasing ΔT shifts the magnetization decay data from the curve corresponding to the 1000 min curve to the 30 min curve. Thus the reheating appears to reinitialize the aging process.

Figure 5 right shows the results for a negative temperature cycle. The important feature here is that a temporary decreasing of the temperature leads to a 'freezing' of the relaxation. In other words the effect of the time

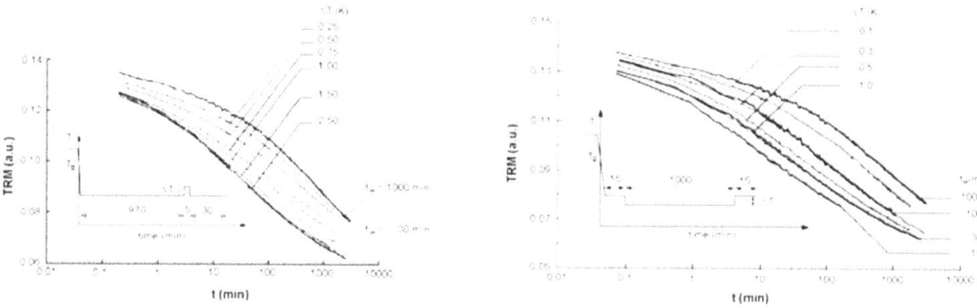

Fig. 5. Effect of a positive (left) and negative (right) temperature cycle on the thermoremanent magnetization (thin lines). The bold lines are reference curves without temperature cycling. The procedure is shown in the inset. These experimental results are taken from Vincent *et al.* [41]

spent at the lower temperature diminishes and eventually disappears as ΔT decreases.

3 Basic spin glass models

Starting point of our modelling are Ising spin glass models [42]. As an example let us consider a short range Edwards-Anderson Hamiltonian on a square grid with periodic boundary conditions of size L x L

$$\mathcal{H} = \sum_{<i,j>} J_{ij}\, s_i\, s_j - \sum_i H s_i \qquad , \tag{1}$$

where the sum is to be performed over all pairs of neighboring Ising spins s_i which can only take the values $+1$ or -1. H denotes a weak external magnetic field which can be applied to the spin glass according to the experimental setup. The interaction constants J_{ij} are uniformly distributed with a zero mean and standard deviation normalized to unity. This assumption defines the units of energy.

The interaction constants are randomly chosen using a random number generator. Each set of interaction constants is one spin glass realization. To obtain reliable data, which is not dependent on a single choice of interaction constants, we have to perform an average over many spin glass realizations. Such an analysis heavily depends on the availability of appropriate computing power.

The number of states is 2^N if N is the number of spins and grows exponentially with the system size, a feature common to the systems we are interested in. Thus either only small systems can be considered or a coarse graining procedure needs to be applied in order to reduce the number of states. The latter approach leads directly to the concept of an energy landscape with the thermal relaxation process painted as a random hopping process.

4 Energy landscapes

4.1 Branch and bound

In a first step we need to map out the state space of our system, and due to the excessively large number of states we restrain ourselves to the energetically low lying part of the state space, which is important for the low temperature regime. We use a recursive Branch-and-Bound algorithm to determine all energetically low-lying states up to a given cut-off energy E_{cutoff} [32,33,43]. The main idea of this method is to search a specific binary tree of all states. The search can be restricted by finding lower bounds for the minimal reachable energy inside of a subtree (branch). If this lower bound is higher than the energy of a suboptimal state already found, it is not necessary to examine the corresponding subtree. A first good suboptimal state can be found by recursively solving smaller subproblems.

We start with the smallest possible subsystem: one spin. This system has only two states with zero energy, corresponding to the spin up or down. In the next step we consider one additional spin. Now there exist four configurations, each with a certain energy. Thereafter spins are added one at a time, and every time the number of configurations is doubled.

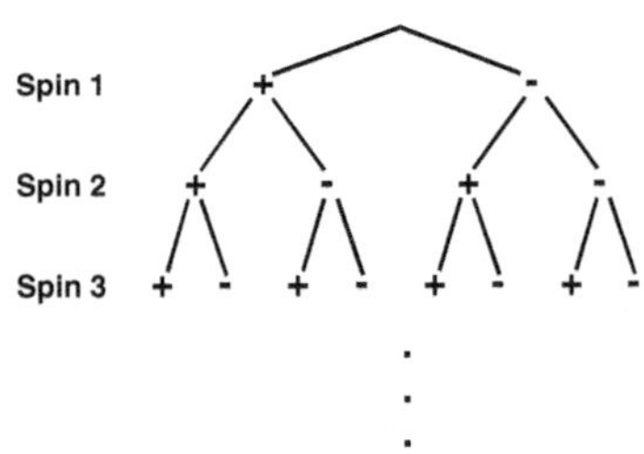

Fig. 6. Configuration tree for the Branch-and-Bound algorithm

This can be visualized by a so-called configuration tree (see Fig. 6). In the first level (from the top) there is only one spin with two configurations. In the next level spin 2 is added and there are four configurations and so on. Each node in the configuration tree has a certain energy corresponding to the interactions between the already considered spins. But in all nodes, except the lowest-level nodes, there are terms in (1) for spins which have not been considered yet. From this knowledge we can derive a lower bound for all nodes which are located in the subtree below. This lower bound is found by assuming that each of the lacking terms in (1) will sum up in a way such that (1) is minimized. If all states below a cut-off energy should be found, then only those subtrees are excluded from the further consideration, whose lower bound of the reachable minimum energy is larger than the energy of the known local minimum plus the cut-off energy. Due to the exclusion of large subtrees from consideration, the Branch-and-Bound algorithm provides a very efficient way to find the low-energy region of a discrete state-space.

Several interesting features can be observed in the data obtained. A first important result of this Branch-and-Bound algorithm is that the number of states grows sub-exponentially with the cut-off energy. A further careful analysis showed that the deviation from an exponential increase can be parameterized in different ways, for details see [32,34].

4.2 Thermal relaxation dynamics

We now turn to the dynamics in the enumerated state space. The states α of our model system are defined by the configuration of all spins $\{s_i\}$, and each state has its energy $E(\alpha)$ as given by the Hamiltonian (1). Neighboring states are obtained from each other by flipping one of the spins, and $N(\alpha)$ will denote the set of neighbors of a state α.

The thermal relaxation in contact with a heat bath at temperature T can be modelled as a discrete time Marcov process. The thermally induced hopping process induces a probability distribution in the state space. The time development of $P_\alpha(k)$, the probability to be in state α at step k, can then be described by a master equation [44]

$$P_\alpha(k+1) = \sum_\beta \Gamma_{\alpha\beta}(T) P_\beta(k). \tag{2}$$

The transition probabilities $\Gamma_{\alpha\beta}(T)$ depend on the temperature T. They have to insure that the stationary distribution is the Boltzmann distribution $P_\alpha^{eq}(T) = g_\alpha \exp(-E_\alpha/T)/Z$, where $Z = \sum_\alpha P_\alpha^{eq}$ is the partition function and g_α is the degeneracy of state α. The latter is needed if the states α already represent quantities which include more than one micro state. Possible choices for the transition probabilities are the Glauber dynamics [45], and the Metropolis dynamics [46]. Here we use:

$$\Gamma_{\beta\alpha} = \begin{cases} \Pi_{\beta\alpha} \exp(-\Delta E/T) & \text{if } \Delta E > 0,\ \alpha \neq \beta \\ \Pi_{\beta\alpha} & \text{if } \Delta E \leq 0,\ \alpha \neq \beta \\ 1 - \sum_{\xi \neq \alpha} \Gamma_{\xi\alpha} & \text{if } \alpha = \beta, \end{cases} \tag{3}$$

where $\Delta E = E(\beta) - E(\alpha)$, $\Pi_{\beta\alpha}$ equals $1/|N(\alpha)|$ if the states α and β are neighbors and equals zero otherwise, and $|N(\alpha)|$ is the number of neighbors of state α. The Boltzmann factor in (3) makes the transition over an energy barrier a slow process compared to the relaxation within a valley of the energy landscape.

5 Cluster models

5.1 Structural coarse graining

The number of states found by the branch-and-bound algorithm grows nearly exponentially with the cut-off energy [32]. In order to handle larger systems we thus need to coarse grain our description by collecting sets of microscopic states into larger clusters. Our aim is to obtain a reduced model such that the macroscopic properties of the dynamics will not be altered. In order to obtain a good approximation for the dynamical properties of the system on macroscopic time scales it is important that the inner relaxation in a cluster is faster than the interaction with the surrounding clusters. To fulfill this condition for low temperatures a cluster must not contain any energy barriers. The idea of such a coarse graining was advanced already in 1988 [24] and a practical implementation of the above mentioned criteria is described in [31,34].

The clustering algorithm which we refer to as the *NB-clustering* (No-Barrier-clustering) in the following proceeds as follows:

```
1. Sort all states according to ascending energy
2. Start with one of the lowest-energy states
   -- create a cluster to which this state belongs
   -- the reference energy of the cluster is the energy of this state
   -- create a new valley to which the cluster and the state belong
3. Consider one of the states with equal energy, or if not present, the state
   with next higher energy
4. If the new state is
   4.1 not neighbored to states considered yet ⇒
       -- create a new cluster to which the new state belongs
       -- the reference energy of the cluster is the energy of the new state
       -- create a new valley to which the new cluster and the new state
          belong
   4.2 neighbored to states which belong to different valleys (such states we
       call barrier states) ⇒
       -- link the connected valleys to one new large valley
       -- create a new cluster to which the new state belongs
       -- the new cluster belongs to the new large valley
   4.3 neighbored to states which belong to one valley and
     4.3.1 one cluster ⇒ the state is added to this cluster
     4.3.2 different clusters ⇒ the state is added to the cluster with the
             highest reference energy
5. go to step 3 until all states have been considered
```

This algorithm produces clusters without any internal barrier. In Fig. 7 (left) this structural coarse graining has been applied to a small subsystem with 28 states for demonstration purposes. The subsystem shown is actually a low energy subset of the states of an Ising spin glass. The resulting coarse grained system is shown in Fig. 7 (right), where the numbers inside a cluster are the numbers of the states which are lumped into the cluster. The figure shows two kinds of clusters: *local minimum clusters* are those which contain a local minimum of the energy, *energy barrier clusters* are those which connect two energetically lower clusters.

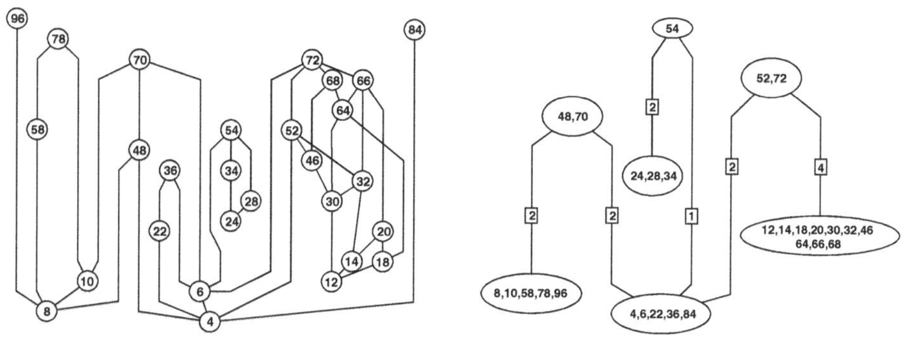

Fig. 7. Left: Microscopic states with connections. Right: Coarse grained state space obtained from the microscopic states using the algorithm. The number in a square at a connection is the number of the corresponding microscopic connections

In order to gain an understanding of the structure of the energy landscape and the resulting coarse grained clusters we studied [34] several different features, of which we here report only a few: First we analysed the number of local minimum clusters $N_{\mathrm{lmc}}^{\mathrm{gs}}$, which are accessible from the ground state without exceeding an energy barrier of E_{b}. This data is obtained by applying a lid energy which is initially set to the energy of the global minimum and is subsequentially raised until the cutoff energy is reached. For each lid energy we search the data base for accessible local minimum clusters without climbing above the given lid energy. The summed-up results for 88 spin glass realizations of size 8 x 8 are shown in Fig. 8. One sees that the number of local minimum clusters accessible from the ground state $N_{\mathrm{lmc}}^{\mathrm{gs}}$ increases fast with the energy, but not exponentially as a function of the barrier energy. This behavior agrees well with the subexponential growth in the density of states seen for such systems [32]. Interestingly a power law $N_{\mathrm{lmc}}^{\mathrm{gs}} \propto E_{\mathrm{b}}^{3.9}$ fits the resulting curve quite well.

While the subexponential increase of $N_{\mathrm{lmc}}^{\mathrm{gs}}$ indicates that a binary tree with its exponential increase does not reflect exactly the results from the enumerated state space of our spin glass Hamiltonian the increase in $N_{\mathrm{lmc}}^{\mathrm{gs}}$ seems strong enough to warrant the use of the binary trees as simple models. But based on our investigation one can now take a step towards a more realistic tree concept.

To obtain more information about the Edwards-Anderson state space structure we refined the analysis of accessible local minimum clusters by

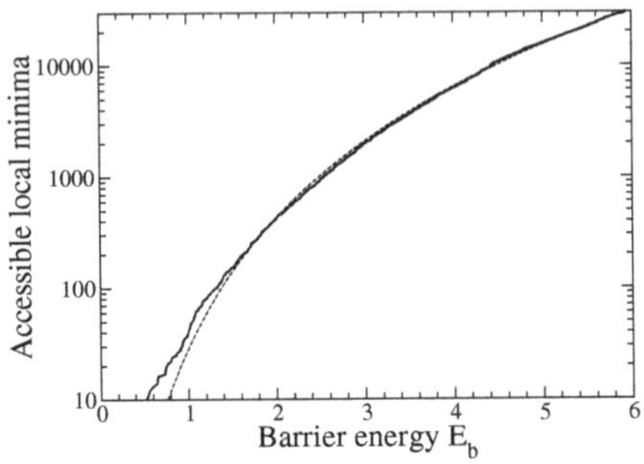

Fig. 8. The number of local minimum clusters $N_{\mathrm{lmc}}^{\mathrm{gs}}$ in the NB-clustered state space which are accessible from the global minimum without crossing energy barriers higher than E_{b}. The dashed line corresponds to a power fit which is $\propto E_{\mathrm{b}}^{3.9}$

distinguishing between the local minimum clusters at different energies. We start again in the global minimum but we count only those destination local minimum clusters with an energy which is lower than 1, 2, 3 or 4, respectively (Fig. 9).

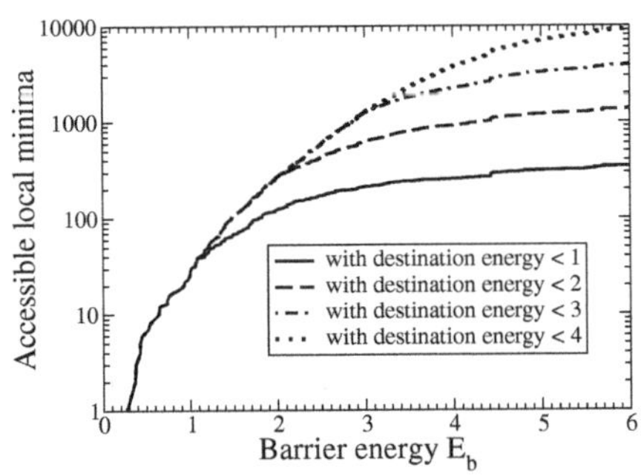

Fig. 9. The number of local minimum clusters in the NB-clustered state space which are accessible from the global minimum without crossing energy barriers higher than E_{b}. For the different curves only those local minimum clusters are counted which have an energy difference to the global minimum less than 1,2,3 or 4, respectively

Again we find that at a given barrier energy the number of accessible local minimum clusters increases (sub-)exponentially with energy. Interestingly most of the barriers lead into energetically high lying local minima, and paths from the global minimum over a high energy barrier back to an energetically low lying local minimum cluster are relatively rare.

Further information is obtained by studying the dependence of the number of accessible local minimum clusters $N_{\mathrm{lmc}}^{\mathrm{lm}}$ when starting in a local minimum cluster. We counted the number of all accessible local minimum clusters when climbing a certain barrier energy relative to the starting point. In Fig. 10 we show the number of accessible local minimum clusters $N_{\rm lmc}^{\rm lm}(E_{\rm rb})$ as a function of the relative barrier energy $E_{\rm rb}$ summed over 88 spin glass realizations. Here each pair of local minimum clusters contributed twice to this number, but (generically) at two different barrier energies depending on the depth of the local minima on each side. We find that $N_{\rm lmc}^{\rm lm}(E_{\rm rb})$ grows slower than exponentially, but it still shows a strong increase.

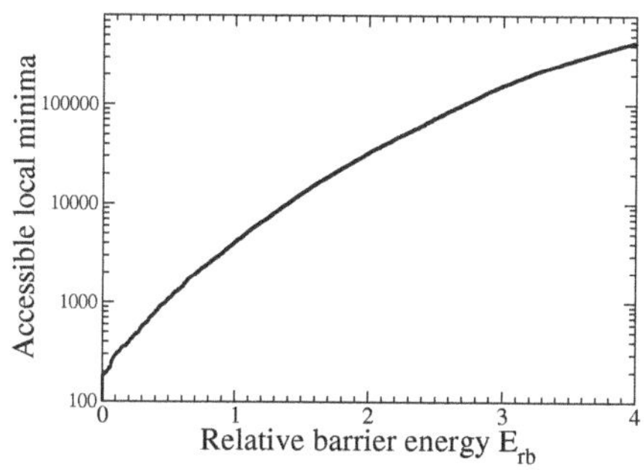

Fig. 10. The number of local minimum clusters $N_{\rm lmc}^{\rm lm}$ accessible from another local minimum by crossing energy barriers which are not higher than $E_{\rm rb}$. The energy is counted relative to the energy of the starting point (local minimum). The data is taken from each path between two local minima in 88 spin glass realizations of size 8x8

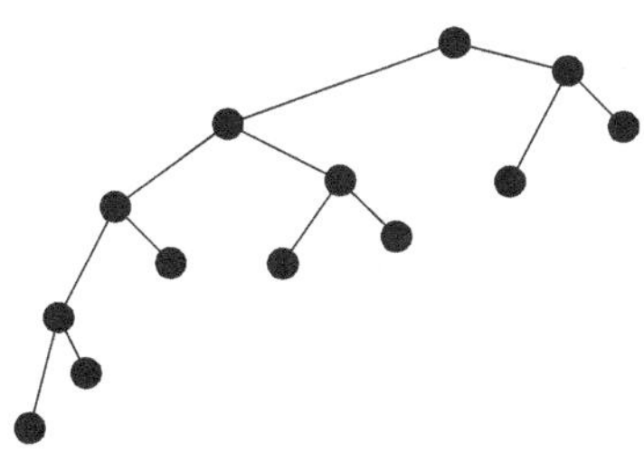

Fig. 11. Possible coarse grained state space structure compatible with the data obtained from *NB-clustering*. Note the similarity to the LS-tree shown in Fig. 17 (right)

The results found in our investigation indicate, that a hierarchical structure depicted in Fig. 11 is compatible with the structure of the enumerated state space under NB-clustering. It is quite similar to a modified LS-tree, which is discussed below, where the subtrees below the short edges are smaller than those below the long edges. This picture captures the sub-exponential growth of accessible local minima with the barrier energy as well as the significant growth of the number of local minimum clusters with increasing energy.

5.2 Dynamical coarse graining

Consider now the thermal hopping on the complex energy landscape. Rather than calculating the full time dependence of the probability distribution in the enumerated state space, we can choose to monitor the presence or absence from a cluster of the coarse grained system. We have hereby defined a stochastic process, which in general will not be a Marcov process [47], because the induced transition probability from cluster to cluster might depend on the internal (microscopic) distribution within one cluster. However, it turns out that inside a coarse grained area very quickly a kind of local equilibrium distribution is established, which then makes the coarse grained relaxation process (at least approximately) marcovian. The result is that Marcov processes on the coarse grained systems are good modelling tools for the thermal relaxation of complex systems [23, 24, 48].

Using this insight, the proper transition rates of the coarse grained system can be determined. To find the structure of this rates let us start with the exact calculation of the transition probability between two neighboring clusters based on the microscopic picture. The probability flux from the states belonging to cluster C_ν to the states belonging to cluster C_μ is given by

$$J_{\mu\nu} = G_{\mu\nu} P_\nu = \sum_{\alpha \in C_\mu, \beta \in C_\nu} \Gamma_{\alpha\beta} p_\beta \ , \tag{4}$$

where P_ν is the total probability to be in cluster C_ν, i.e. the sum of the probabilities of all states in cluster C_ν, and $G_{\mu\nu}$ is the transition rate from cluster C_ν to cluster C_μ. Again we assume that the internal relaxation inside the clusters is fast compared to the relaxation between different clusters. For the time scale of interest all clusters are in internal equilibrium, i.e. $p_\beta \propto \exp(-E_\beta/T)$. As the microscopic transition rates have the form $\Gamma_{\alpha\beta} \propto \exp(-\max(E_\alpha - E_\beta, 0)/T)$, the coarse grained transition rate is

$$G_{\mu\nu} = \frac{\sum_{\alpha \in C_\mu, \beta \in C_\nu} \Pi_{\alpha\beta} \exp(-\max(E_\alpha, E_\beta)/T)}{\sum_{\alpha \in C_\nu} \exp(-E_\alpha/T)} \ . \tag{5}$$

The roughest simplification (here referred to as procedure A) would be to consider all states of a cluster as one state with a certain energy $\hat{E}_\mu$ which is chosen as the mean energy of the microscopic states. Following this idea the sums in (5) can be simplified to

$$\hat{G}_{\mu\nu} = \frac{\hat{T}_{\mu\nu} \min(\exp(-(\hat{E}_\mu - \hat{E}_\nu)/T), 1)}{\hat{n}_\nu} \ , \tag{6}$$

where $\hat{T}_{\mu\nu}$ is related to the number of connections between cluster C_ν and cluster C_μ, and $\hat{n}_\nu$ is the number of states collected in cluster C_ν. A better approximation can be achieved if each cluster is modelled by a two-level system, which is in internal equilibrium (procedure B). For a detailed description of these procedures see Klotz et al. [31].

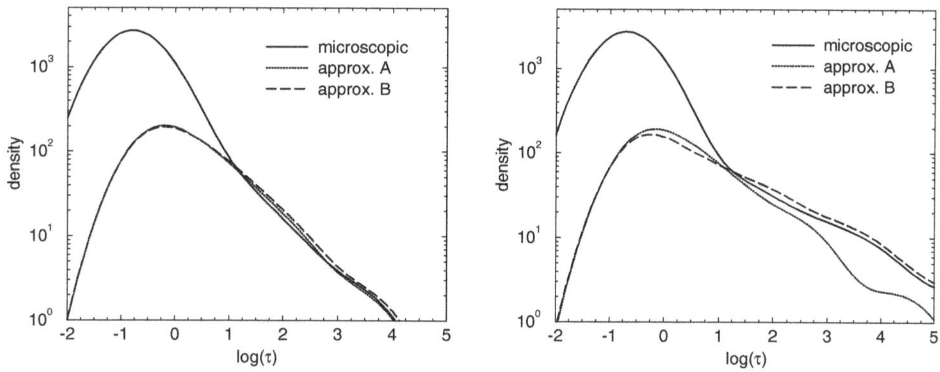

Fig. 12. Smoothed relaxation time densities versus the logarithm of the relaxation time for $T = 1$ (left) and $T = 0.25$ (right) for the microscopic system and the two procedures

In order to check the quality of the approximations made by the coarse graining of the energy landscape one can for instance look at the density of relaxation times. Fig. 12 shows this density of relaxation times for $T = 1$ and $T = 0.25$ respectively. The spectra have been computed with a resolution of 0.2 on the logarithmic τ-scale. In the case of high temperatures (Fig. 12 (left)) we see a good agreement of the two procedures compared to the original microscopic system in the range of large relaxation times, while for lower temperature (Fig. 12 (right)) procedure B provides the better approximation. For short times the microscopic system has many more eigenvalues, which are neglected in the coarse grained system. Thus the dynamics in the coarse grained state space is a good approximation of the dynamics in the microscopic system for slow processes which are the important ones for the analysis of low temperature relaxation phenomena.

In a more detailed analysis [34] of the kinetic factors $T_{\mu\nu}$ we required that the coarse grained dynamics reflects the microscopic one at a given temperature. Then we find that the connections with a small energy difference have (on average) a larger kinetic factor. This is especially important for the dynamics following a temperature quench. Then the occupation probability of the system moves to lower energy states. As the connections with low energy differences are faster, the system will preferably end up in a local minimum with a comparably high energy. The equilibration towards the global minimum states is then driven by the slow modes of relaxation.

The above algorithm for the coarse graining of complex state spaces shows in which way the idea of an energy landscape can help in modelling complex systems. This technique is independent of the model and can not only be used for Ising spin-glass models as demonstrated, but also for other complex systems such as Lennard-Jones systems or proteins.

6 Aging in the enumerated state space

We now turn to the modelling of the aging experiments. If the concepts presented above are valid, then one should be able to simulate the aging behavior based on the coarse grained dynamics. First we investigate a simple ZFC experiment where the spin glass is quenched below its glass temperature with no external magnetic field. We perform the analysis on the low-energy part of an enumerated and coarse grained state space derived from our spin glass Hamiltonian. In this approach we directly access the properties of the state space (e.g., the energy and magnetization of the clusters) and perform a dynamics following the Metropolis algorithm.

To model experiments with an external magnetic field we set $H \neq 0$ in the second term of the Hamiltonian (1). While in principle the external field changes the state space structure and the coarse graining procedure should be redone for the state space in the external field, we here assume that the magnetic field is weak and that it changes only slightly the structure of the state space. Thus we just change the energies of the already coarse grained system

$$\hat{E}_\mu^{\mathrm{H}} = \hat{E}_\mu - H\hat{M}_\mu \ , \tag{7}$$

for all Clusters p. $\hat{M}_\mu$ is the effective magnetization of the cluster C_μ, which is computed from the magnetizations of the micro-states in this cluster and the assumption of internal equilibration at the given temperature, and

$$\hat{E}_\mu = -T\ln\left(\frac{1}{\hat{n}_\mu}\sum_{\alpha\in C_\mu} \mathrm{e}^{-E_\alpha/T}\right) . \tag{8}$$

Technically we solve the eigenvalue problem of the corresponding master equation determining eigenvalues and eigenvectors. Then macroscopic properties of the spin glass can be obtained by performing the thermal average and the average over many spin glass realizations (ensemble average). We simulated the ZFC procedure in 500 spin glass realizations with altogether more than 17 million micro-states. They can be reduced to coarse grained state spaces with about 276.000 clusters. Our choice for the external magnetic field was $H = 0.1$ and for the temperature $T = 0.2$.

The averaged results of these calculations are shown in Fig. 13. The magnetization increases after the application of the magnetic field, since the spins tend to align with the direction of the external field. States with positive magnetization become energetically favored. However, we find that the magnetization increases slower in the curves with a larger waiting time. This system clearly shows aging features. The corresponding curves of the relaxation rate $S(t) = \mathrm{d}M(t)/\mathrm{d}\log(t)$ in Fig. 14 display maxima at times which are roughly equal to the waiting time.

This remarkable result indicates that the considered coarse grained state space still represents all ingredients necessary for aging effects. We find that

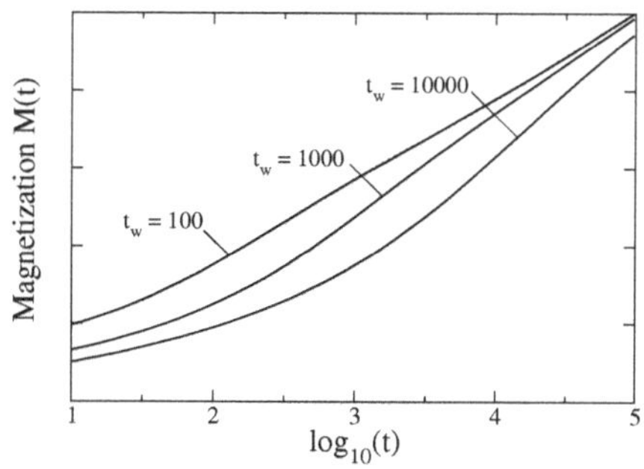

Fig. 13. Magnetization of a simple ZFC procedure averaged over 500 spin glass realizations. The probes are cooled to a low temperature T=0.2 with no external field. After the waiting time $t_{\rm w}$ a weak external field $H = 0.1$ is applied. We clearly find an aging behavior, the measured magnetization depends on the waiting time

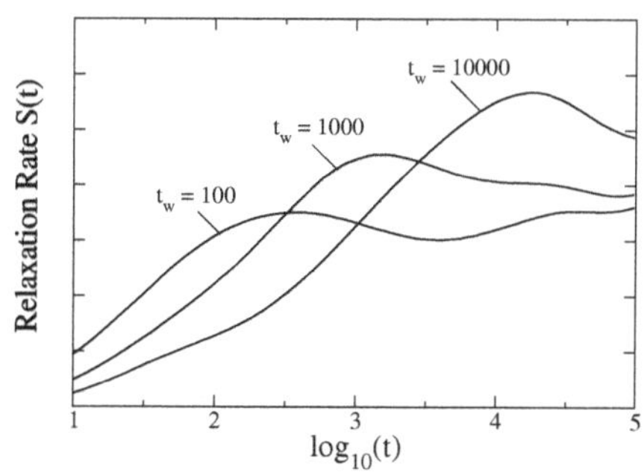

Fig. 14. Relaxation rate $S(t) = \mathrm{d}M(t)/\mathrm{d}\log(t)$ for the magnetization shown in Fig. 13. As in the experiment we find maxima in the relaxation rate at measurement times which correspond to the waiting time

the appropriate dynamics on the coarse grained state space is able to reproduce the effects observed in spin glass experiments and also in the dynamics of model state spaces.

The second set of experiments we simulated were the ZFC experiment with temperature step [13], in which the temperature during the waiting time $t_{\rm w}$ is ΔT lower than during the measurement time t. The experimental results show that in the curves with a lower temperature during the waiting time the maximum of the relaxation rate is shifted towards smaller times. In our simulations we find excellent agreement with the experimental results: the maxima of the relaxation rate are shifted towards earlier times for increasing size of the temperature step (Fig. 15). This behaviour is easily understood. The systems relax slower during the waiting time due to the lower temperature which leads to decreased transition probabilities over the energy barriers. Thus the effect of the waiting time is reduced.

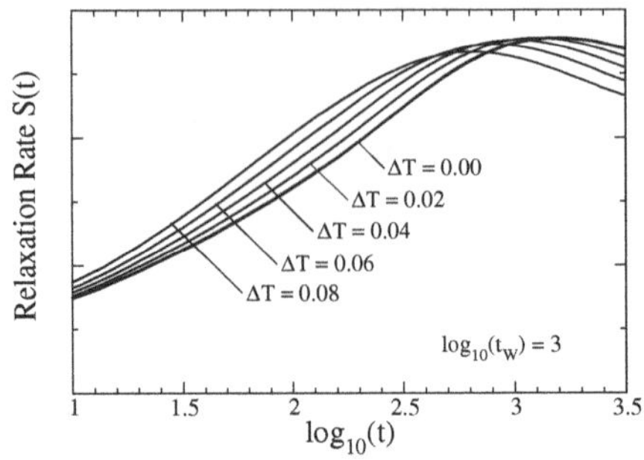

Fig. 15. Relaxation rate $S(t) = \mathrm{d}M(t)/\mathrm{d}log(t)$ of the procedure with temperature step, where the temperature during the waiting time is ΔT lower than during the measurement time. The measurement temperature is $T = 0.2$, the external magnetic field is $H = 0.1$ and the waiting time for all curves is $t_{\mathrm{w}} = 1000$. The ensemble average is performed over the same 500 spin glass realizations used in the previous figure

Using the fact that the measurement time at which the relaxation rate is maximum equals the waiting time for the simple ZFC-experiment we can introduce the apparent waiting time of a curve by the position of the maximum of their relaxation rate on the time axis. In Fig. 16 we plot the logarithm of the apparent waiting time as a function of the temperature step ΔT. These results are in a good agreement with the experiments [13] for different waiting times and also with simulations on heuristical models [26].

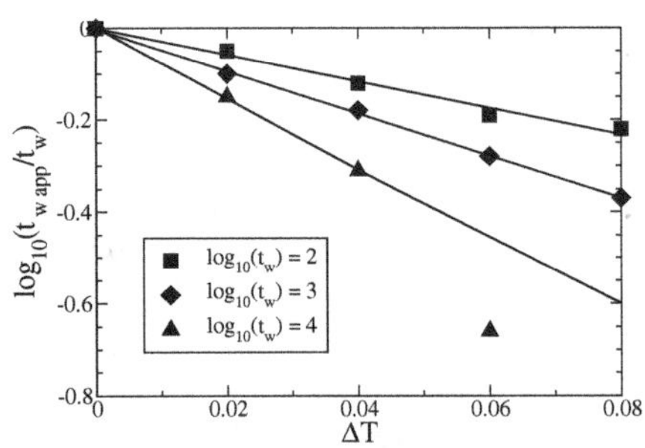

Fig. 16. The logarithm of the apparent waiting time as a function of the temperature step ΔT (symbols) for different waiting times t_{w}. The apparent waiting time is defined by the maximum of the relaxation rate, which is shifted to lower times in experiments with temperature step. The lines are linear fits. In the case of $t_{\mathrm{w}} = 10000$ the fit is restricted to $\Delta T \leq 0.04$

We find increasing deviations from the linear behavior for large temperature steps for the longest waiting time. They are due to the coarse grained dynamics used, which does depend on the temperature for which it is determined.

We here used the measurement temperature $T = 0.2$ in the coarse graining procedure, which we also used for the dynamics during the waiting time due to computational restrictions. This approximation is only valid for small ΔT and leads to the observed deviations for large ΔT.

7 Tree models

The above reported results on enumerated state spaces suggest that hierarchical structures like the one shown in Fig. 11 could be good models for aging dynamics. Historically it was already shown long before detailed studies of spin glass state spaces were performed that tree models show typical features of spin glasses [26, 49–51]. One basic reason for their success is that they efficiently model a hierarchy of time scales by a sequence of energy barriers at increasing heights [24, 25], which allows to reproduce the aging effects observed over many magnitudes of time, i.e. these tree models possess the many relaxation time scales seen in aging.

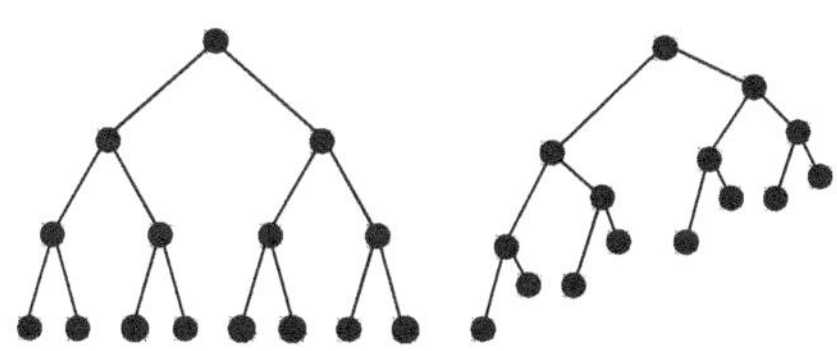

Fig. 17. Tree models for the state space structure of spin glasses

Early work dealt with the simple symmetric tree model shown on the left in Fig. 17. Experimental results as shown in Fig. 18 can be reproduced successfully [25, 26] using such symmetric tree models. However, those simple model state-spaces cannot reproduce the age-reinitialization and freezing effects observed in the temperature-cycling experiments of Vincent et al [12]. For this reason a so-called LS-tree as displayed in Fig. 17 (right) was introduced [27, 28, 50].

All nodes but those at the lowest level have two 'daughters' connected to their mother by a 'Long' and a 'Short' edge (which explains the name LS-tree), such that the energy differences become $\Delta E = L$ and $S < L$ respectively. The dynamics is given by a nearest neighbor random walk on this structure. The nondiagonal elements of the transition matrix $\Gamma_{\mu\nu}$ are zero except for states connected by an edge, in which case they can all be expressed in terms of up and down rates along the edge:

$$\Gamma_{\text{up}} = f_j \kappa_j \mathrm{e}^{-\Delta E_j / T} \qquad \Gamma_{\text{down}} = f_j \ . \tag{9}$$

The index j distinguishes between L- and S-edges, and κ_j is the ratio between the degeneracy of a node and that of the corresponding daughter node. The

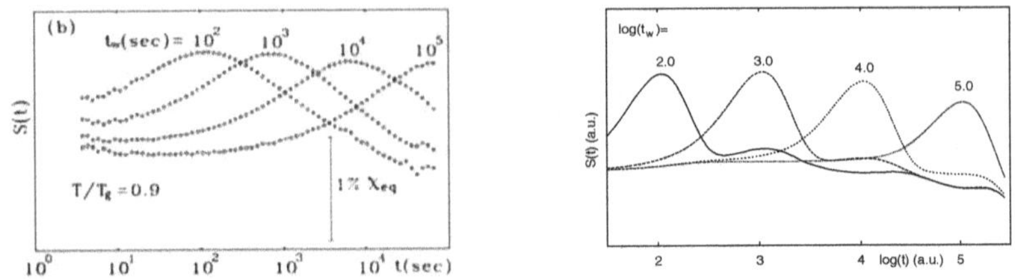

Fig. 18. A comparison between the experimental data for ZFC-experiments and the tree model data. Note that the latter reproduces very well the maxima in the relaxation rate at times which correspond to the waiting times

diagonal elements of the transition matrix are given by the condition that each column sum vanish to ensure conservation of probability.

The most important feature of this model are the kinetic factors f_j controlling the relaxation speed along each edge. As detailed balance only prescribes the ratios of the hopping rates between any two neighbors the f_j can be freely chosen without affecting the equilibrium properties of the model. The exponents of the slow algebraic relaxation [50] are not affected by any arbitrary choice of these parameters, as numerically demonstrated by Uhlig *et al.* [27]. Nevertheless, non-uniform kinetic factors have a decisive effect on the dynamics following temperature steps, which destroy local equilibrium on short time scales [52].

To qualitatively understand how the competition comes about, consider the extreme case where the system, initially at high temperature, is quenched to zero temperature. Since upward moves are forbidden, the probability flows downwards through the system, splitting at each node in the ratio $f_{\mathrm{L}} : f_{\mathrm{S}}$, independently of energy differences. Thus, if f_{S} is larger than f_{L}, the probability is preferably funnelled through the short edges, and ends up mainly in high lying metastable states. If the system is heated up even so slightly after the quench, thermal relaxation sets in and redistributes the probability. Eventually the distribution of probability becomes independent of the values of f_{L} and f_{S}, and the low energy states are favored.

The above description is now easily extended to a many level tree. An initial quench creates a strongly non-equilibrium situation mainly determined by the relaxation speeds, whereupon the slow relaxation takes over. At any given time subtrees of a certain size will have achieved internal equilibration, while larger ones will have not.

A temperature cycle – that is temperature increase followed by a decrease of the same size or vice versa – always destroys this internal equilibration as it induces a fast redistribution of probability. In positive temperature cycles probability is pumped up to energetically higher nodes which leads to a partial reset when the temperature is lowered. By way of contrast negative temperature cycles push probability down the fast edges, a process which is readily reversed when the temperature is raised again.

In the following we compare our model predictions to the thermoremanent magnetization experiments of Vincent *et al.* [41], which we discussed in Sect. 2.3. The important feature was that increasing ΔT shifts the magnetization decay data from the curve corresponding to the 1000 min curve to the 30 min curve. The corresponding effect in our model is shown in the left part of the Fig. 19. The parallels between model and experiment are obvious. In both cases the reheating appears to reinitialize the aging process.

Figure 5 (right) shows the results for a negative temperature cycle. The important feature here is that a temporary decreasing of the temperature leads to a 'freezing' of the relaxation. This is seen in the model data as well as, see Fig. 19 (right).

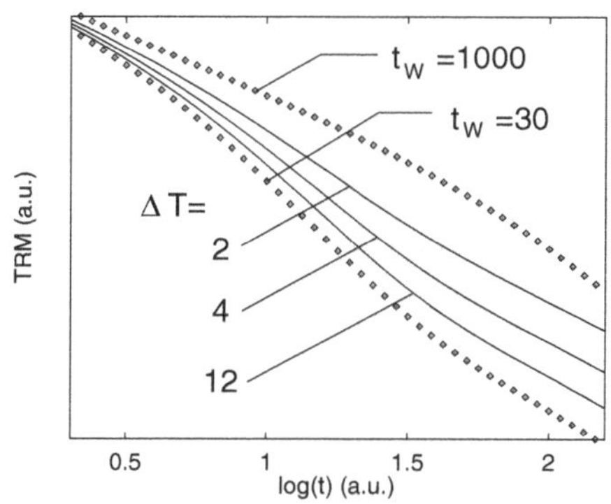

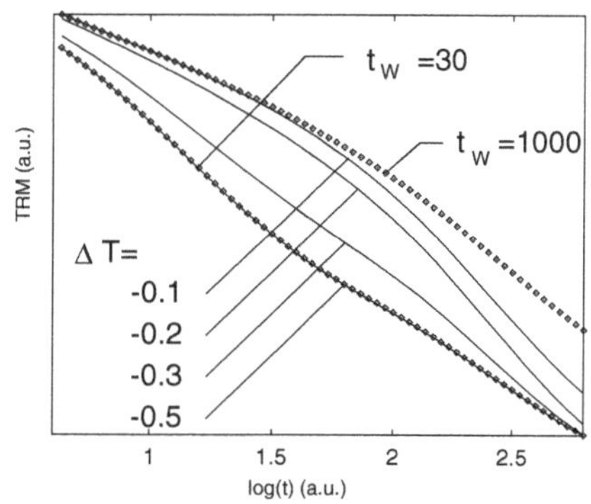

Fig. 19. Model results corresponding to the experimental results above

8 Summary and outlook

Our aim was to model spin glass aging experiments starting from a microscopic spin glass Hamiltonian. We enumerated the energetically low lying parts of the state space of an Edwards-Anderson spin glass. We obtained the states as well as their connectivity which allowed us to analyse features such as density of states. A coarse graining procedure reduced the vast amount of states such that a reduced description became possible without loosing the main features of the relaxation dynamics. The NB-clustering scheme produced coarse grained nodes (clusters) which have no internal barriers thus being effectively very close to thermal equilibrium at all times. Based on a direct simulation of the aging experimental set-up we could show that our approach captures the essential features of the experiments. Finally we showed that even coarser models such as the LS-tree already model the quite complicated re-initialization experiments successfully.

This research program was only possible because we could make use of the compute power available to us. This rested partly on the the hardware in form of parallel machines as well as compute clusters but to an even larger extent on the efficient algorithms and their implementation we developed over the years.

Acknowledgements

KHH thanks Paolo Sibani for numerous discussions about aging experiments and their modelling and our continued collaboration over the last two decades.

References

1. L. Lundgren, P. Svedlindh, P. Nordblad, and P. Beckman. Dynamics of the relaxation-time spectrum in a cumn spin-glass. *Phys. Rev. Lett.*, 51(10):911–914, 1983.
2. M. Ocio, H. Bouchiat, and P. Monod. Observation of $1/f$ magnetic fluctuations in a spin glass. *J. Phys. Lett. France*, 46:647–652, 1985.
3. P. Nordblad, P. Svedlindh, J. Ferré, and M. Ayadi. $Cd_{0.6}Mn_{0.4}Te$, a semiconducting spin glass. *Journal of Magnetism and Magnetic Materials*, 59(3–4):250–254, 1986.
4. Ph. Refregier, M. Ocio, J. Hammann, and E. Vincent. Nonstationary spin glass dynamics from susceptibility and noise measurements. *J. Appl. Phys.*, 63(8):4343–4345, 1988.
5. P. Svedlindh, P. Granberg, P. Nordblad, L. Lundgren, and H. S. Chen. Relaxation in spin glasses at weak magnetic fields. *Phys. Rev. B*, 35(1):268–273, 1987.
6. J. Hamman, M. Lederman, M. Ocio, R. Orbach, and E. Vincent. Spin-glass dynamics – relation between theory and experiment – a beginning. *Physica A*, 185(1–4):278–294, 1992.
7. A. G. Schins, E. M. Dons, A. F. M. Arts, H. W. de Wijn, E. Vincent, L. Leylekian, and J. Hammann. Aging in two-dimensional ising spin glasses. *Phys. Rev. B*, 48(22):16524–16532, 1993.
8. C. Djurberg, K. Jonason, and P. Nordblad. Magnetic relaxation phenomena in a CuMn spin glass. *Eur. Phys. J. B*, 10(1):15–21, 1999.
9. K. Jonason, P. Nordblad, E. Vincent, J. Hammann, and J.-P. Bouchaud. Memory interference effects in spin glasses. *Eur. Phys. J. B*, 13(1):99–105, 2000.
10. R. Mathieu, P. Jönsson, D. N. H. Nam, and P. Nordblad. Memory and superposition in a spin glass. *Phys. Rev. B*, 63(09):092401/1–092401/4, 2001.
11. R. Mathieu, P. E. Jönsson, P. Nordblad, H. Aruga Katori, and A. Ito. Memory and chaos in an ising spin glass. *Phys. Rev. B*, 65(1):012411/1–012411/4, 2002.
12. E. Vincent, J. Hammann, and M. Ocio. Slow dynamics in spin glasses and other complex systems. In D. H. Ryan, editor, *Recent Progress in Random Magnets*, page 207. World Scientific, Singapore, 1992.
13. P. Granberg, L. Sandlund, P. Norblad, P. Svedlindh, and L. Lundgren. Observation of a time-dependent spatial correlation length in a metallic spin glass. *Phys. Rev. B*, 38(10):7097–7100, 1988.

14. D. S. Fisher and D. A. Huse. Nonequilibrium dynamics of spin glasses. *Phys. Rev. B*, 38:373–385, 1988.
15. H. G. Katzgraber, M. Palassini, and A. P. Young. Monte carlo simulations of spin glasses at low temperatures. *Phys. Rev. B*, 63:184422, 2001.
16. M. Palassini and A. P. Young. Nature of the spin glass state. *Phys. Rev. Lett.*, 85:3017, 2000.
17. P. E. Jönsson, H. Yoshino, Nordblad P., Aruga Katori H., and Ito A. Domain growth by isothermal aging in 3d ising and heisenberg spin glasses. *Phys. Rev. Lett.*, 88:257204, 2002.
18. G. Parisi. Infinite number of order parameters for spin-glasses. *Phys. Rev. Lett.*, 43(23):1754–1756, 1979.
19. G. Parisi, R. Ranieri, F. Ricci-Tersenghiz, and J.J. Ruiz-Lorenzo. Mean field dynamical exponents in finite-dimensional Ising spin glass. *J. Phys. A: Math. Gen.*, 30(20):7115–7131, 1997.
20. A Montanari and F. Ricci-Tersenghi. On the nature of the low-temperature phase in discontinuous mean-field spin glasses. *Eur. Phys. J. B*, 33:339, 2003.
21. I. R. Pimentel, T. Temesvari, and C. De Dominicis. Spin glass transition in a magnetic fieldi: a renormalization group study. *Phys. Rev. B*, 65:224420, 2003.
22. Andrew T. Ogielski and D. L. Stein. Dynamics on ultrametric spaces. *Phys. Rev. Lett.*, 55(15):1634–1637, 1985.
23. K. H. Hoffmann, S. Grossmann, and F. Wegner. Random walk on a fractal: Eigenvalue analysis. *Z. Phys. B*, 60:401–414, 1985.
24. K. H. Hoffmann and P. Sibani. Diffusion in hierarchies. *Phys. Rev. A*, 38(8):4261–4270, 1988.
25. P. Sibani and K. H. Hoffmann. Hierarchical models for aging and relaxation of spin glasses. *Phys. Rev. Lett.*, 63(26):2853–2856, 1989.
26. C. Schulze, K. H. Hoffmann, and P. Sibani. Aging phenomena in complex systems: A hierarchical model for temperature step experiments. *Europhys. Lett.*, 15(3):361–366, 1991.
27. C. Uhlig, K. H. Hoffmann, and P. Sibani. Relaxation in self similar hierarchies. *Z. Phys. B*, 96:409–416, 1995.
28. K. H. Hoffmann, S. Schubert, and P. Sibani. Age reinitialization in hierarchical relaxation models for spin-glass dynamics. *Europhys. Lett.*, 38(8):613–618, 1997.
29. T. Klotz and S. Kobe. Exact low-energy landscape and relaxation phenomena in Ising spin glasses. *Acta Physica Slovaca*, 44:347, 1994.
30. T. Klotz and S. Kobe. "valley structures" in the phase space of a finite 3d ising spin glass with +or-i interactions. *J. Phys. A: Math. Gen.*, 27(4):L95–L100, 1994.
31. T. Klotz, S. Schubert, and K. H. Hoffmann. Coarse graining of a spin-glass state space. *J. Phys.: Condens. Matter*, 10(27):6127–6134, 1998.
32. T. Klotz, S. Schubert, and K. H. Hoffmann. The state space of short-range Ising spin glasses: the density of states. *Eur. Phys. J. B*, 2(3):313–317, 1998.
33. S. Schubert and K. H. Hoffmann. Aging in enumerated spin glass state spaces. *Europhys. Lett.*, 66(1):118–124, 2004.
34. S. Schubert and K. H. Hoffmann. The structure of enumerated spin glass state spaces. *Comp. Phys. Comm.*, 174:191–197, 2006.
35. M. Alba, M. Ocio, and J. Hammann. Ageing process and response function in spin glasses: an analysis of the thermoremanent magnetization decay in ag:mn (2.6%). *Europhys. Lett.*, 2:45, 1986.

36. M. Alba, E. Vincent, J. Hammann, and M. Ocio. Field effect on aging and relaxation of the thermoremanent magnetization in spin glasses (low-field regime). *J. Appl. Phys.*, 61(8):4092–4094, 1987.
37. M. Alba, J. Hammann, M. Ocio, Ph. Refregier, and H. Bouchiat. Spin-glass dynamics from magnetic noise, relaxation, and susceptibility measurements. *J. Appl. Phys.*, 61(8):3683–3688, 1987.
38. N. Bontemps and R. Orbach. Evidence for differing short- and long-time decay behavior in the dynamic response of the insulating spin-glass $eu_0.4sr_0.6s$. *Phys. Rev. B*, 37(9):4708–4713, 1988.
39. C. Rossel, Y. Maeno, and I. Morgenstern. Memory effects in a superconducting y-ba-cu-o single crystal: A similarity to spin-glasses. *Phys. Rev. Lett.*, 62(6):681–684, 1989.
40. K. Biljakovic, J. C. Lasjaunias, and P. Monceau. Aging effects and nonexponential energy relaxations in charge-density-wave systems. *Phys. Rev. Lett.*, 62(13):1512–1515, 1989.
41. E. Vincent, J. Hammann, and M. Ocio. Slow dynamics in spin glasses and other complex systems. Saclay Internal Report SPEC/91-080, Centre D'Etudes de Saclay, Orme des Merisiers, 91191 Gif-sur-Yvette Cedex, France, October 1991. also in *Recent Progress in Random Magnets*, D.H. Ryan editor.
42. K. H. Fischer and J. A. Hertz. *Spin Glasses*. Cambridge University Press, 1991.
43. A. Hartwig, F. Daske, and S. Kobe. A recursive branch-and-bound algorithm for the exact ground state of ising spin-glass models. *Comp. Phys. Comm.*, 32(2):133–138, 1984.
44. N. G. van Kampen. *Stochastic Processes in Physics and Chemistry*. Elsevier, Amsterdam, 1997.
45. K. Binder and D. W. Heermann. *Monte Carlo Simulation in Statistical Physics*, volume 80 of *Springer Series in Solid-State Sciences*. Springer-Verlag, 1992.
46. N. Metropolis, A. W. Rosenbluth, M. N. Rosenbluth, A. H. Teller, and E. Teller. Equation of state calculations by fast computing machines. *J. Chem. Phys.*, 21:1087–1091, 1953.
47. B. Andresen, K.H. Hoffmann, K. Mosegaard, J. Nulton, J.M. Pedersen, and P. Salamon. On lumped models for thermodynamic properties of simulated annealing problems. *J. Phys. France*, 49:1485–1492, 1988.
48. P. Sibani. Anomalous diffusion and low-temperature spin-glass susceptibility. *Phys. Rev. B*, 35(16):8572–8578, 1987.
49. K. H. Hoffmann and P. Sibani. Relaxation and aging in spin glasses and other complex systems. *Z. Phys. B*, 80:429–438, 1990.
50. P. Sibani and K. H. Hoffmann. Relaxation in complex systems: Local minima and their exponents. *Europhys. Lett.*, 16(5):423, 1991.
51. K. H. Hoffmann, T. Meintrup, C. Uhlig, and P. Sibani. Linear-response theory for slowly relaxing systems. *Europhys. Lett.*, 22(8):565–570, 1993.
52. K. H. Hoffmann and J. C. Schön. Kinetic features of preferential trapping on energy landscapes. *Foundations of Physics Letters*, 18(2):171–182, 2005.

Random Walks on Fractals

Astrid Franz[1], Christian Schulzky[2], Do Hoang Ngoc Anh[3], Steffen Seeger[3], Janett Balg[3], and Karl Heinz Hoffmann[3]

1 Philips Research Laboratories,
Röntgenstr. 24-26,
22315 Hamburg, Germany
astrid.franz@philips.com
2 zeb/information.technology
Hammer Str. 165,
48153 Münster, Germany
cschulzky@zeb.de
3 Technische Universität Chemnitz, Institut für Physik
09107 Chemnitz, Germany
{anh.do, seeger, hoffmann}@physik.tu-chemnitz.de,
janett.balg@s2001.tu-chemnitz.de

1 Introduction

Porous materials such as aerogel, porous rocks or cements exhibit a fractal structure for a range of length scales [1]. Diffusion processes in such disordered media are widely studied in the physical literature [2, 3]. They exhibit an anomalous behavior in terms of the asymptotic time scaling of the mean square displacement of the diffusive particles,

$$\langle r^2(t) \rangle \sim t^\gamma, \tag{1}$$

where $r(t)$ is the distance of the particle from its origin after time t. In porous media the diffusion exponent γ is less than one, describing a slowed down diffusion compared to the linear time behavior known for normal diffusion. The random walk dimension is defined via equation (1) as

$$d_\mathrm{w} = 2/\gamma, \tag{2}$$

which on fractals has a value greater than 2.

This kind of diffusion can be modelled by random walks on fractal latices, as for instance the Sierpinski carpet family. Here three efficient methods are represented for calculating the random walk dimension on Sierpinski carpets: First, for finitely ramified regular fractals a resistance scaling algorithm can be used yielding a resistance scaling exponent. This exponent is related to the random walk dimension via the Einstein relation, using analogies between

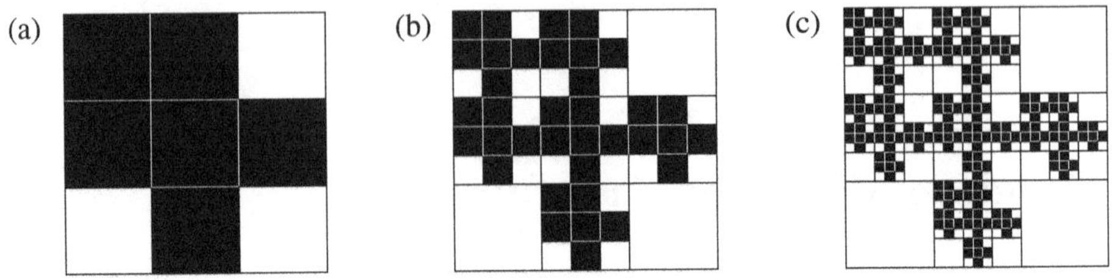

Fig. 1. Example of a Sierpinski carpet generator (a) and the result of the second (b) and third (c) construction step of the resulting carpet

random walks on graphs and resistor networks. Secondly, random walks are simulated. Thirdly, the master equation describing the time evolution of the probability distribution is iterated. The last two methods can also be applied on random fractals, which are a more realistic model for real materials.

At the end we shortly discuss differential equation approaches to anomalous diffusion. When going from the time-discrete random walks to time-continuous diffusion processes, different kinds of differential equations describing such processes have been investigated. Fractional diffusion equations are a bridging regime between irreversible normal diffusion and reversible wave propagation. For this regime a counter-intuitive behavior of the entropy production is found, and ways for solving this paradox are shown.

2 Sierpinski carpets

Sierpinski carpets are determined by a so-called generator, i.e. a square, which is divided into $n \times n$ subsquares, and m of the subsquares are black and the remaining $n^2 - m$ are white. The construction procedure for a regular Sierpinski carpet described by this generator is as follows: Starting with a square in the plane, divide it into $n \times n$ smaller congruent squares and erase all the squares corresponding to the white squares in the generator. Every one of the remaining smaller squares is again divided into $n \times n$ smaller subsquares, and again the ones marked white in the generator are erased (see Fig. 1 for an example). This procedure is continued ad infinitum resulting in a fractal called Sierpinski carpet with a fractal dimension $d_{\mathrm{f}} = \ln(m)/\ln(n)$ [4]. The result of finitely many construction steps is called a pre-carpet or pre-fractal.

Sierpinski carpets can be finitely ramified or infinitely ramified. For finitely ramified carpets, every part can be separated from the rest by cutting a finite number of connections. This property can be checked in the generator: The carpet is finitely ramified, if the first and last row of the generator coincide in exactly one black subsquare, and the same holds true for the first and last column. Sierpinski carpets, which are not finitely ramified, are called infinitely ramified.

The diffusion process on Sierpinski carpets can be modelled by random walks on pre-carpets. A random walker is at every time step on one of the black subsquares. In the next time step it can either move to one of the neighboring black squares, where subsquares are called neighbors if they coincide in one edge, or it stays on the spot. The transition probabilities depend on whether the 'blind ant' or 'myopic ant' algorithm is used [2]. The blind ant can choose each direction with the same probability. If the chosen direction is 'forbidden', i.e. there is no black square in this direction, it stays on its position. In contrast the myopic ant chooses the direction with equal probability from the permitted ones for each time step.

Equivalently to the description by squares, we can assign a graph to the pre-carpet by placing the vertices at the midpoints of the black squares and connecting vertices corresponding to neighboring squares by an edge, and perform random walks on this graph.

For random walks on such graph structures, a variety of methods for determining the random walk dimension d_{w} will be presented.

3 Resistance scaling

Random walks on graph structures and the current flow through an adequate resistor network are strongly connected [5]. Assigning a unit resistance to every edge of the graph representation of a Sierpinski pre-carpet we get the corresponding resistor network. For finitely ramified Sierpinski carpets the Einstein relation [3, 6–8]

$$d_{\mathrm{w}} = d_{\mathrm{f}} + \zeta \tag{3}$$

holds, where ζ is the scaling exponent of the resistance R with the linear length L of the network: $R \sim L^{\zeta}$.

Since the fractal dimension d_{f} of the Sierpinski carpet is known (see Sect. 1) it remains to determine the resistance scaling exponent ζ in order to get the random walk dimension d_{w} via equation (3). To achieve this, we developed an algorithm, which

1. converts the resistor network corresponding to the Sierpinski carpet generator into a triangular or rhomboid network,
2. replaces the nodes of the generator network with these triangles or rhombi,
3. converts the resulting network again into a triangle or rhombus,
4. repeats steps 2 and 3, until convergence is reached, i.e. successive networks differ by a constant scaling factor.

Whether the resistor network will be converted into a triangular or rhomboid network depends on how many contact points a resistor network has. That are squares in the carpet, where one generator network may be connected to a neighboring one. Every resistor network with three or four contact points can be converted into a triangular or rhomboid network by the use of Kirchhoff's laws.

Further details of this algorithm and its implementation using computer algebra methods can be found in [9, 10]. This algorithm yields the resistance scaling exponent and hence the random walk dimension for finitely ramified Sierpinski carpets with arbitrary accuracy. Therefore it is a powerful tool for investigating random walk properties on finitely ramified fractals.

Similar methods can give other scaling exponents for fractals as the chemical dimension, which describes the scaling of the shortest path between two points with the linear distance [11]. The pore structure of finitely and even infinitely ramified Sierpinski carpets can be described by a hole-counting polynomial, from which scaling exponents for the distribution of holes with different areas and perimeters can be determined [12]. Furthermore, we can compute the fractal dimension of the external boundary and the boundaries of internal cavities [13]. Such exponents are important parameters to characterize the fractal properties and may have a decisive influence on the anomalous diffusion behavior.

4 Effective simulation of random walks

The direct way of studying the time behavior of the mean square displacement of random walkers on pre-fractals is the simulation of the random walks itself. This method is not restricted to finitely ramified fractals. The asymptotic scaling behavior is reached when the log-log-plot of the mean square displacement over time reaches a straight line. According to equation (1) the slope of the curve is equivalent to γ and can be approximated by linear regression. Normally long times and hence large pre-fractals have to be considered in order to reach this linear behavior.

For finitely ramified Sierpinski carpets we developed an efficient storing scheme, which only takes the generator as input and represents the actual walker position by a hierarchical coordinate notation (see [14] for a detailed description of this scheme). In this way we are able to perform long random walks on effectively unbounded pre-fractals. In order to get good statistics, the number of walkers has to be sufficiently large, too. Since the memory requirements for storing the Sierpinski pre-carpet are negligible, the simplest way for parallelizing the random walk algorithm is to start random walkers on each available CPU separately and collect the results after a given number of time steps.

Real materials often exhibit a fractal structure for a certain range of length scales only, they appear uniform at larger scales. In order to get more realistic models for porous materials, we investigated repeated Sierpinski carpets, i.e. we applied the construction procedure for Sierpinski carpets over a few iterations (the number of applied iterations is referred to as stage of the carpet) and repeated these pre-carpets periodically in order to construct a homogeneous structure at large length scales [15]. In the log-log-plot in Fig. 2 a cross-over can be observed from the anomalous diffusion regime at small length scales

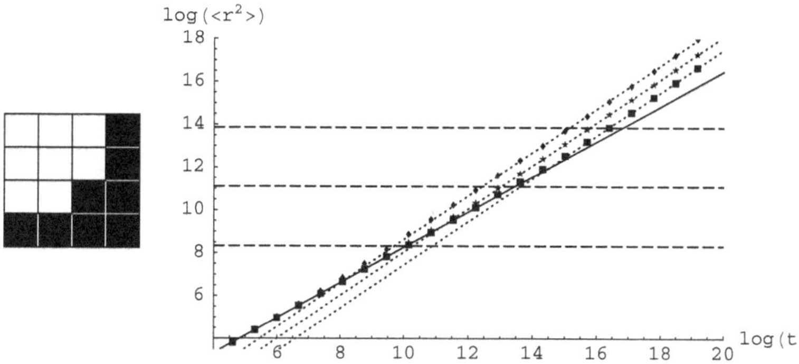

Fig. 2. Mean square displacement for the myopic ant random walk on repeated Sierpinski carpet resulting from the show generator for stages 3 (diamonds), 4 (stars) and 5 (squares), together with linear fits (solid and dotted lines) and the theoretical crossover values (dashed lines)

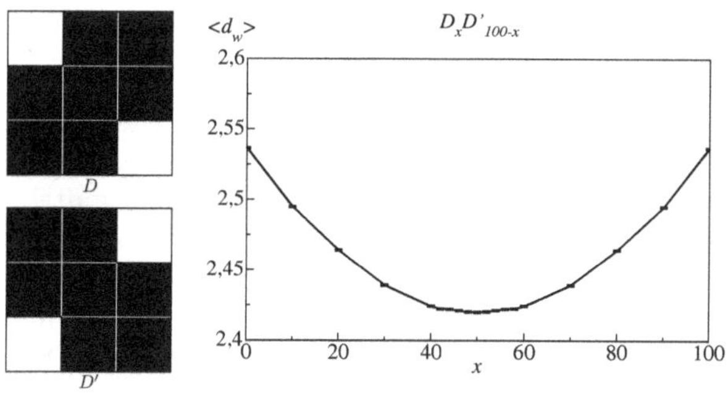

Fig. 3. $\langle d_{\mathrm{w}} \rangle$ vs. percentage x for the mixture $D_x D'_{100-x}$ with error bars indicating the standard deviation $\sigma_{\langle d_{\mathrm{w}} \rangle}$ of the average d_{w}. Although D and D' have the same d_{f} and d_{w} we can observe a decrease of $\langle d_{\mathrm{w}} \rangle$ with increasing disorder

to the normal diffusion regime at large length scales. Furthermore, we investigated the diffusion constant D, which is an additional quantity to characterize the speed of diffusive particles. One important result of this research was, that carpets with same d_{w} and same d_{f} might have quite different values of D, depending on the repeated carpet stage of the considered iterator and also on the diffusion constant of the starting regime [15].

A step further in the direction of modelling real disordered materials is shown in [16]. There we constructed random fractals by combining several (not necessarily finitely ramified) Sierpinski carpet generators. Our simulations show that random walkers on such mixed structures are also characterized by a specific random walk dimension $\langle d_{\mathrm{w}} \rangle$. We noticed that even if the different generators have the same d_{w} and d_{f} there are variations in their observed effective $\langle d_{\mathrm{w}} \rangle$. We found that increasing disorder leads to a slower diffusion in several cases. But on the other hand for some carpet configura-

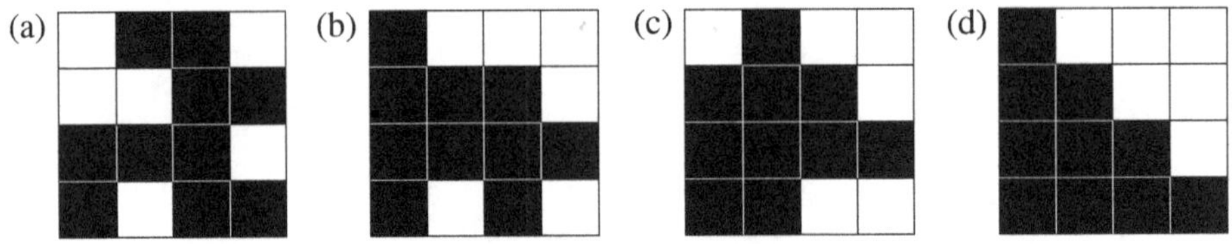

Fig. 4. Four examples of Sierpinski carpet generators

tions a decrease of the random walk dimension (corresponding to an enhanced diffusion speed) can be observed with increasing disorder (see Fig. 3), contrary to the expected behavior for regular structures like crystals.

5 Master equation approach

The random walk investigated in the previous section can be characterized by a master equation

$$P(t+1) = \Gamma \cdot P(t), \tag{4}$$

where $P(t)$ denotes the probability distribution that a walker is at a certain position at time $t = 0, 1, 2, \ldots$ and Γ is the transition matrix. Compared to performing random walks, iterating (4) has the advantage that a statistical average over a large number of walkers is not necessary any more. But of course now memory is needed for every position in the pre-fractal covered by a non-vanishing probability. In the case of finitely ramified Sierpinski carpets we used a dynamic storing scheme in order to keep the memory requirements as small as possible [17].

For more general fractals we investigated possibilities for a parallelization on a compute cluster to circumvent memory restrictions. The communication requirements increase with ongoing time, hence the connectivity between parts of the pre-fractal has to be carefully associated with the communication between the compute nodes in order to ensure efficiency.

The master equation approach yields the whole probability distribution for every time step, which contains much more information than just the scaling behavior of the mean square displacement over time. Thus the results of the master equation algorithm may be a starting point for the investigation of the scaling properties of the probability distribution itself.

We applied all three methods (resistance scaling, random walk, master equation) to get estimates for the random walk dimension of the four 4×4 carpets investigated in [18] (see Fig. 4). For random walks and the master equation approach we calculated 10,000 time steps, in addition we averaged over 10,000,000 walkers. The resulting data for the carpet pattern given in Fig. 4(a) are shown in Fig. 5.

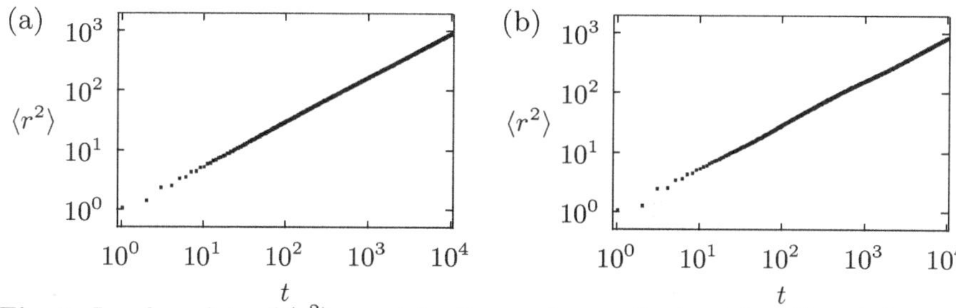

Fig. 5. Log-log-plots of $\langle r^2 \rangle$ over t for the random walk algorithm (i) and the master equation iteration (ii) for carpet pattern shown in Fig. 4(a)

Table 1 shows the resulting d_w from the regression together with confidence intervals of 95%, the theoretical values resulting from the resistance scaling method and in the last column the values of [18].

Table 1. Estimates for d_w for the four carpet patterns of Fig. 4 resulting from our three methods and from [18]

	Random walk	Master equation	Resistance	Dasgupta et. al.
a	2.68 ± 0.02	2.71 ± 0.05	2.66	2.538 ± 0.002
b	2.52 ± 0.02	2.52 ± 0.06	2.58	2.528 ± 0.002
c	2.47 ± 0.02	2.49 ± 0.05	2.49	2.524 ± 0.002
d	2.47 ± 0.02	2.50 ± 0.09	2.51	2.514 ± 0.002

The data from the resistance scaling method can be taken as reference values as they have been computed to very high accuracy. As can be seen from Table 1, the values for d_w can be really different although all four example carpets have the same fractal dimension. We remark that the small oscillations showing up in the master equation data (Fig. 5) can be reduced by an additional average over starting points, which is included in the random walk data.

6 Corresponding differential equations

Going from the time-discrete processes considered in the previous sections to a time-continuous description gives us the transition from random walks to diffusion processes. Many suggestions [19–21] have been advanced to generalize the well-known Euclidean diffusion equation

$$\frac{\partial}{\partial t} P(r,t) = \frac{\partial^2}{\partial r^2} P(r,t) \qquad (5)$$

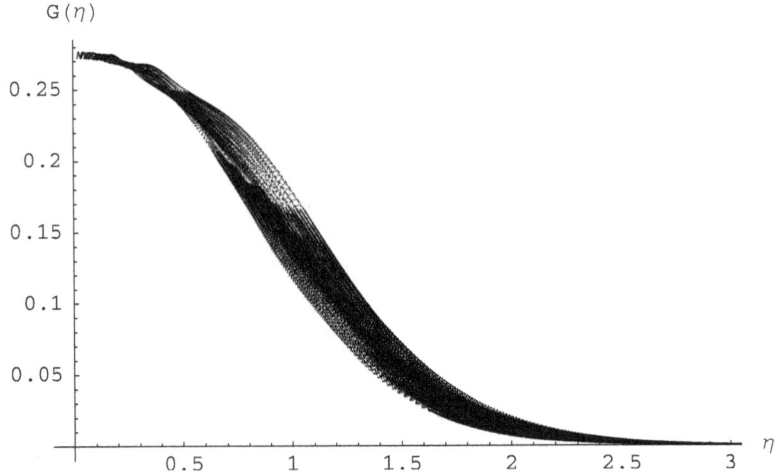

Fig. 6. The cloud $G(\eta)$ for the diffusion on a Sierpinski gasket. It is generated by taking the data $P(r,t)$ for many times t and transforming them using equation (6)

to anomalous diffusion. All these approaches are only partially successful, either they give a good approximation for small or for large r. The difficulty rests with the fact that fractals, by definition, are 'spiky' or 'rough'. This leads to the problem that they do not lend themselves to description by differential or integral equations.

A more promising starting point for describing diffusion processes on fractals is the use of the natural similarity group [22, 23]

$$P(r,t) = t^{-d_\mathrm{w}/d_\mathrm{f}} G(\eta), \tag{6}$$

where $\eta = rt^{-1/d_\mathrm{w}}$ is the similarity variable and the function $G(\eta)$ is the natural invariant representation of the probability density function $P(r,t)$. By using (6) we investigated structures in $G(\eta)$ on the Sierpinski gasket, which we called 'clouds', 'fibres' and 'echoes' [24]. The clouds accrue as a result of plotting the $G(\eta)$ density function (GDF) against η seen in Fig. 6 and they exhibit the multivalued and even self-similar character of these functions.

This structure appears to be the assembly of a number of curves, which we named fibres. All of them can be explained in terms of sets of 'echo points', where every fibre belongs to a certain 'echo class' (see Fig. 7). Using these echo classes we are able, at least in principle, to produce smooth GDF [24]. In case of the Koch curve we derived an ordinary differential equation describing random walks on it as a representative of one of the simplest fractal structures [25].

Furthermore, the time fractional diffusion equation

$$\frac{\partial^\gamma}{\partial t^\gamma} P(r,t) = \frac{\partial^2}{\partial r^2} P(r,t) \tag{7}$$

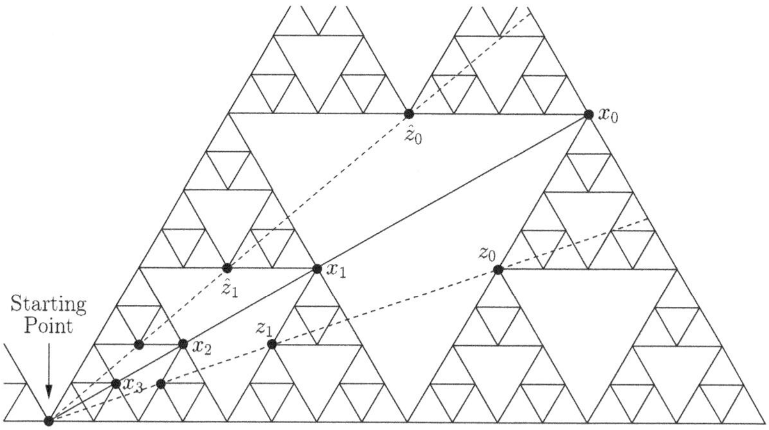

Fig. 7. A few members of three different classes of echo points $x_k, z_k, \hat{z}_k$ are represented on a gasket schematic. The x_k are situated along a symmetry line, while the points z_k and $\hat{z}_k$ are reflections of each other about that line

can be considered, which for $\gamma = 1$ coincides with the normal diffusion equation (5) describing an irreversible process and for $\gamma = 2$ corresponds to the reversible wave propagation. Hence the regime $1 < \gamma < 2$ forms a bridge between irreversible and reversible processes, which should be visible in the entropy production. We found a counter-intuitive increase of the entropy production as γ rises towards 2 [22]. This holds true not only for the Shannon entropy, but also for the Tsallis and Rényi entropies [26]. Other bridging schemes between the normal diffusion equation and the wave equation can be described by the space fractional diffusion equation or the telegraphers equation [27] and show, at least partially, the same counter-intuitive behavior. Looking at the microscopic random processes behind all these equations, the entropy production paradox can be partially explained by the fact, that the walkers in the diffusion equation move with an infinite speed (Brownian motion), while the walkers in the wave equation move with a finite speed. Hence the entropy time derivative is not a suited measure for comparing different γ regimes, but the basis for comparison should be the first moment [22].

7 Conclusions

Fractals are used as a simple model for porous media in order to describe diffusive processes. In contrast to uniform media, the mean square displacement of diffusive particles, modelled by random walkers, does not scale linearly with time t, but with t^{2/d_w}, where the random walk dimension d_w is usually greater than 2. For a better understanding of anomalous diffusion encountered in a number of experimental contexts, different methods for investigating random walks on fractals were studied. We determined the scaling exponent of the mean square displacement of the walkers with time as an important quantity to characterize diffusion properties. Furthermore we developed techniques to calculate further properties of diffusion on fractals, i.e. the resistance scaling exponent, chemical dimension or the pore structure.

Some of the methods additionally yield the whole probability distribution for every time step, containing much more information than just the scaling behavior of the mean square displacement. Going to a time-continuous description, the time evolution of the probability distribution can be described by differential equations. In case of fractional differential equations, they are a bridging scheme between the irreversible normal diffusion equation and the reversible wave equation. However, the entropy production is no adapted quantity characterizing this regime.

Acknowledgment

We want to thank C. Essex, M. Davison, X. Li, and S. Tarafdar for helpful discussions during our extended collaborations.

References

1. B. B. Mandelbrot. *Fractals - Form, Chance and Dimension.* W. H. Freeman, San Francisco, 1977.
2. S. Havlin and D. Ben-Avraham. Diffusion in disordered media. *Adv. Phys.*, 36(6):695–798, 1987.
3. A. Bunde and S. Havlin, editors. *Fractals and Disordered Systems.* Springer, Berlin, Heidelberg, New-York, 2nd edition edition, 1996.
4. K. J. Falconer. *Techniques in fractal geometry.* John Wiley & Sons Ltd, Chichester, 1997.
5. P. Tetali. Random walks and the effective resistance of networks. *J. Theor. Prob.*, 4(1):101–109, 1991.
6. J. A. Given and B. B. Mandelbrot. Diffusion on fractal lattices and the fractal Einstein relation. *J. Phys. B: At. Mol. Opt. Phys.*, 16:L565–L569, 1983.
7. M. T. Barlow, R. F. Bass, and J. D. Sherwood. Resistance and spectral dimension of Sierpinski carpets. *J. Phys. A: Math. Gen.*, 23(6):L253–L238, 1990.
8. A. Franz, C. Schulzky, and K. H. Hoffmann. The Einstein relation for finitely ramified Sierpinski carpets. *Nonlinearity*, 14(5):1411–1418, 2001.
9. C. Schulzky, A. Franz, and K. H. Hoffmann. Resistance scaling and random walk dimensions for finitely ramified Sierpinski carpets. *SIGSAM Bulletin*, 34(3):1–8, 2000.
10. A. Franz, C. Schulzky, S. Seeger, and K. H. Hoffmann. Diffusion on fractals – efficient algorithms to compute the random walk dimension. In J. M. Blackledge, A. K. Evans, and M. J. Turner, editors, *Fractal Geometry: Mathematical Methods, Algorithms, Applications*, IMA Conference Proceedings, pages 52–67. Horwood Publishing Ltd., Chichester, West Sussex, 2002.

11. A. Franz, C. Schulzky, and K. H. Hoffmann. Using computer algebra methods to determine the chemical dimension of finitely ramified Sierpinski carpets. *SIGSAM Bulletin*, 36(2):18–30, 2002.
12. A. Franz, C. Schulzky, S. Tarafdar, and K. H. Hoffmann. The pore structure of Sierpinski carpets. *J. Phys. A: Math. Gen.*, 34(42):8751–8765, 2001.
13. P. Blaudeck, S. Seeger, C. Schulzky, K. H. Hoffmann, T. Dutta, and S. Tarafdar. The coastline and lake shores of a fractal island. *J. Phys. A: Math. Gen.*, 39:1609–1618, 2006.
14. S. Seeger, A. Franz, C. Schulzky, and K. H. Hoffmann. Random walks on finitely ramified Sierpinski carpets. *Comp. Phys. Comm.*, 134(3):307–316, 2001.
15. S. Tarafdar, A. Franz, S. Schulzky, and K. H. Hoffmann. Modelling porous structures by repeated Sierpinski carpets. *Physica A*, 292(1-4):1–8, 2001.
16. D. H. N. Anh, K. H. Hoffmann, S. Seeger, and S. Tarafdar. Diffusion in disordered fractals. *Europhys. Lett.*, 70(1):109–115, 2005.
17. A. Franz, C. Schulzky, S. Seeger, and K. H. Hoffmann. An efficient implementation of the exact enumeration method for random walks on Sierpinski carpets. *Fractals*, 8(2):155–161, 2000.
18. R. Dasgupta, T. K. Ballabh, and S. Tarafdar. Scaling exponents for random walks on Sierpinski carpets and number of distinct sites visited: A new algorithm for infinite fractal lattices. *J. Phys. A: Math. Gen.*, 32(37):6503–6516, 1999.
19. B. O'Shaughnessy and I. Procaccia. Analytical solutions for diffusion on fractal objects. *Phys. Rev. Lett.*, 54(5):455–458, 1985.
20. M. Giona and H. E. Roman. Fractional diffusion equation for transport phenomena in random media. *Physica A*, 185:87–97, 1992.
21. R. Metzler, W. G. Glockle, and T. F. Nonnemacher. Fractional model equation for anomalous diffusion. *Physica A*, 211(1):13–24, 1994.
22. K. H. Hoffmann, C. Essex, and C. Schulzky. Fractional diffusion and entropy production. *J. Non-Equilib. Thermodyn.*, 23(2):166–175, 1998.
23. C. Schulzky, C. Essex, M. Davison, A. Franz, and K. H. Hoffmann. The similarity group and anomalous diffusion equations. *J. Phys. A: Math. Gen.*, 33(31):5501–5511, 2000.
24. M. Davison, C. Essex, C. Schulzky, A. Franz, and K. H. Hoffmann. Clouds, fibres and echoes: a new approach to studying random walks on fractals. *J. Phys. A: Math. Gen.*, 34(20):L289–L296, 2001.
25. C. Essex, M. Davison, C. Schulzky, A. Franz, and K. H. Hoffmann. The differential equation describing random walks on the Koch curve. *J. Phys. A: Math. Gen.*, 34(41):8397–8406, 2001.
26. C. Essex, C. Schulzky, A. Franz, and K. H. Hoffmann. Tsallis and Rényi entropies in fractional diffusion and entropy production. *Physica A*, 284(1-4):299–308, 2000.
27. X. Li, C. Essex, M. Davison, K. H. Hoffmann, and C. Schulzky. Fractional diffusion, irreversibility and entropy. *J. Non-Equilib. Thermodyn.*, 28(3):279–291, 2003.

Lyapunov Instabilities of Extended Systems

Hong-liu Yang and Günter Radons

Technische Universität Chemnitz, Institut für Physik
09107 Chemnitz, Germany
hongliu.yang@physik.tu-chemnitz.de
radons@physik.tu-chemnitz.de

1 Introduction

One of the most successful theories in modern science is statistical mechanics, which allows us to understand the macroscopic (thermodynamic) properties of matter from a statistical analysis of the microscopic (mechanical) behavior of the constituent particles. In spite of this, using certain probabilistic assumptions such as Boltzmann's *Stosszahlansatz* causes the lack of a firm foundation of this theory, especially for non-equilibrium statistical mechanics. Fortunately, the concept of chaotic dynamics developed in the 20th century [1] is a good candidate for accounting for these difficulties. Instead of the probabilistic assumptions, the dynamical instability of trajectories can make available the necessary fast loss of time correlations, ergodicity, mixing and other dynamical randomness [2]. It is generally expected that dynamical instability is at the basis of macroscopic transport phenomena and that one can find certain connections between them. Some beautiful theories in this direction were already developed in the past decade. Examples are the escape-rate formalism by Gaspard and Nicolis [3, 4] and the Gaussian thermostat method by Nosé, Hoover, Evans, Morriss and others [5, 6], where the Lyapunov exponents were related to certain transport coefficients.

Very recently, molecular dynamics simulations on hard-core systems revealed the existence of regular collective perturbations corresponding to the smallest positive Lyapunov exponents (LEs), named hydrodynamic Lyapunov modes [7]. This provides a new possibility for the connection between Lyapunov vectors, a quantity characterizing the dynamical instability of trajectories, and macroscopic transport properties. A lot of work [8–14] has been done to identify this phenomenon and to find out its origin. The appearance of these modes is commonly thought to be due to the conservation of certain quantities in the systems studied [8–12]. A natural consequence of this expectation is that the appearance of such modes might not be an exclusive feature of hard-core systems and might be generic to a large class of Hamiltonian systems. However, until very recently, these modes have only been identified in the computer simulations of hard-core systems [8, 14].

Here we review our current results on Lyapunov spectra and Lyapunov vectors (LVs) of various extended systems with continuous symmetries, especially on the identification and characterization of hydrodynamic Lyapunov modes. The major part of our discussion is devoted to the study of Lennard-Jones fluids in one- and two-dimensional spaces, wherein the HLMs are, for the first time, identified in systems with soft-potential interactions [15, 16]. By using the newly introduced LV correlation functions, we demonstrate that the LVs with $\lambda \approx 0$ are highly dominated by a few components with low wave numbers, which implies the existence of hydrodynamic Lyapunov modes (HLMs) in soft-potential systems. Despite the wave-like character of the LVs, no step-like structure exists in the Lyapunov spectrum of the systems studied here, in contrast to the hard-core case. Further numerical simulations show that the finite-time Lyapunov exponents fluctuate strongly. Studies on dynamical LV structure factors conclude that HLMs in Lennard-Jones fluids are propagating. We also briefly outline our current results on the universal features of HLMs in a class of spatially extended systems with continuous symmetries. HLMs in Hamiltonian and dissipative systems are found to differ both in respect of spatial structure and in the dynamical evolution. Details of these investigations can be found in our publications [17–19].

2 Numerical method for determining Lyapunov exponents and vectors

2.1 Standard method

The equations of motion for a many-body system may always be written as a set of first order differential equations $\dot{\Gamma}(t) = F(\Gamma(t))$, where Γ is a vector in the D-dimensional phase space. The tangent space dynamics describing infinitesimal perturbations around a reference trajectory $\Gamma(t)$ is given by

$$\delta\dot{\Gamma} = M(\Gamma(t)) \cdot \delta\Gamma \tag{1}$$

with the Jacobian $M = \frac{\mathrm{d}F}{\mathrm{d}\Gamma}$. The time averaged expansion or contraction rates of $\delta\Gamma(t)$ are given by the Lyapunov exponents [1]. For a $D-$dimensional dynamical system there exist in total D Lyapunov exponents for D different directions in tangent space. The orientation vectors of these directions are the Lyapunov vectors $e^{(\alpha)}(t)$, $\alpha = 1,\cdots,D$.

For the calculation of the Lyapunov exponents and vectors the offset vectors have to be reorthogonalized periodically, either by means of Gram-Schmidt orthogonalization or QR decomposition [20, 21]. To obtain scientifically useful results, one needs large particle numbers and long integration times for the calculation of certain long time averages. This enforces the use of parallel implementations of the corresponding algorithms. It turns out that the repeated reorthogonalization is the most time consuming part of the algorithm.

2.2 Parallel realization

As parallel reorthogonalization procedures we have realized and tested several parallel versions of Gram-Schmidt orthogonalization and of QR factorization based on blockwise Householder reflection. The parallel version of classical Gram-Schmidt (CGS) orthogonalization is enriched by a reorthogonalization test which avoids a loss of orthogonality by dynamically using iterated CGS. All parallel procedures are based on a 2-dimensional logical processor grid and a corresponding block-cyclic data distribution of the matrix of offset vectors. Row-cyclic and column-cyclic distributions are included due to parameterized block sizes, which can be chosen appropriately. Special care was also taken to offer a modular structure and the possibility for including efficient sequential basic operations, such as that from BLAS [22], in order to efficiently exploit the processor or node architecture. For comparison we consider the standard library routine for QR factorization from ScaLAPACK [23].

Performance tests of parallel algorithms have been done on a Beowulf cluster, a cluster of dual Xeon nodes, and an IBM Regatta p690+. Results can be found in [24]. It is shown that by exploiting the characteristics of processors or nodes and of the interconnections network of the parallel hardware, an efficient combination of basic routines and parallel orthogonalization algorithms can be chosen, so that the computation of Lyapunov spectra and Lyapunov vectors can be performed in the most efficient way.

3 Correlation functions for Lyapunov vectors

In previous studies, certain smoothing procedures in time or space were applied to the Lyapunov vectors, in order to make the wave structure more obvious. For a 1d hard-core system with only a few particles, these procedures have been shown to be quite useful in identifying the existence of hydrodynamic Lyapunov modes [13]. For soft-potential systems, the smoothing procedures are no longer helpful in detecting the hidden regular modes and can even damage them [14]. Here we will introduce a new technique based on a spectral analysis of LVs, which enables us to identify unambiguously the otherwise hardly identified HLMs and to characterize them quantitatively.

In the spirit of molecular hydrodynamics [25], we introduced in [15, 16] a dynamical variable called *LV fluctuation density*,

$$u^{(\alpha)}(r,t) = \sum_{j=1}^{N} \delta x_j^{(\alpha)}(t) \cdot \delta(r - r_j(t)), \tag{2}$$

where $\delta(z)$ is Dirac's delta function, $r_j(t)$ is the position coordinate of the j-th particle, and $\{\delta x_j^{(\alpha)}(t)\}$ is the coordinate part of the α-th Lyapunov vector at time t. The spatial structure of LVs is characterized by the *static LV structure factor* defined as

$$S_u^{(\alpha\alpha)}(k) = \int \langle u^{(\alpha)}(r,0)u^{(\alpha)}(0,0)\rangle e^{-ik\cdot r}\mathrm{d}r, \tag{3}$$

which is simply the spatial power spectrum of the LV fluctuation density. Information on the dynamics of LVs can be extracted via the *dynamic LV structure factor*, which is defined as

$$S_u^{(\alpha\alpha)}(k,\omega) = \int\int \langle u^{(\alpha)}(r,t)u^{(\alpha)}(0,0)\rangle e^{-ik\cdot r}e^{i\omega t}\mathrm{d}r\mathrm{d}t. \tag{4}$$

With the help of these quantities the controversy [7,8] about the existence of hydrodynamic Lyapunov modes in soft-potential systems has been successfully resolved [15].

4 Numerical results for 1d Lennard-Jones fluids

4.1 Models

The Lennard-Jones system studied has the Hamiltonian

$$H = \sum_{j=1}^{N} m v_j^2/2 + \sum_{j<l} V(x_l - x_j). \tag{5}$$

where the interaction potential among particles $V(r) = 4\epsilon\left[(\frac{\sigma}{r})^{12} - (\frac{\sigma}{r})^6\right] - V_c$ if $r \le r_c$ and $V(r) = 0$ otherwise with $V_c = 4\epsilon\left[(\frac{\sigma}{r_c})^{12} - (\frac{\sigma}{r_c})^6\right]$. Here the potential is truncated in order to lower the computational burden.

The system is integrated using the velocity form of the Verlet algorithm with periodic boundary conditions [26]. In our simulations, we set $m = 1$, $\sigma = 1$, $\epsilon = 1$ and $r_c = 2.5$. All results are given in reduced units, i.e., length in units of σ, energy in units of ϵ and time in units of $(m\sigma^2/48\epsilon)^{1/2}$. The time step used in the molecular dynamics simulation is $h = 0.008$. The standard method invented by Benettin et al. and Shimada and Nagashima [20,21] is used to calculate the Lyapunov characteristics of the systems studied. The time interval for periodic re-orthonormalization is $30h$ to $100h$. Throughout this paper, the particle number is typically denoted by N, the length of the system by L and the temperature by T.

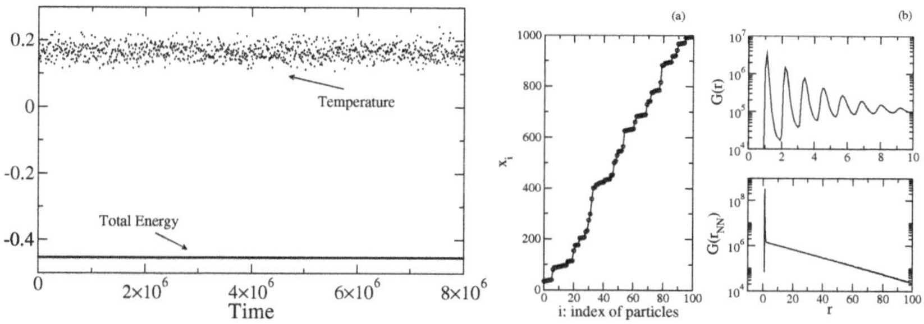

Fig. 1. Left: Time evolution of temperature $T \equiv \langle mv^2 \rangle$ and total energy. Right: a) Snapshot of the particle positions x_i vs. index of particles i and b) pair distribution function $G(r)$ obtained from the distances between all particles (upper panel) and from nearest neighbors only (lower panel) for the stationary state shown in the left part (see [25] for the definition of $G(r)$). The sharp peaks in $G(r)$ imply that the state is a broken-chain state [27]. The parameter setting used here is: $N = 100$, $L = 1000$ and $T = 0.2$

4.2 The stationary state

The time evolution of state variables like temperature T and total energy for a case with parameter setting $N = 100$, $L = 1000$ and $T = 0.2$ is shown in Fig. 1. At the beginning of our molecular dynamics simulation, the particles are placed randomly in the interval $[0, L]$. Their velocities are chosen randomly from a Boltzmann distribution. In order to equilibrate the system, it is coupled to a stochastic heat bath with the given temperature T. In Fig. 1, the stage with thermal bath is omitted and only the part of the evolution with constant total energy is shown. The almost constant value of the temperature means that the system has already reached a stationary state and one can start the calculation of the Lyapunov instability of the system.

The pair distribution function $G(r)$ shown in Fig. 1 tells us that the stationary state for $T = 0.2$ is a broken-chain state with short range order. This is generic for 1d Lennard-Jones systems with not too high density [27].

4.3 Smooth Lyapunov spectrum with strong short-time fluctuations

The Lyapunov spectrum for the case $N = 100$, $L = 1000$ and $T = 0.2$ is shown in Fig. 2. Only half of the spectrum is shown here, since all LEs of Hamiltonian systems come in pairs according to the conjugate-pairing rule. In the enlargement shown in the inset of Fig. 2 for the part near $\lambda^{(\alpha)} \approx 0$, one can not see any step-wise structure in the Lyapunov spectrum, in contrast to the case of hard-core systems [8]. This is the typical result obtained for our soft potential system.

The fluctuations in local instabilities of trajectories are demonstrated by means of the distribution of finite-time LEs. By definition, finite-time

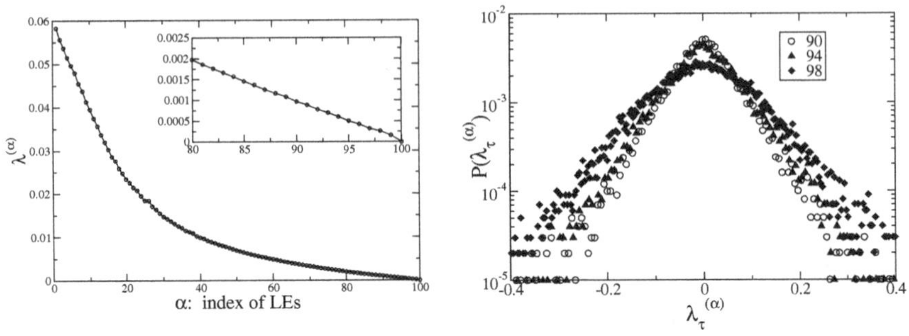

Fig. 2. Left: Lyapunov spectrum of the state shown in Fig. 1. The enlargement of the part in the regime $\lambda^{(\alpha)} \approx 0$ shows that no step-wise structure exists here in contrast to the case of hard-core systems. This is the typical result for our soft potential systems. Right: Distribution of the finite-time Lyapunov exponent $\lambda_\tau^{(\alpha)}$ where τ is equal to the period of re-orthonormalization. The strong fluctuations of $\lambda_\tau^{(\alpha)}$ are one of the possible reasons for the disappearance of the step-wise structures in the Lyapunov spectrum

Lyapunov exponents λ_τ measure the expansion rate of trajectory segments of the duration τ. In Fig. 2, such distributions are presented for some LEs in the regime $\lambda \approx 0$. Fluctuations of the finite time Lyapunov exponents are quite large compared to the difference between their mean values, i.e., $\sigma(\lambda_\tau^{(\alpha)}) \equiv \sqrt{\langle {\lambda_\tau^{(\alpha)}}^2 \rangle - \langle \lambda_\tau^{(\alpha)} \rangle^2} \gg |\lambda^{(\alpha)} - \lambda^{(\alpha+1)}|$. Here, $\langle \cdots \rangle$ means time average. The strong fluctuations in local instabilities constitute one of the possible reasons for the disappearance of the step-wise structures in the Lyapunov spectra. They could also cause the mixing of nearby Lyapunov vectors. The mixing may be at the basis of the intermittency observed in the time evolution of the spatial Fourier transformation of LVs (see Sect. 4.4).

4.4 Spatial structure of LVs with $\lambda^{(\alpha)} \approx 0$

LV fluctuation density

Another quantity used to characterize the local instability of trajectories are Lyapunov vectors $\delta\Gamma^{(\alpha)}$, which represent expanding or contracting directions in tangent space. In the study of hard-core systems, Posch et al. found that the coordinate part of the Lyapunov vectors corresponding to $\lambda \approx 0$ are of regular wave-like character [7, 8]. They are referred to as *hydrodynamic Lyapunov modes.* Here, we are searching for the counterpart of these modes in our soft-potential system.

Remember that each of the LVs consists of two parts: the displacement δx_i in coordinate space and δv_i in momentum space. In past studies of hydrodynamic Lyapunov modes in hard-core systems, only the coordinate part δx_i was considered. This is due to an interesting feature of hydrodynamic Lyapunov

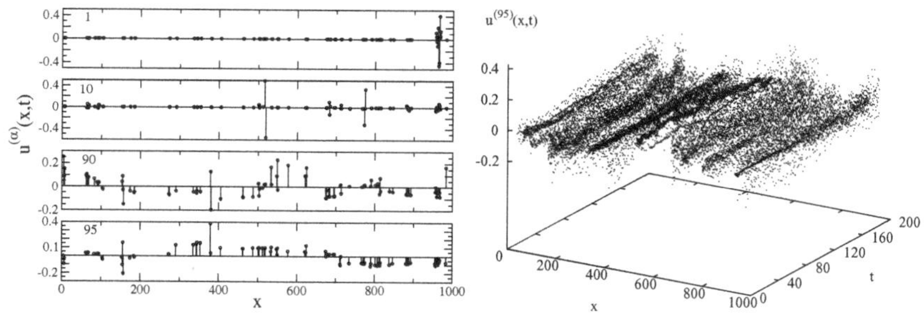

Fig. 3. Left: $u^{(\alpha)}(x,t)$ for LVs with index $\alpha = 1$, 10, 90, and 95 respectively. Note that the LVs with $\alpha = 1$ and 10 are more localized while those with $\alpha = 90$ and 95 are more distributed. Right: Time evolution of $u^{(95)}(x,t)$ for the same parameters as in Fig. 3. No clear wave structure can be detected

modes found in [10], which states that the angles between the coordinate part and the momentum part are always small, i.e, the two vectors are nearly parallel. Therefore, it is sufficient to use only δx_i for the study of $\delta \Gamma$. For our soft potential systems, we find that the angles between the coordinate part and the momentum part are no longer as small as in the hard-core systems. However, we will still follow the tradition and study the coordinate part of LV first, before coming to the momentum part.

For the one-dimensional Lennard-Jones fluids treated here, the LV fluctuation density defined in (2) is reduced to

$$u^{(\alpha)}(x,t) = \sum_{j=1}^{N} \delta x_j^{(\alpha)}(t) \cdot \delta(x - x_j(t)) \tag{6}$$

where $\delta x_j^{(\alpha)}(t)$ constitutes the coordinate component of the Lyapunov vector with index α, and $x_j(t)$ represents the instantaneous position of the j-th particle.

The profiles of $u^{(\alpha)}(x,t)$ for some typical LVs of the 1d Lennard-Jones system are presented in Fig. 3. It can be seen that $u^{(\alpha)}(x,t)$ for LVs corresponding to the largest Lyapunov exponents are highly localized, for example $u^{(1)}(x,t)$ and $u^{(10)}(x,t)$, while those for LV_{90} and LV_{95} are more distributed. The temporal evolution of $u^{(95)}(x,t)$ is also shown in Fig. 3, in order to make the possibly existing wave-like structure more evident. A wave structure, however, cannot unambiguously be detected here with the naked eye.

Intermittency in time evolution of instantaneous static LV structure factors

Based on the spatial Fourier transformation of $u^{(\alpha)}(x,t)$

$$\tilde{u}_k^{(\alpha)}(t) = \int u^{(\alpha)}(x,t) e^{-ikx} \mathrm{d}x = \sum_{j=1}^{N} \delta x_j^{(\alpha)} \cdot e^{-ik \cdot x_j(t)} \tag{7}$$

we introduce a quantity called *instantaneous static LV structure factor*, which reads

$$s_{uu}^{(\alpha)}(k,t) \equiv |\tilde{u}_k^{(\alpha)}(t)|^2. \tag{8}$$

It is nothing but the instantaneous spatial power spectrum of $u^{(\alpha)}(x,t)$. The quantity will be used to characterize the dynamical evolution of Lyapunov vectors. The long time average (and ensemble average) of $s_{uu}^{(\alpha)}(k,t)$ recovers the static LV structure defined in (3). We expect that in $S_{uu}^{(\alpha)}(k) \equiv \langle s_{uu}^{(\alpha)}(k,t)\rangle$ the contribution of stochastic fluctuations will be averaged out while the information on the collective modes will remain and accumulate. The following results show that this technique is quite successful in detecting the vague collective modes.

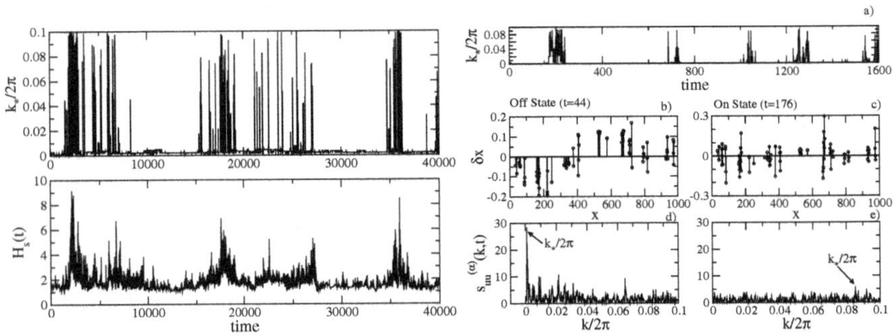

Fig. 4. Left: Intermittent behaviors of the peak wave-number k_* and spectral entropy $H_s(t)$ for the spatial Fourier spectrum of $u^{(95)}(x,t)$. Right: a) Variation of the peak wave number k_* with time. b),c) Two typical snapshots of LV_{95}, *off* and *on* state at $t = 44$ and 176 respectively. d),e) their spatial Fourier transform. The spectrum for the off state has a sharp peak at small k_*, while that for the on state has no dominant peak

The time evolution of the instantaneous static LV structure factor $s_{uu}^{(95)}(k,t)$ for Lyapunov vector No. 95 is shown in Fig. 4 as an example. Two quantities are recorded as time goes on. One is the peak wave-number k_*, which marks the position of the highest peak in the spectrum $s_{uu}^{(\alpha)}(k,t)$ (see Fig. 4). The other is the spectral entropy $H_s(t)$ [28], which measures the distribution property of the spectrum $s_{uu}^{(\alpha)}(k,t)$. It is defined as:

$$H_s(t) = -\sum_{k_i} s_{uu}^{(\alpha)}(k_i,t) \ln s_{uu}^{(\alpha)}(k_i,t). \tag{9}$$

A smaller value of $H_s(t)$ means that the spectrum $s_{uu}^{(\alpha)}(k,t)$ is highly concentrated on a few values of k, i.e., these components dominate the behavior of the LV. Both of these quantities behave intermittently, as shown in Fig. 4. Large intervals of nearly constant low values (*off state*) are interrupted by short period of bursts (*on state*) where they have large values. Details of typical *on* and *off* states are shown in the right part of Fig. 4. One can see that the off state is dominated by low wave-number components (see the sharp peak at low wave-number k_*), while the on state is more noisy and there are no significant dominant components. This intermittency in the time evolution of the instantaneous static LV structure factors is a typical feature of soft potential systems. It is conjectured that this is a consequence of the mixing of nearby LVs caused by the wild fluctuations of local instabilities. Due to the mutual interaction among modes, the hydrodynamic Lyapunov modes in the soft potential systems are only of finite life-time. In the dynamic Lyapunov structure function estimated, the peak representing the propagating (or oscillating) Lyapunov modes is of finite width. This is support for our conjecture that the hydrodynamic Lyapunov modes are of finite life-time.

Dispersion relation of hydrodynamic Lyapunov modes

Now, we consider the static LV structure factor $S_{uu}^{(\alpha)}(k)$, which is the long-time average of the instantaneous quantity $s_{uu}^{(\alpha)}(k)$. Two cases with $L = 1000$ and 2000 are shown in Fig. 5. It is not hard to recognize the sharp peak at $\lambda \approx 0$ in the contour plot of the spectrum. With increasing Lyapunov exponents, the peak shifts to the larger wave number side. A dashed line is plotted to make clear how the wave number of the peak $k_{\max}$ changes with $\lambda^{(\alpha)}$. To further demonstrate this point, the value of the Lyapunov exponent $\lambda^{(\alpha)}$ is plotted versus $k_{\max}$ of corresponding LVs in Fig. 6. We call this the *dispersion relation* of the hydrodynamical Lyapunov modes. The numerical fitting of the data shows that for $\lambda \approx 0$, $\lambda^{(\alpha)} \sim k_{\max}^{\gamma}$ with the exponent $\gamma \approx 1.2$. We presume that a linear dispersion relation $\lambda^{(\alpha)} \sim k_{\max}$ may be obtained as the thermodynamic limit is approached and the deviation from the linear function of the data shown in Fig. 6 could be due to finite-size effects.

In order to show that the peak in $S_{uu}^{(\alpha)}(k)$ is not a result of the highly regular packing of particles in the broken-chain state, the static structure function [25]

$$S(k) \equiv \int G(r) e^{-ikr} \mathrm{d}r \tag{10}$$

for the case $L = 2000$ is plotted in the same figure as $S_{uu}^{(\alpha)}(k)$, where $G(r)$ is the pair correlation function shown in Fig. 1. Obviously, $S(k)$ is nearly constant in the regime $k \approx 0$, the place where a sharp peak was observed in $S_{uu}^{(\alpha)}(k)$. The regular packing of particles causes the formation of a peak at $k/2\pi \approx 0.9$ in $S(k)$. This corresponds to a tiny peak at the same k-value in $S_{uu}^{(\alpha)}(k)$ for those LVs with $\lambda \approx 0$. These facts show clearly that the collective modes observed in the LVs are not caused by the regular packing of particles.

All of our results shown above provide strong evidence to the fact that the Lyapunov vectors corresponding to the smallest positive LEs in our 1d

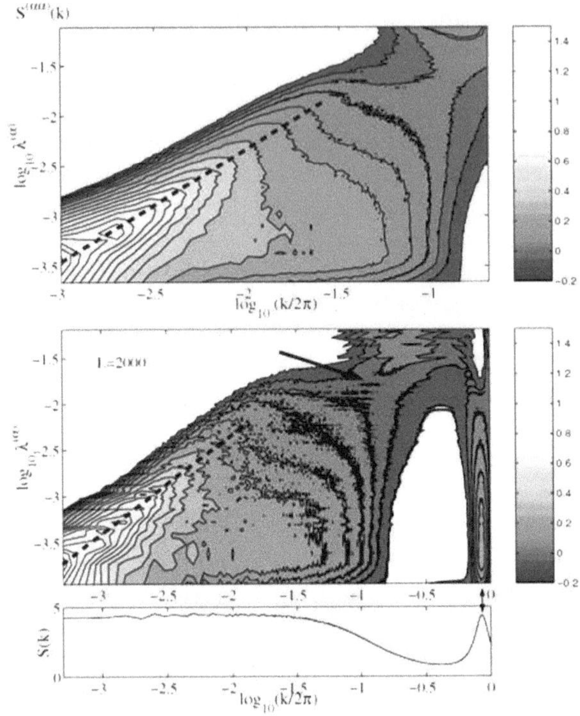

Fig. 5. Contour plot of the spectra $S_{uu}^{(\alpha)}(k)$ for $L = 1000$ and 2000. A ridge structure can easily be recognized in the regime $k \approx 0$ and $\lambda \approx 0$. To guide the eyes, a dashed line is plotted to show how the peak wave-number $k_{\max}$ changes with λ. The sudden jump in $k_{\max}$ is marked with an arrow

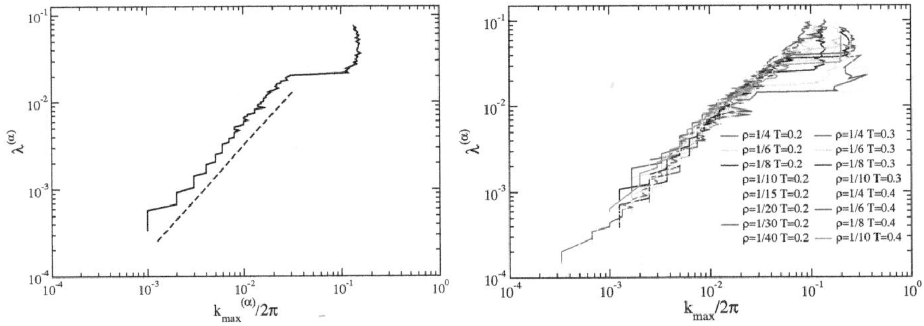

Fig. 6. Left: $\lambda^{(\alpha)}$ vs. $k_{\max}$ for a case with $L = 1000$ and $T = 0.2$. The dashed line has the form $\lambda^{(\alpha)} \sim k_{\max}^{1.2}$. Right: Dispersion relation $\lambda^{(\alpha)}$ vs. $k_{\max}$ for cases with various densities and temperatures. Note that in the regime $\lambda \approx 0$, the data from all simulations collapse to a single curve. Fitting the low wave-number part to a power-law function $\lambda^{\alpha} \sim k_{\max}^{\gamma}$ gives $\gamma \approx 1.2 \pm 0.1$. Here, $N = 100$

Lennard-Jones system are highly dominated by a few components with small wave numbers, i.e, they are similar to the Hydrodynamic Lyapunov modes found in hard-core systems. The wave-like character becomes weaker and weaker as the value of the LE is increased gradually from zero.

Influence of density and temperature

To study how the change in density influences the behavior of LVs, we increase the length L of the system from 200 to 4000, while the particle number N remains fixed at 100. From the time evolution of k_* shown in Fig. 7, one can see that, with increasing the density $\rho = N/L$, the occurrence of the *on*-state becomes more frequent, i.e., the domination of low wave-number components is much weaker. The spatial Fourier spectra for LVs with LEs in the regime $\lambda^{(\alpha)} \simeq 0$, however, are always dominated by certain low wave-number components irrespective of the density (see Fig. 7).

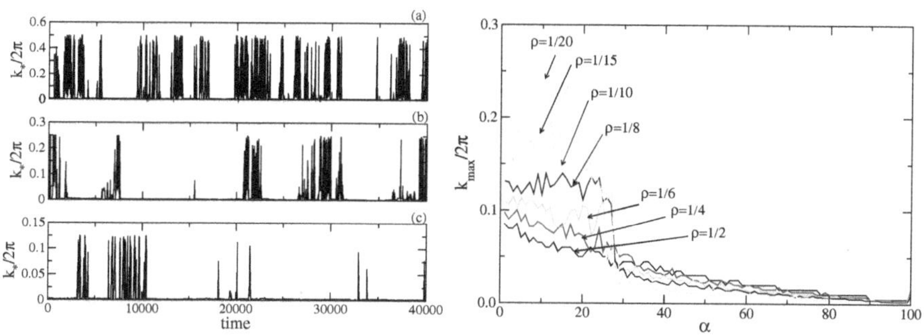

Fig. 7. Left: Time evolution of k^* as shown in Fig. 4, but with different density (a) $\rho = 1/2$, (b) 1/4 and (c) 1/8 respectively. Right: $k_{\max}$ vs. α for simulation with different densities. Here, $T = 0.2$ and the particle number $N = 100$

An important point to note is the collapse of data of dispersion relations from simulations with various densities and temperatures to a single curve (see Fig. 7). For hydrodynamic Lyapunov modes in our system this means that the dispersion function $\lambda_\alpha(k)$ is universal for the particle densities and the system temperatures studied. Fitting the data to a power-law function $\lambda_\alpha \sim k_{\max}^{\gamma}$ states that the value of the exponent γ is 1.2 ± 0.1 which is not far from the expected linear dispersion. Since our simulations are limited to cases with relatively low density, the possibility of a density-dependence of the dispersion relation can not be ruled out for high densities.

Searching for HLMs in momentum components of LVs

We will now turn to investigations on the spatial Fourier spectrum of the momentum part of LVs. Unfortunately, all the spectra are more or less homogeneously distributed over all wave-numbers. For all the cases tested, no

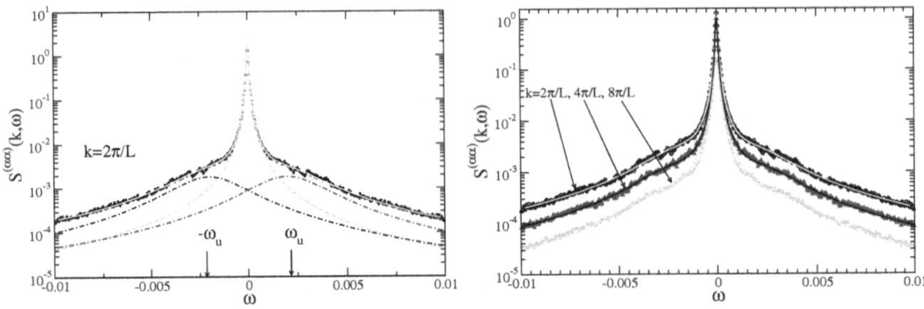

Fig. 8. Left: Dynamic LV structure factor $S_u^{(\alpha\alpha)}(k,\omega)$ for $\alpha = 96$ and $k = 2\pi/L$. The full line results from a 3-pole fit. The corresponding decomposition into three Lorentzians is also shown. Right: A comparison of such fits for $k = 2\pi/L$, $4\pi/L$, and $8\pi/L$

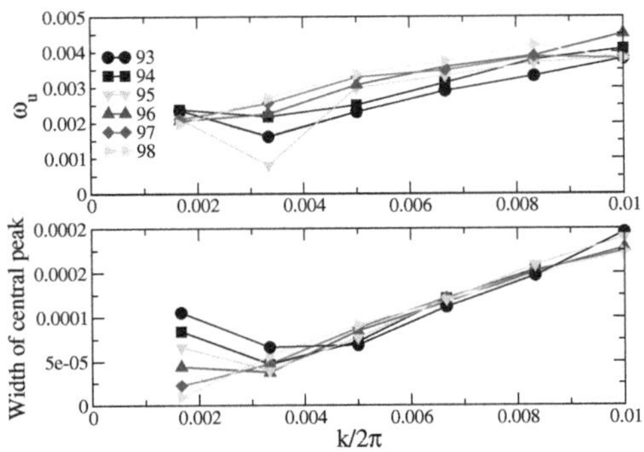

Fig. 9. Dispersion relations $\omega^{(\alpha)}(k)$ (top) and the $k-$dependence of the width of the central peak (bottom) obtained from 3-pole approximations as in Fig. 8

wave-like structure as in the coordinate part can be identified. One may wonder why no mode-like collective motion is observed in the momentum part. There are two possibilities: one is that the momentum part does contain information similar to the coordinate part but because of the strong noise it is too weak to be detected. The other is that there is no similarity between the two parts at all and regular long wave-length modes exist only in the coordinate part. The results of our current investigation on simple model systems support the former possibility [18].

4.5 Dynamic LV structure factors

More detailed information about the dynamical evolution of Lyapunov vectors can be obtained from the dynamic LV structure factors $S_u^{(\alpha\alpha)}(k,\omega)$, which encode in addition to the structural also the temporal correlations. As usual, the equal time correlations can be recovered by a frequency integration $S_u^{(\alpha\alpha)}(k) = \int S_u^{(\alpha\alpha)}(k,\omega)\mathrm{d}\omega$. In Fig. 8 we show typical examples for

$S_u^{(\alpha\alpha)}(k,\omega)$. It consist of a central "quasi-elastic" peak with shoulders resulting from dynamical excitations quite similar to the dynamic structure factor $S(k,\omega)$ of fluids [25]. In order to extract the dynamical information we use a 3-pole approximation for $S_u^{(\alpha\alpha)}(k,\omega)$, which amounts to fitting the latter by a superposition of three Lorentzians, one central peak at $\omega = 0$ and two symmetric peaks located at $\omega = \pm\omega_u(k)$. The fits are also shown in Fig. 8. They describe the frequency dependence of $S_u^{(\alpha\alpha)}(k,\omega)$ quite well. Such a dependence arises naturally e.g. from continued fraction expansions based on Mori-Zwanzig projection techniques [25], which may also be applied to this problem. These fits allow us to extract the dispersion relations $\omega^{(\alpha)}(k)$ for each of the hydrodynamic Lyapunov modes with index α. The results are shown in Fig. 9 for several of the Lyapunov modes. Clearly, this tells us that a Lyapunov mode corresponding to exponent λ is characterized, apart from the dominating wave number $k(\lambda)$, by a typical frequency $\omega(k(\lambda))$. Because $\frac{d\omega}{dk}$ is non-vanishing, this implies propagating wave-like excitations. The origin of the characteristic frequency $\omega(k(\lambda))$ is not yet fully understood. Probably, it reflects the rotational motion of the orthogonal frame formed by LVs around its reference trajectory [29]. The full LV dynamics of the soft-potential system treated here, however, is more complex than that of the hard-core systems. For instance, the peaks in $S_u^{(\alpha\alpha)}(k,\omega)$ are of finite width (see Fig. 8). This fact is consistent with our observation that several quantities characterizing the dynamical aspect of Lyapunov vectors evolve erratically in time (see Sec. 4.4), which implies that the coherent wave-like motion is switched on and off intermittently. These facts suggest that the hydrodynamic Lyapunov modes in soft-potential systems have finite life-times.

5 Lyapunov modes in 2D Lennard-Jones fluids

In isotropic fluids with $d > 1$ the static LV structure factor $S_u^{(\alpha\alpha)}(\mathbf{k})$ becomes a second rank tensor. Cartesian components $S_{\mu\nu}^{(\alpha\alpha)}(\mathbf{k})$ of $S_u^{(\alpha\alpha)}(\mathbf{k})$ can be expressed in terms of longitudinal and transverse correlation functions $S_\mathrm{L}^{(\alpha\alpha)}$ and $S_\mathrm{T}^{(\alpha\alpha)}$ as $S_{\mu\nu}^{(\alpha\alpha)}(\mathbf{k}) = \hat{k}_\mu \hat{k}_\nu S_\mathrm{L}^{(\alpha\alpha)}(k) + (\delta_{\mu\nu} - \hat{k}_\mu \hat{k}_\nu) S_\mathrm{T}^{(\alpha\alpha)}(k)$ with $\hat{k}_\mu = (\mathbf{k}/k)_\mu$. As an example, we present in Fig. 10 the contour plot of the two correlation functions S_L and S_T for LV No. 140 of a two-dimensional Lennard-Jones system with $N = 100$, $T = 0.8$ and $L_x \times L_y = 20 \times 20$. The difference between the two components is quite obvious. However, as can be seen from Fig. 11, $S_\mathrm{L}^{(\alpha\alpha)}(k)$ and $S_\mathrm{T}^{(\alpha\alpha)}(k)$ for two-dimensional cases behave similar to the one-dimensional case shown in Fig. 5. The fact implies the existence of hydrodynamic Lyapunov modes also in two-dimensional cases. In addition, both the longitudinal and transverse components are characterized by a linear dispersion relation, which has been found to be typical for Hamiltonian systems [18,19]. Further numerical simulations show that the transverse modes are non-propagating, in contrast to the longitudinal components.

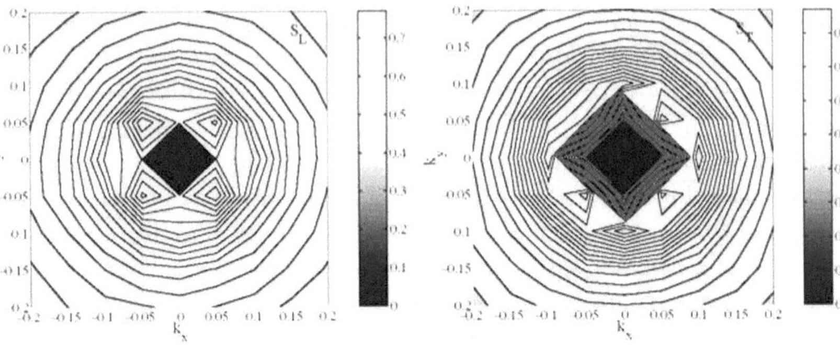

Fig. 10. Contour plots of $S_{\mathrm{L}}^{(\alpha\alpha)}(\mathbf{k})$ and $S_{\mathrm{T}}^{(\alpha\alpha)}(\mathbf{k})$ of a LV with $\alpha = 140$ in a 2D system with $N = 100$, $T = 0.8$ and $L_x \times L_y = 20 \times 20$. Obviously, the longitudinal and transverse components behave differently

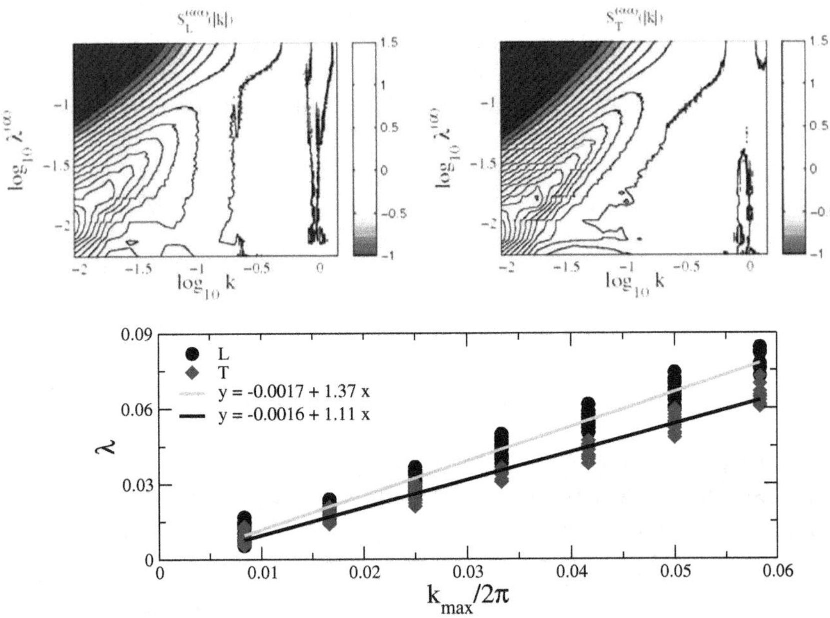

Fig. 11. Contour plots of $S_{\mathrm{L}}^{(\alpha\alpha)}(k)$ and $S_{\mathrm{T}}^{(\alpha\alpha)}(k)$ (upper row). As in Fig. 5 the corresponding dispersion relation $\lambda(k_{\max})$ (lower row) of the hydrodynamic Lyapunov modes in a 2D system is extracted ($N = 100$, $T = 0.8$ and $L_x \times L_y = 5 \times 120$)

6 Universal features of Lyapunov modes in spatially extended systems with continuous symmetries

Relying on the LV correlation function method, we have up to now successfully identified the existence of HLMs in the following spatially extended systems:

Coupled map lattices (CMLs) with either Hamiltonian or dissipative local dynamics

$$v_{t+1}^l = (1-\gamma)v_t^l + \epsilon[f(u_t^{l+1} - u_t^l) - f(u_t^l - u_t^{l-1})] \tag{11}$$

$$u_{t+1}^l = u_t^l + v_{t+1}^l \tag{12}$$

and

$$u_{t+1}^l = u_t^l + \epsilon[f(u_t^{l+1} - u_t^l) - f(u_t^l - u_t^{l-1})] \tag{13}$$

where $f(z)$ is a nonlinear map, t is the discrete time index, $l = \{1, 2, \cdots, L\}$ is the index of the lattice sites and L is the system size. We set the damping coefficient $\gamma = 0$ and use periodic boundary conditions $\{u_t^0 = u_t^L, u_t^{L+1} = u_t^1\}$, unless it is explicitly stated otherwise.

Dynamic XY model [30] with the Hamiltonian

$$H = \sum_i \dot{\theta}_i + \epsilon \sum_{ij} [1 - cos(\theta_j - \theta_i)]. \tag{14}$$

Kuramoto-Sivashinsky equation [31]

$$h_t = -h_{xx} - h_{xxxx} - h_x^2. \tag{15}$$

A common feature of these systems is that they all hold certain continuous symmetries and conserved quantities, which have been shown to be essential for the occurrence of Lyapunov modes [18]. Our numerical simulations and analytical calculations indicate that these systems fall into two groups with respect to the nature of hydrodynamic Lyapunov modes. To be precise, the dispersion relations are characterized by $\lambda \sim k$ and $\lambda \sim k^2$ in Hamiltonian and dissipative systems respectively, as Fig. 12 indicates. Moreover, the HLMs in Hamiltonian systems are propagating, whereas those in dissipative systems show only diffusive motion. Examples of dynamic LV structure factors for two CMLs are presented in the right row of Fig. 12. In a), each spectrum has two sharp symmetric side-peaks located at $\pm\omega_u$. Furthermore, $\omega_u \simeq \pm c_u k$ for $k \geq 2\pi/L$. These facts suggest that the HLMs in coupled standard maps are propagating. The spectrum of coupled circle maps in b) has only a single central peak and can be well approximated by a Lorentzian curve [18], which implies that the HLMs in this system fluctuate diffusively. In addition, no step structures in Lyapunov spectra have been found in contrast to the hard-core systems. The quantities characterizing the dynamical evolutions of LVs in these systems exhibit intermittent behaviors.

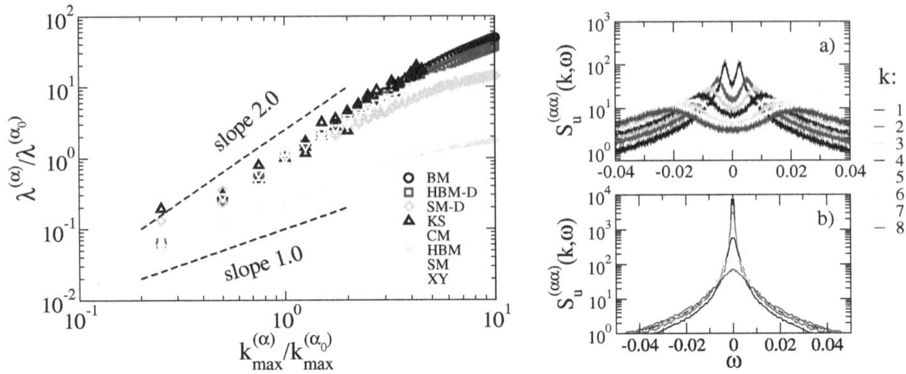

Fig. 12. Left: The λ-k dispersion relations for various extended systems with continuous symmetries. The normalized data for different systems collapse on two master curves. These results strongly support our conjecture that there are two classes of systems with $\lambda \sim k$ and $\lambda \sim k^2$ respectively. Systems in the group with $\lambda \sim k$ include (11) with $f(z) = \frac{1}{2\pi}\sin(2\pi z)$ (SM), (11) with $f(z) = 2z \pmod 1$ (HBM) and the 1d XY model (XY) [30]. Systems belonging to class $\lambda \sim k^2$ are (13) with $f(z) = \frac{1}{2\pi}\sin(2\pi z)$ (CM), (13) with $f(z) = 2z \pmod 1$ (BM), (11) with $f(z) = \frac{1}{2\pi}\sin(2\pi z)$ and $\gamma = 0.7$ (SM-D), (11) with $f(z) = 2z \pmod 1$ and $\gamma = 0.7$ (HBM-D) and the 1d Kuramoto-Sivashinsky equation (KS). Right: Dynamic LV structure factors $S_u^{(\alpha\alpha)}(k,\omega)$ for a) coupled standard maps, (11) with $\epsilon = 1.3$; b) coupled circle maps, (13) with $\epsilon = 1.3$

7 Conclusion and discussion

We have presented numerical results for the Lyapunov instability of Lennard-Jones systems. Our simulations show that the step-wise structures found in the Lyapunov spectrum of hard-core systems disappear completely here. This is presumed to be the result of the strong fluctuations in the finite-time LEs [8]. A new technique based on the spatial Fourier spectral analysis is employed to reveal the vague long wave-length structure hidden in LVs. In the resulting spatial Fourier spectrum of LVs with $\lambda \simeq 0$, a significantly sharp peak with low wave-number is found. This serves as strong evidence for the fact that hydrodynamic Lyapunov modes do exist in soft-potential systems [32]. The disappearance of the step-structures and the survival of the hydrodynamic Lyapunov modes show that the latter are more robust and essential than the former. Studies on dynamical LV structure factors conclude that longitudinal HLMs in Lennard-Jones fluids are propagating. Going beyond many-particle systems, we have shown that, for a large class of extended systems, HLMs of Hamiltonian and dissipative cases are different both in respect of spatial structure and in the dynamical evolution.

The intermittency in the temporal evolution of certain quantities characterizing LVs indicates that the coherent wave-like excitations in LVs switch on and off erratically, which suggests that the HLMs in soft-potential systems have finite life-times. This finding is consistent with the observation that in dynamic LV structure factors the side-peaks representing the dynamical excitations have finite widths. This may also be related to the strong fluctuations in the finite-time LEs.

Until now, only the coordinate part of LVs is used in most of the studies on hydrodynamic Lyapunov modes. For the case of hard-core systems, this is reasonable due to an interesting feature of those LVs corresponding to near-zero LEs found in [10], namely the fact that the angles between the coordinate part and the momentum part are always small, i.e, the two vectors are nearly parallel. For our soft potential systems, we find that the angles between the coordinate part and the momentum part are no longer as small as in hard-core systems. However, we failed to detect any long wave-length structures in the momentum part of LVs. This could be explained by our current results on the simple model system of coupled map lattices (CMLs) [18].

The standard method of [20] was employed to calculate Lyapunov exponent and vectors quantities for our many-particle ($N = 100 - 1000$) Lennard-Jones system in d dimension. A set of $2dN \times 2dN$ linear and $2dN$ nonlinear ordinary differential equations have to be integrated simultaneously in order to obtain the dynamics of $2dN$ offset vectors in tangent space and the reference trajectory in phase space, respectively. This enforces the use of parallel implementations of the corresponding algorithms. Performance tests of our parallel algorithms were executed on clusters with different hardware properties. An efficient combination of basic routines and parallel orthogonalization algorithms make our exploration of the challenge field of Lyapunov instabilities of large dynamical systems feasible with the state art of computer power and yields the reported interesting results.

Acknowledgments

We thank W. Just, W. Kob, A. Latz, A. S. Pikovsky and H. A. Posch for fruitful discussions. Special thanks go to W. Kob for providing us with the code of molecular dynamics simulations and to G. Rünger and M. Schwind for the help on the parallel algorithms.

References

1. J.-P. Eckmann and D. Ruelle. Ergodic theory of chaos and strange attractors. *Rev. Mod. Phys.*, 57:617, 1985; E. Ott. *Chaos in Dynamical Systems.* Cambridge University Press, Cambridge, 1993.
2. N. S. Krylov. *Works on the Foundations of Statistical Mechanics.* Princeton University Press, Princeton, 1979.
3. P. Gaspard. *Chaos, Scattering, and Statistical Mechanics.* Cambridge University Press, Cambridge, 1998.
4. J. P. Dorfman. *An Introduction to Chaos in Nonequilibrium Statistical Mechanics.* Cambridge University Press, Cambridge, 1999.

5. D. J. Evans and G. P. Morriss. *Statistical Mechanics of Nonequilibrium Liquids.* Academic, New York, 1990.
6. Wm. G. Hoover. *Time Reversibility, Computer Simulation, and Chaos.* World Scientific, Singapore, 1999.
7. H. A. Posch and R. Hirschl. Simulation of blliards and of hard-body fluids. In D. Szasz, editor, *Hard Ball Systems and the Lorenz Gas*, Springer, Berlin, 2000.
8. C. Forster, R. Hirschl, H. A. Posch, and Wm. G. Hoover. Perturbed phase-space dynamics of hard-disk fluids. *Physics D*, 187:294, 2004.
9. J.-P. Eckmann and O. Gat. Hydrodynamic Lyapunov modes in translation-invariant systems. *J. Stat. Phys.*, 98:775, 2000.
10. S. McNamara and M. Mareschal. Origin of the hydrodynamic Lyapunov modes. *Phys. Rev. E,* 64:051103, 2001.
11. A. de Wijn and H. van Beijeren. Goldstone modes in Lyapunov spectra of hard sphere systems. *Phys. Rev. E,* 70:016207, 2004.
12. T. Taniguchi and G. P. Morriss. Stepwise structure of Lyapunov spectra for many-particle systems using a random matrix dynamics. *Phys. Rev. E,* 65:056202, 2002.
13. T. Taniguchi and G. P. Morriss. Boundary effects in the stepwise structure of the Lyapunov spectra for quasi-one-dimensional systems. *Phys. Rev. E,* 68:026218, 2003.
14. Wm. G. Hoover, H. A. Posch, C. Forster, C. Dellago and M. Zhou. Lyapunov modes of two-Dimensional many-body systems; soft disks, hard disks, and rotors. *J. Stat. Phys.,* 109:765, 2002.
15. H.-L. Yang and G. Radons. Lyapunov instabilities of Lennard-Jones fluids. *Phys. Rev. E,* 71:036211, 2005, see also arXiv:nlin.CD/0404027.
16. G. Radons and H.-L. Yang. Static and dynamic correlations in many-particle Lyapunov vectors. arXiv:nlin.CD/0404028.
17. H.-L. Yang and G. Radons. Universal features of hydrodynamic Lyapunov modes in extended systems with continuous symmetries. *Phys. Rev. Lett.,* 96:074101, 2006.
18. H.-L. Yang and G. Radons. Hydrodynamic Lyapunov modes in coupled map lattices. *Phys. Rev. E,* 73:016202, 2006.
19. H.-L. Yang and G. Radons. Dynamical behavior of hydrodynamic Lyapunov modes in coupled map lattices. *Phys. Rev. E,* 73:016208, 2006.
20. G. Benettin, L. Galgani and J. M. Strelcyn. Kolmogorov entropy and numerical experiments. *Phys. Rev. A,* 14:2338, 1976.
21. I. Shimada and T. Nagashima. A numerical approach to ergodic problem of dissipative dynamical systems. *Prog. Theor. Phys.,* 61:1605, 1979.
22. J. Dongarra, J. Du Croz, I. Duff and S. Hammarling. A set of Level 3 Basic Linear Algebra Subprograms. *ACM Trans. Math. Soft.,* 16:1, 1990.
23. L. S. Blackford, et al.. *ScaLAPACK Users' Guide.* Society for Industrial and Applied Mathematics, Philadelphia, 1997.
24. G. Radons, G. Rünger, M. Schwind and H.-L. Yang. Parallel algorithms for the determination of Lyapunov characteristics of large nonlinear dynamical systems. Proceedings of PARA04, WORKSHOP ON STATE-OF-THE-ART IN SCIENTIFIC COMPUTING, Lyngby, June 20-23, 2004, Lecture Notes of Computer Science, vol. 3272, 1131, Springer, Berlin, 2005.
25. J. P. Boon and S. Yip. *Molecular Hydrodynamics.* McGraw-Hill, New York, 1980.

26. W. Kob and H. C. Andersen. Testing mode-coupling theory for a supercooled binary Lennard-Jones mixture I: The van Hove correlation function. *Phys. Rev. E,* 51:4626, 1995.
27. F. H. Stillinger. Statistical mechanics of metastable matter: superheated and stretched liquids. *Phys. Rev. E,* 52:4685, 1995.
28. R. Livi, M. Pettini, S. Ruffo, M. Sparpaglione and A. Vulpiani. Equipartition threshold in nonlinear large Hamiltonian systems: The Fermi-Pasta-Ulam model. *Phys. Rev. A,* 31:1039, 1985.
29. J.-P. Eckmann, C. Forster, H. A. Posch and E. Zabey. Lyapunov modes in hard-disk systems. *J. Stat. Phys.,* 118:795, 2005.
30. D. Escande, H. Kantz, R. Livi and S. Ruffo. Self-consistent check of the validity of Gibbs calculus using dynamical variables. *J. Stat. Phys.,* 76:605, 1994; M. Antoni and S. Ruffo. Clustering and relaxation in Hamiltonian long-range dynamics. *Phys. Rev. E,* 52:2361, 1995.
31. Y. Kuramoto and T. Tsuzuki. Persistent propagation of concentration waves in dissipative media far from thermal equilibrium. *Prog. Theor. Phys.,* 55:356, 1976; G. I. Sivashinsky. *Acta Astron.,* 4:1177, 1977.
32. C. Forster and H. A. Posch. Lyapunov modes in soft-disk fluids. *New J. Phys.,* 7:32, 2005, see also arXiv:nlin.CD/0409019

The Cumulant Method for Gas Dynamics

Steffen Seeger[1], Karl Heinz Hoffmann[2], and Arnd Meyer[3]

[1] seeger@physik.tu-chemnitz.de
[2] Technische Universität Chemnitz, Institut für Physik
09107 Chemnitz, Germany
hoffmann@physik.tu-chemnitz.de
[3] Technische Universität Chemnitz, Fakultät für Mathematik
09107 Chemnitz, Germany
arnd.meyer@mathematik.tu-chemnitz.de

1 Introduction

Characterizing fluid flow by the ratio of mean free path and a characteristic flow length (the KNUDSEN number Kn) we have two extremes: dense gases ($Kn \ll 1$) where modeling by EULER or NAVIER-STOKES equations is valid and rarefied gases ($Kn \gg 1$) for which modeling by the BOLTZMANN equation is necessary. Developing models for the intermediate transition regime is subject to active current research because despite the tremendously growing increase in computational and algorithmic computing performance, numerical simulation of flows in the transition regime remains a challenging problem. Thus there is a considerable gap in the ability to model flows where mean free path and characteristic flow lengths are comparable. However, efficient methods for simulating transition regime flows will be an important design tool for micro-scale machinery, where dense gas models become invalid.

As can be learned from the tremendous advances of computational methods for the NAVIER-STOKES equations, an important key for the efficient solution is the use of (structured) adaptivity and parallel methods of solution. For an efficient parallel implementation with good speedup, the resulting numerical scheme should require only local operations and require only moderate communication. Adaptivity, however, has so far mainly been used for grid refinement to achieve the accuracy goals for numerical solution. To obtain the necessary gain in efficiency for the transition regime flows, it seems natural to extend the concept of adaptivity also to physical modelling (that is, the 'detail' of the equations of motion to be solved in a certain flow regime). This requires a consistent theory of how these equations are related to each other, and how the coupling could be achieved consistently.

The derivation of reduced, but nevertheless consistent equations of motion for fluid flow is therefore an interesting question of current research. The goal is do develop a description of fluid flow which can be adjusted in detail between that of the BOLTZMANN equation and that of EULER equations. Here we present some results from our work on physical models for these flows, trying to shed more light on the question how properly chosen physical models might allow for such adaptivity not only in numerical solution but also in physical modelling.

Two schemes that try to successively approximate the BOLTZMANN equation to include more and more detail are moment methods [1–3] and the cumulant method [4–6]. Commonly a rather small number of moments or cumulants is used and thus these methods do not capture the full information available with the phase space density, but focus on physically interesting macroscopic quantities, such as density, flow velocity, mean kinetic energy, shear stress and heat flux.

This work is intended to give a review of the theory and results for the cumulant method, in order to serve as a starting point for the interested reader. The first part of the review gives a short introduction to kinetic theory, namely the BOLTZMANN equation, the interaction model, and the space-homogeneous case. The second part discusses the basic concepts and differences of the various moment methods known from the literature. The last part presents the cumulant method, results on properties of the resulting equations, a simple numerical scheme to solve them and possible boundary conditions. A summary and outlook to possible future research closes this review.

2 Kinetic theory

Considering flow of an inert mixture of gases we assume the fluid is composed of N_s different species, enumerated by the set $\mathcal{N}_s$ with each species having its own set of associated properties. For a species s, these are the particle mass m_s and the laws of interaction with particles of any other species r. We let the phase space density $f_s(t, \underline{x}, \underline{c}_s)$ denote the density of particles of species s at time t and position $\underline{x}$, moving with absolute velocity $\underline{c}$. The f_s are normalized such that

$$n_s(t, \underline{x}) = \int \mathrm{d}\underline{c}\ f_s(t, \underline{x}, \underline{c}) \tag{1}$$

are the partial particle number densities of the various species (integrals are taken over the whole velocity space $\mathcal{R}^d$).

2.1 The BOLTZMANN equation

The f_s have to satisfy the BOLTZMANN equation. Considering a sufficiently dilute gas in an inertial frame, where accounting only for binary collisions is sufficient, this integro-differential equation in the f_s takes the form [7]

$$\partial_t f_s + \underline{c} \cdot \partial_{\underline{x}} f_s + \underline{a}_s \cdot \partial_{\underline{c}} f_s = \sum_{r \in \mathcal{N}_s} S_{rs}[f_r, f_s]\,; \qquad \forall s \in \mathcal{N}_s \tag{2}$$

where S_{rs} denotes the collision operator and $\underline{a}_s$ denotes the acceleration of a s-particle due to external forces as a function of time t, particle position $\underline{x}$ and particle velocity $\underline{c}$. We regard forces exerted on particles due to particle-particle interaction as internal and any other (e.g. gravity) as external. The functional $S_{rs}[f_r, f_s]$ occurring in equation (2) describes the change of f_s due to interaction of particles of species r with particles of species s. By restriction to the description of a sufficiently dilute gas it may be assumed that: I) contributions by other than binary collisions may be neglected; II) the range of interaction is much less than the mean free path; III) particle trajectories before and after the collision are approximately rectilinear; IV) the distribution functions are constant over the range of interaction; V) particles about to collide are not correlated. With these assumptions S_{rs} can be written as the integral operator

$$S_{rs}[f_r, f_s] = \sum_r \int \mathrm{d}\underline{c}_r \, \mathrm{d}\underline{n} \; \sigma_{rs} \, \|\underline{c}_{rs}\| \, (\hat{f}_r \hat{f}_s - f_r f_s) \tag{3}$$

with scattering cross section $\sigma_{rs} = \sigma_{rs}(\underline{n}, \|\underline{c}_{rs}\|)$, relative collision velocity $\underline{c}_{rs} = \underline{c}_r - \underline{c}_s$, $\|\underline{c}_{rs}\| = \sqrt{\underline{c}_{rs} \cdot \underline{c}_{rs}}$ denoting the usual quadratic vector norm and the collision parameter vector $\underline{n}$. The integral over $\mathrm{d}\underline{n}$ covers the d-dimensional sphere $\|\underline{n}\| = 1$. $\hat{f}_r$ and $\hat{f}_s$ are the phase-space densities evaluated at the velocities $\hat{\underline{c}}_r$ and $\hat{\underline{c}}_s$ after the collision.

For a gas at standard conditions, these assumptions are justified: with a typical molecule diameter of about 0.3 nm and a particle density of $n \approx 2.7 \cdot 10^{25}\,\mathrm{m}^{-3}$ the average distance of the particles is about ten times the molecule diameter, making the first assumption hold. The mean free path is $\approx 10\,\mathrm{nm}$, so it is about 10 times longer than the range of interaction, which makes the second assumption reasonable. This allows to abstract from the particular particle trajectories during an encounter to states 'before' and 'after' a collision. With intermolecular forces generally several powers of ten larger than external forces (e.g. gravity) the third assumption allows to neglect the effect of external forces during a collision. The fourth assumption is certainly valid if there are no steep gradients in density or kinetic temperature. The last assumption does not have any obvious justification but is postulated to hold for dilute gases (where particles participating in a collision stem from regions a few mean free path lengths apart), but it is supported by excellent agreement of the results of kinetic theory of dilute gases with known experiments.

2.2 The Enskog equation of change

Describing the evolution of mixtures of gases on a macroscopic scale we are interested in partial quantities, such as the partial pressure, density, etc. of each species. Given a phase space density $f_s(t, \underline{x}, \underline{c})$, the density of a partial macroscopic thermodynamic quantity $\overline{(\Phi)}_s(t, \underline{x})$ can be obtained as the mean of an associated microscopic function $\Phi(t, \underline{x}, \underline{c})$

$$\overline{(\Phi)}_s(t,\underline{x}) = \int \mathrm{d}\underline{c}\ \Phi(t,\underline{x},\underline{c})\, f_s(t,\underline{x},\underline{c}) \ . \tag{4}$$

Multiplying (2) with Φ and integrating over $\underline{c}$ we obtain [8] a balance equation for the quantity $\overline{(\Phi)}_s$ (also known as ENSKOG's general equation of change)

$$\partial_t \overline{(\Phi)}_s + \partial_{\underline{x}} \cdot \overline{(\underline{c}\,\Phi)}_s = \overline{\left(\partial_t \Phi + \partial_{\underline{x}} \cdot \underline{c}\Phi + \partial_{\underline{c}} \cdot \underline{a}_s \Phi\right)}_s + \sum_{r\in N_s} \int \mathrm{d}\underline{c}\ \Phi\, S_{rs} \ , \tag{5}$$

where we made use of partial integration and the property $f_s \to 0$ for $\|\underline{c}\| \to \infty$ due to (1).

2.3 The MAXWELL gas model

Throughout this work we will assume the particular interaction model of so-called MAXWELL molecules, for which the particles repel each other with a force inversely proportional to the $(2d-1)$th power of their distance. As has been known for quite a long time [9], this simplifies the collision integral considerably. Since the BOLTZMANN-type collision operator (3) is similar to a convolution integral, the simplifications are even greater when employing a FOURIER-transformation with regard to particle velocity. This method of transformation of the BOLTZMANN equation into an equation for the characteristic function [10] has been proposed and extensively studied by BOBYLEV et al. [11] with a review of the method and results given in [12]. It can be applied also for other interaction models (i.e. hard spheres and the BGK approximation) but here we restrict our considerations to MAXWELL interaction.

Setting $\Phi = \frac{1}{(2\,\pi)^{d/2}}\, e^{\mathrm{i}\underline{\chi}\cdot\underline{c}}$ we find the associated $\overline{(\Phi)}_s$ to be the characteristic function $\varphi_s(t,\underline{x},\underline{\chi})$ with the equation of motion

$$\begin{aligned}
\partial_t \varphi_s + \partial_{\underline{x}} \cdot \partial_{\mathrm{i}\underline{\chi}} \varphi_s &= \Gamma_s + \sum_{r\in N_s} \Xi_{rs}[\varphi_r,\varphi_s]\,, \\
\text{collision term}\quad \Xi_{rs}[\varphi_r,\varphi_s] &= \tfrac{1}{(2\pi)^{d/2}} \int \mathrm{d}\underline{c}\ S_{rs}\, e^{\mathrm{i}\underline{\chi}\cdot\underline{c}} \\
\text{and force term}\quad \Gamma_s[\varphi_s] &= \tfrac{1}{(2\,\pi)^{d/2}} \int \mathrm{d}\underline{c}\ f_s\, \partial_{\underline{c}} \cdot \underline{a}_s\, e^{\mathrm{i}\underline{\chi}\cdot\underline{c}} \ .
\end{aligned} \tag{6}$$

After a straightforward calculation following the idea of BOBYLEV [11] we find

$$\begin{aligned}
\Xi^{2\mathrm{D}}_{rs}[\varphi_r,\varphi_s] &= \sqrt{\tfrac{2\,\kappa_{rs}}{\mu_{rs}}\,(2\pi)^2} \int \mathrm{d}\varepsilon\ \Omega[\varphi_r,\varphi_s] \\
\Xi^{3\mathrm{D}}_{rs}[\varphi_r,\varphi_s] &= \sqrt{\tfrac{2\,\kappa_{rs}}{\mu_{rs}}\,(2\pi)^3} \int \mathrm{d}\varepsilon\,\mathrm{d}\varphi\ \varepsilon\, \Omega[\varphi_r,\varphi_s]
\end{aligned} \tag{7}$$

for the 2D and 3D MAXWELL gas respectively. Here ε is a dimensionless collision parameter, φ (without any index) is the tilt of the scattering plane and

κ_{rs} is the parameter controlling the strength of interaction between particles of species r and s. The integral kernel $\Omega[\varphi_r, \varphi_s]$ is given by

$$\Omega[\varphi_r, \varphi_s] = \varphi_r(\underline{\chi} \cdot \underline{\underline{D}}_+^+)\, \varphi_s(\underline{\chi} \cdot \underline{\underline{D}}_+^-) - \varphi_r(\underline{0})\, \varphi_s(\underline{\chi}) \tag{8}$$

with

$$\underline{\underline{D}}_+^+ = \left[\tfrac{1+\Delta_{rs}}{2}\, \underline{\underline{1}} - \tfrac{1+\Delta_{rs}}{2}\, \underline{\underline{S}}^{-1}\right] \qquad \underline{\underline{D}}_+^- = \left[\tfrac{1-\Delta_{rs}}{2}\, \underline{\underline{1}} + \tfrac{1+\Delta_{rs}}{2}\, \underline{\underline{S}}^{-1}\right], \tag{9}$$

where $\underline{\underline{S}}^{-1}$ denotes the *inverse* of the scattering matrix $\underline{\underline{S}}$ that *rotates* $\underline{c}_{rs}$ to $\hat{\underline{c}}_{rs}$. Note that this choice is a bit different from the common notation in kinetic theory, where the mapping from $\underline{c}_{rs}$ to $\hat{\underline{c}}_{rs}$ is chosen as an unitary, symmetric matrix. The common notation makes the mapping between $\underline{c}_{rs}$ and $\hat{\underline{c}}_{rs}$ self-inverse, but here we prefer the given notation for easier calculation of the integrals occurring in the production terms. Δ_{rs} and μ_{rs} are calculated from m_r and m_s, the masses of r- and s-particles, as

$$\mu_{rs} = \frac{m_r\, m_s}{m_r + m_s} \qquad\qquad \Delta_{rs} = \frac{m_r - m_s}{m_r + m_s}. \tag{10}$$

From the actual kinematics of a collision we find in 2D

$$\vartheta^{2\mathrm{D}}(\varepsilon) = \pi\left(1 - \frac{\varepsilon}{\sqrt{1+\varepsilon^2}}\right) \quad \text{and} \quad \underline{\underline{S}}(\varepsilon) = \begin{pmatrix} \cos\vartheta(\varepsilon) & -\sin\vartheta(\varepsilon) \\ \sin\vartheta(\varepsilon) & \cos\vartheta(\varepsilon) \end{pmatrix}. \tag{11}$$

with the collision angle ϑ given as a function of the dimensionless collision parameter ε.

For convenient numerical implementation we use a system of units such that the relations and equations given remain unchanged but with $k_B = 1$ and the atomic mass unit $m_u = 1$. That is, choosing a reference temperature, pressure and volume as that of 1 mol of an ideal gas at the ice point of water and using the atomic mass as unit for particle masses, we have determined the units for length, time, energy etc. by the condition $k_B = 1$. We can establish such a system of units for the 2D-case considered here as well and give any numerical results, plots etc. in dimensionless form.

2.4 Spatially homogeneous Boltzmann equation

Let us now examine the spatially homogeneous case [13] for equation (2). All points $\underline{x}$ in space are assumed equal, so $f_s(t, \underline{x}, \underline{c})$ does not depend on $\underline{x}$ and further the effect of external forces is neglected ($\underline{a} = 0$). Thus we are concerned with the temporal relaxation of a set of N_s phase space densities $f_s(t, \cdot, \underline{c}) : \mathcal{R}_+ \times \mathcal{R}^d \to \mathcal{R}_+$ from initial non-equilibrium densities $f_s(0, \cdot, \underline{c})$ to equilibrium densities $f_s^{\mathrm{eq}}(\underline{c}) = f(\infty, \cdot, \underline{c})$. According to (2) and (6), this relaxation is determined by the equation

$$\partial_t f_s = \sum_r S_{rs}[f_r, f_s] \qquad \text{or} \qquad \partial_t \varphi_s = \sum_r \Xi_{rs}[\varphi_r, \varphi_s]\,. \tag{12}$$

In general quite many different solutions to the non-linear equation (12) can be studied [12] regarding their properties but they are all expressed in the form of converging series, whose coefficients are calculated recurrently. However, it appears that there exists a (so far unique) non-trivial (e.g. non-equilibrium) solution to the non-linear equations (12) that can be given in finite, analytic form using elementary functions. Reported by BOBYLEV [12, 14] and KROOK and WU [15, 16] this solution can be found by assuming that a dimensionless time scale

$$\tau = 1 - \theta\, e^{-\lambda\, t} \tag{13}$$

can be introduced and is common to all species s. λ denotes a relaxation rate to be determined. The arbitrary parameter $\theta \in [0, \frac{2}{d+2}]$ allows to control the initial deviation of f from equilibrium ($\theta = 0$), and is limited by the requirement that $f_s(t, \cdot, \underline{c}) \geq 0$ for all species, times t and particle velocities $\underline{c}$. In addition to the common time scale the mean kinetic energies are assumed to be the same for each species, so the specific energies $\varepsilon_s = \frac{d}{2} \frac{k_B}{m_s} T$ are given by a common temperature T. Inserting equations (7), (8), (9), the ansatz

$$\varphi_s(t, \cdot, \underline{\chi}) = n_s \frac{\exp\left(-\tau \frac{\varepsilon_s}{d} \chi^2\right)}{(2\pi)^{d/2}} \left(p_s(\tau) + q_s(\tau)\, \tau \frac{\varepsilon_s}{d} \chi^2\right) \tag{14}$$

with $\chi = \|\underline{\chi}\|$ into (12) and equating coefficients for equal powers of χ allows to derive three conditions on $p_s(\tau)$, $q_s(\tau)$ and λ: an ordinary differential equation for $p_s(\tau)$ from the $\underline{\chi}^0$ term; another ordinary differential equation for $p_s(\tau)$ and $q_s(\tau)$ from the $\underline{\chi}^2$ term and finally an algebraic relation from the $\underline{\chi}^4$ term. From the ordinary differential equations a solution can be determined as

$$\varphi_s(t, \cdot, \underline{\chi}) = n_s \frac{\exp\left(-\tau \frac{\varepsilon_s}{d} \chi^2\right)}{(2\,\pi)^{d/2}} \left(1 + (\tau - 1) \frac{\varepsilon_s}{d} \chi^2\right) \tag{15}$$

$$f_s(t, \cdot, \underline{c}) = n_s \frac{\exp\left(-\frac{1}{\tau} \frac{d}{2\,\varepsilon_s} \frac{c^2}{2}\right)}{(\pi\, \tau\, 4\, \varepsilon_s / d)^{d/2}} \left[1 + \frac{\tau - 1}{\tau} \left(\frac{d}{2} - \frac{1}{\tau} \frac{d}{2\,\varepsilon_s} \frac{c^2}{2}\right)\right] \tag{16}$$

with particle density n_s, particle mass m_s and specific energy ε_s. The algebraic relation obtained from the $\underline{\chi}^4$ term determines λ. Fig. 1 depicts the initial ($t = 0$) and final ($t = \infty$) phase space density, and real parts of the first and second characteristic function (also known as the cumulant generating function)

$$\kappa_s(\underline{\chi}) = \ln\left((2\,\pi)^{d/2}\, \varphi_s(\underline{\chi})\right) \tag{17}$$

for a mono-atomic gas with $\kappa_{11} = 1$, $m_1 = 1$, $\varepsilon_1 = 1$ and $\theta = \frac{1}{2}$.

While the phase-space-density is a strictly non-negative function, the characteristic function, though real-valued for real $\underline{\chi}$, appears to have a zero in

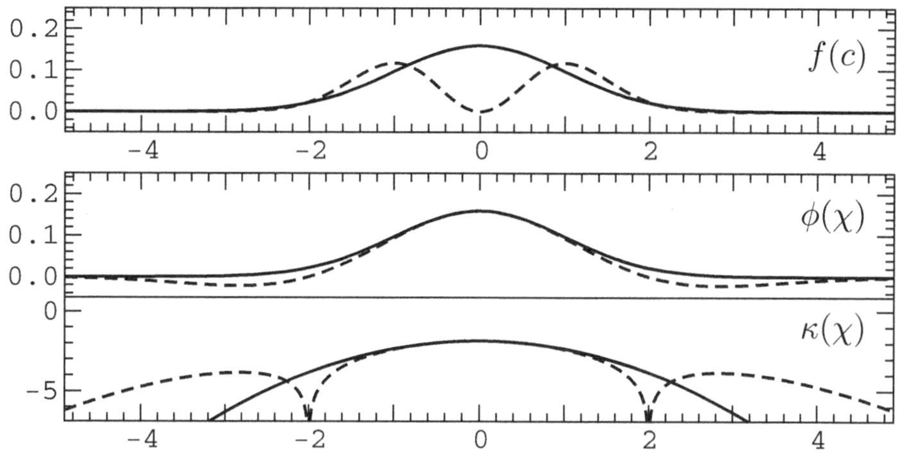

Fig. 1. Radial cross sections of the initial (dashed) and final (solid) phase space density (top), the characteristic function (center) and the real part of the second characteristic function (bottom) for the 2D case. The phase space density is non-negative, the characteristic function, however, has two zeros which move to larger χ with time. These zeros result in two poles for the second characteristic function and a non-continuous imaginary part

the radial part, which results in a singularity for the real part and a non-continuous, but piecewise constant imaginary part of the cumulant generating function (we restrict our considerations to the main branch of ln). With time advancing, the zero moves toward larger χ converging to the equilibrium solution.

In the following we will restrict our considerations to the 2D-case, for which we obtain

$$\lambda = \sum_r n_r \sqrt{\frac{2\,\kappa_{rs}}{\mu_{rs}}}\, \frac{1 - {\Delta_{rs}}^2}{2}\, \frac{\omega_2 - {\Delta_{rs}}^2\,(\omega_2 - 4\,\omega_1)}{4} \tag{18}$$

where the ω_n are constants given by

$$\omega_n = \int_{-\infty}^{+\infty} \mathrm{d}\varepsilon\ 1 - \cos\left(n\,\vartheta^{2\mathrm{D}}(\varepsilon)\right)\ . \tag{19}$$

In general relation (18) is not satisfied for all species s if we allow for an arbitrary choice of the species parameters μ_{rs}, Δ_{rs} and κ_{rs}. Thus it should be read as one rule for determination of λ and $(N_s - 1)$ constraints on the species parameters.

3 Mesoscopic fluid modelling

For simulating transition regime flows it becomes essential to extend the well known macroscopic models to include a more detailed description of the fluid flow. This is because models for rather dense gases, such as the EULER or NAVIER-STOKES equations become invalid and methods of solving the BOLTZMANN equation become too expensive in this regime. Such flow conditions are mainly characterized by the mean free path being in the order of a characteristic flow length. Thus the flow conditions for the transition regime are between dense gases ($Kn \ll 1$) where modeling by EULER or NAVIER-STOKES equations is valid and rarefied gases ($Kn \gg 1$) for which the BOLTZMANN equation is an adequate description.

3.1 Moment equations

By repeated differentiation of (6) with regard to $\underline{\chi}$ and setting $\underline{\chi} = 0$ afterwards we find an infinite system of coupled, nonlinear balance equations

$$\begin{aligned} \partial_t M_s^0 + \partial_{\underline{x}} \cdot M_s^1 &= G_s^0 + \textstyle\sum_r P_{rs}^0 \\ \partial_t M_s^1 + \partial_{\underline{x}} \cdot M_s^2 &= G_s^1 + \textstyle\sum_r P_{rs}^1 \\ &\vdots \\ \partial_t M_s^\alpha + \partial_{\underline{x}} \cdot M_s^{\alpha+1} &= G_s^\alpha + \textstyle\sum_r P_{rs}^\alpha \\ &\vdots \end{aligned} \quad \text{with} \quad \begin{aligned} M_s^\alpha &= (2\pi)^{d/2}\, \partial_{\mathrm{i}\underline{\chi}}^\alpha\, \varphi_s \Big|_{\underline{\chi}=0} \\ G_s^\alpha &= (2\pi)^{d/2}\, \partial_{\mathrm{i}\underline{\chi}}^\alpha\, \Gamma_s \Big|_{\underline{\chi}=0} \\ P_{rs}^\alpha &= (2\pi)^{d/2}\, \partial_{\mathrm{i}\underline{\chi}}^\alpha\, \Xi_{rs} \Big|_{\underline{\chi}=0} \end{aligned} \tag{20}$$

for the so-called moments $M_s^\alpha = \overline{(\underline{c}^\alpha)}_s$. For powers of the particle velocity $\underline{c}$ or the partial derivative $\partial_{\mathrm{i}\underline{\chi}}$ with respect to imaginary unit i times inverse velocity $\underline{\chi}$ a scalar exponent α denotes the tensorial power of order α, while a multi-index will denote the appropriate element of such a tensor. Similarly, for the moments M_s^α, force terms G_s^α and productions P_{rs}^α, a scalar superscript α denotes the full tensor (of rank α) and a multi-index the corresponding element.

Moments have been of particular interest, because the low order moments are related to particle number density, momentum and energy density, all of which are conserved quantities. And further, if one assumes the phasespace density f_s describes a local equilibrium state (where equilibrium parameters may vary with t and $\underline{x}$), the low order moment equations give actually the EULER equations.

The advection term for equation α couples moments of order α and $\alpha+1$, as the flux in one equation appears as density in the equation of next higher order and vice versa. In general we must assume that production terms may couple between any orders α, but for the particular interaction model considered here we observe only coupling towards lower orders. Currently it appears to be not even known whether the quantities appearing in equation (20) are well-defined functions for *every* solution of (2). E.g. boundedness of the appearing integrals or proper differentiability of the moments are by no means obvious for non-equilibrium solutions. Anyhow, moments have been shown to be well-defined at least for the spatially homogeneous [17] and the nearly homogeneous [18] case, so it might be more generally true.

3.2 The closure problem

One promising approach to modelling mesoscopic fluid flow is to consider finite subsets of (20) as approximations of (2) by taking only equations of order $\alpha = 0, \ldots, N_\alpha$, where the 'level of detail' can be adjusted by N_α. Truncation of (20) at some order N_α imposes the so-called closure problem, which consists in expressing the moments M_s^α, the force terms G_s^α and the productions P_s^α as a function of some set of variables (traditionally the moments themselves) so that the finite subsystem is closed. This is achieved by making an ansatz – or imposing physical principles to determine a suitable ansatz – for the phase space density f_s or the characteristic function φ_s with some ansatz parameters. Given the functional form of the ansatz, a relation of the parameters and the moments is determined by the condition that the first N_α moments of the ansatz should equal those of the 'true' (but unknown) solution. Solving this relation for the parameters as a function of the moments allows to express the ansatz for f_s in terms of the moments and in turn to determine $M_s^{N_\alpha+1}$, G_s^α and P_s^α as a function of the moments so that (20) is closed.

There are various criteria proposed in the literature on how to achieve this closure. It turns out that moment method(s), regardless of the particular closure employed, give results in good agreement with kinetic theory [19–22], where phenomenological descriptions of gases (like NAVIER-STOKES) fail. Nevertheless equations (20) are in general hard to obtain – even for states close to equilibrium – and become quickly very complicated with increasing order N_α. In the following we give a short overview of the most important proposals. There are more closure procedures discussed in the literature (see, for instance [23], [24]) but for the scope of this work we would like to restrict to the following approaches which are conceptually different.

In the method proposed by GRAD [1], f_s is factored into a (local) equilibrium part and a non-equilibrium part, the latter being expanded in a series of HERMITE polynomials ortho-normalized with regard to the $(t, \underline{x})$-local equilibrium part as weight function. From a slightly different point view, the method of GRAD may be viewed as an expansion of the phase space density in terms of HERMITE functions. As the HERMITE functions, however, are the eigenfunctions of the FOURIER-transform this means that we may as well view GRAD's approach as an expansion of the *first* characteristic function in terms of HERMITE-functions. Truncating the expansion at some arbitrary order N_α, this approach allows to determine the closure exactly but it appears to be difficult to give useful expressions for the entropy density, its flux or production rate.

WALDMANN [25] observed that – for states close to thermal equilibrium where a linear approximation of the collision operator holds – expressing f_s as a *sum* of a (local) equilibrium and a small deviation-from-equilibrium part leads to a linear integro-differential equation for the deviation from equilibrium. Assuming a space-homogeneous gas with a solution where the deviation from equilibrium decays exponentially in time leads to an eigenvalue problem for the function describing the deviation from equilibrium. An analytical solution can be given for the special case of a MAXWELL gas where an orthonormal system of eigenfunctions can be constructed as a product of SONINE polynomials and spherical harmonics when using spherical coordinates in velocity space. These correspond to irreducible homogeneous tensors when using cartesian coordinates in velocity space, which is why WALDMANN considers an expansion of the deviation from equilibrium with regard to these eigentensors of the linearized collision operator. Similar to GRAD's method, exact expressions for the entropy density, it's flux or production rate are difficult to obtain.

Existence of a properly defined entropy density is, however, the main emphasis in the framework of Extended Irreversible Thermodynamics (EIT) by MUELLER et al. [2]. There a particular ansatz form is obtained by applying a local formulation of the second law of thermodynamics. Unfortunately, closure is a very hard problem, as the ansatz function is an exponential of a tensor polynomial in $\underline{c}$. Nevertheless entropy density and flux can be given but it appears to be difficult to give an analytic form of the entropy production. Usually one resorts to an expansion of the exponential close to equilibrium, which leads to a closure similar to the one by GRAD [26].

The entropy production is payed more attention to by the modified moment method [27] proposed by EU, where it is regarded a direct measure of energy dissipation. So the ansatz for f_s is chosen such that the entropy production takes a simple form, making the connection of evolution of non-conserved variables and entropy production quite obvious. But again: neither is there a method for exact closure nor is it simple to give an analytic form of the entropy density.

4 The cumulant method

For the cumulant method we make a TAYLOR expansion of the logarithm of the characteristic function, which simplifies the deriviation of the equations satisfied by the ansatz parameters. The resulting equations are considerably simpler than the equations obtained by the Method of GRAD, however, it becomes difficult to give an equivalent procedure of expansion for the phase space density. We find the low order GRAD coefficients to be equivalent to the cumulants [4], but there are additional nonlinear terms in the relation between the ansatz parameters and the moments beginning with order $\alpha = 6$, which we suspect to be the reason for the resulting simplicity of the cumulant equations.

The cumulant method is based on the idea that – being interested in 'macroscopic' quantities – we are also interested in changes on macroscopic (slow) time scales. This allows to assume that fast relaxation processes have (almost) reached their equilibrium state and that their dynamics can be neglected if we are only interested in the slower processes. If we choose cumulants as macroscopic parameters for the description this means that we may assume equilibrium values (which are conveniently zero) for the high-order cumulants.

4.1 Cumulant-Ansatz

Thus our ansatz is a polynomial approximation of the second characteristic function $\kappa_s = \ln\left((2\pi)^{d/2}\,\varphi_s\right)$ so that we have

$$\varphi_s^{CM} = \frac{1}{(2\pi)^{d/2}} \exp\left(\sum_{\alpha=0}^{N_\alpha} \frac{\mathrm{i}^\alpha}{\alpha!} \underline{\chi}^\alpha \cdot C_s^\alpha\right) \tag{21}$$

with some arbitrary truncation number N_α. A closed set of equations of motion – or alternatively the relations between moments M_s^α, productions P_s^α and force terms G_s^α and the cumulants C_s^α – can be directly obtained by (20).

We obtain the following relations between the first order moments and cumulants:

$$\begin{aligned} M^0 &= e^{C^0} & M^{xx} &= e^{C^0}\left(C^x\,C^x + C^{xx}\right) \\ M^x &= e^{C^0} C^x & M^{xy} &= e^{C^0}\left(C^x\,C^y + C^{xy}\right) \\ M^y &= e^{C^0} C^y & M^{yy} &= e^{C^0}\left(C^y\,C^y + C^{yy}\right) \end{aligned}$$

$$\begin{aligned} M^{xxx} &= e^{C^0}\left(C^x\,C^x\,C^x + 3\,C^x\,C^{xx} + C^{xxx}\right) \\ M^{xxy} &= e^{C^0}\left(C^x\,C^x\,C^y + 2\,C^x\,C^{xy} + C^y\,C^{xx} + C^{xxy}\right) = M^{xyx} = M^{yxx} \\ M^{xyy} &= e^{C^0}\left(C^x\,C^y\,C^y + C^x\,C^{yy} + 2\,C^y\,C^{xy} + C^{xyy}\right) = M^{yxy} = M^{yyx} \\ M^{yyy} &= e^{C^0}\left(C^y\,C^y\,C^y + 3\,C^y\,C^{yy} + C^{yyy}\right) \end{aligned} \tag{22}$$

The equations obtained this way from (20) are in balance form. It is clear from their definition that the moments and cumulants of order α have only $\binom{\alpha+d}{d-1}$ linearly independent components, because the product between velocity components is commutative. Denoting the reduced, linear independent variables with a tilde, writing (20) in its reduced from and making use of the chain rule we may derive equations with the cumulants as basic fields, because the JACOBI-matrix $\left(\partial_{\tilde{C}_s} \tilde{M}_s\right)$ is not singular for the reduced variables and so its inverse is defined. The result is a set of equations in convection (or quasi-linear) form, stated for the reduced cumulants as the set of primitive variables

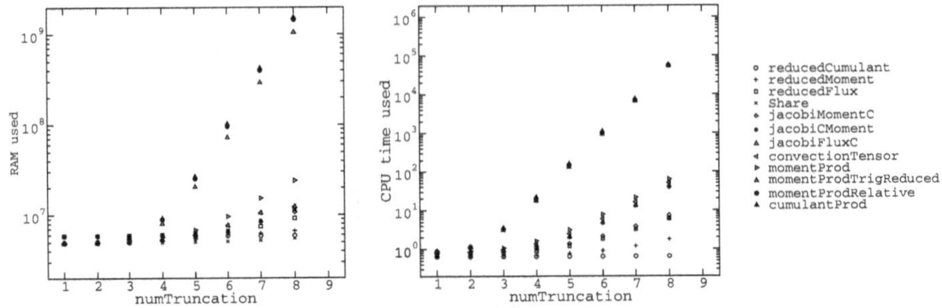

Fig. 2. Scaling of memory (bytes) and CPU time (seconds) requirements to obtain the cumulant equations of a given truncation order **numTruncation** $= N_\alpha$ when run with MATHEMATICA 5.0. Shown are the resources used when the given step is completed. Results are for the most complicated case of an inert mixture, performed on a Intel IA64-Architecture machine (dual 1GHz Itanium2 CPUs and 12GB RAM)

$$\partial_t \tilde{C}_s + \underline{\underline{A_s}} \cdot \partial_{\underline{x}} \tilde{C}_s = \tilde{E}_s + \sum_{r \in \mathcal{N}_s} \tilde{B}_{rs} \tag{23}$$

with

$$\begin{aligned} \text{convection tensor } \underline{\underline{A_s}} &= \left(\partial_{\tilde{C}_s} \tilde{M}_s\right)^{-1} \cdot \left(\partial_{\tilde{C}_s} \underline{\tilde{F}_s}\right), \\ \text{production terms } \tilde{B}_{rs} &= \left(\partial_{\tilde{C}_s} \tilde{M}_s\right)^{-1} \cdot \tilde{P}_{rs}, \\ \text{and force term } \tilde{E}_s &= \left(\partial_{\tilde{C}_s} \tilde{M}_s\right)^{-1} \cdot \tilde{G}_s, \end{aligned} \tag{24}$$

which takes a particularly simple form, namely $\tilde{E}_s = (0 \quad \underline{a}_s \quad 0 \quad 0 \quad \ldots)^{\mathrm{T}}$.

4.2 Symbolic derivation of Cumulant equations

To obtain (23) we mainly have to calculate derivatives of the characteristic function and the collision terms as well as integrate over the collision parameters. Using symbolic formula manipulation systems like MATHEMATICA [28], this process can be automated and equations up to high truncation orders N_α can be obtained. In [29] we give a detailed description of a possible MATHEMATICA implementation. Except for some technicalities the program proceeds in the sequence determined by (24): First the relation between cumulants and moments is derived for both the densities and fluxes in the balance equations (20) (steps `reducedCumulant`, `reducedMoment`, `reducedFlux`). Next the JACOBIan and its inverse, and the convection tensor are calculated (steps `jacobiMomentC`, `jacobiCMoment`, `convectionTensor`). Once the moment productions are known (`momentProd`), integration over the collision parameters can be carried out by simple substitution according to (19) once the powers of trigonometric expressions that stem from (11) have been put in a reduced form (`momentProdTrigReduced`, `momentProdRelative`). As can be seen from figure 2, this step takes most of the time required to obtain the cumulant equations. As a result, one obtains symbolic expressions for the advection tensor, as well as the production terms. However, this calculation has to be carried out only once in order to obtain the equations in symbolic form and to generate code for a numerical solver. Fig. 2 shows the scaling of CPU time and memory requirements to derive equation (23) with the truncation number N_α. The results shown are for the most complicated case (considered here) of an inert mixture. For the single component case, expressions and requirements simplify considerably and equations can be derived up to much higher orders.

4.3 Eigensystem of linearized productions

The production terms $\tilde{B}^\alpha_{rs}(\tilde{C}_r, \tilde{C}_s)$, calculated from (7) by (20) and (24), are in general highly nonlinear functions of the cumulants $\tilde{C}^\alpha_r$ and $\tilde{C}^\alpha_s$. The resulting expressions are much simpler if we rewrite the production terms as $\tilde{B}_{rs} = \tilde{B}_{rs}(\tilde{C}_s + \tilde{C}_{rs}, \tilde{C}_s)$ with the 'relative cumulants' $C^\alpha_{rs} = C^\alpha_r - C^\alpha_s$ and $C^0_{rs} = 0$. For states close to thermodynamic equilibrium the production terms may be linearized by making a TAYLOR-expansion of $\tilde{B}_{s,rs}$ around the equilibrium state. The linearized production terms then read

$$\tilde{B}^{\text{lin}}_{rs} = \underline{\underline{\tilde{B}_s}} \cdot \Delta\tilde{C}_s + \underline{\underline{\tilde{B}_{rs}}} \cdot \Delta\tilde{C}_{rs} \tag{25}$$

with $\Delta\tilde{C} = \tilde{C} - \tilde{C}^{\text{eq}}$,

$$\underline{\underline{\tilde{B}_s}} = \left(\partial_{\tilde{C}_s} \tilde{B}_{s,rs}\right)^{\text{eq}} \quad \text{and} \quad \underline{\underline{\tilde{B}_{rs}}} = \left(\partial_{\tilde{C}_{rs}} \tilde{B}_{s,rs}\right)^{\text{eq}}. \tag{26}$$

For the case of MAXWELL pseudo-molecules both $\underline{\underline{\tilde{B}_s}}$ and $\underline{\underline{\tilde{B}_{rs}}}$ have block-diagonal structure [5], thus for the linearized production terms only cumulants of the same order α are coupled. Performing a JORDAN decomposition $\underline{\underline{\tilde{B}_s}} = \underline{\underline{S}} \cdot \underline{\underline{J}} \cdot \underline{\underline{S}}^{-1}$ with the normal form $\underline{\underline{J}}$ and similarity matrix $\underline{\underline{S}}$ we can calculate the eigenvariables $E^{\alpha ri}$ as components of $\tilde{E} = \underline{\underline{S}}^{-1} \cdot \tilde{C}$ of the linearized production terms for the single component gas [30]:

$$\begin{pmatrix} E^{00}_s \end{pmatrix} = \begin{pmatrix} C^0 \end{pmatrix} \qquad \begin{pmatrix} E^{311}_s \\ E^{312}_s \\ E^{321}_s \\ E^{322}_s \end{pmatrix} = \frac{1}{4}\begin{pmatrix} C^{xxx} + C^{xyy} \\ C^{yyy} + C^{xxy} \\ C^{xxx} - 3\,C^{xyy} \\ C^{yyy} - 3\,C^{xxy} \end{pmatrix}$$

$$\begin{pmatrix} E^{11}_s \\ E^{12}_s \end{pmatrix} = \begin{pmatrix} C^x \\ C^y \end{pmatrix}$$

$$\begin{pmatrix} E^{20}_s \\ E^{211}_s \\ E^{212}_s \end{pmatrix} = \frac{1}{2}\begin{pmatrix} C^{xx} + C^{yy} \\ C^{xx} - C^{yy} \\ 2C^{xy} \end{pmatrix} \qquad \begin{pmatrix} E^{40}_s \\ E^{411}_s \\ E^{412}_s \\ E^{421}_s \\ E^{422}_s \end{pmatrix} = \frac{1}{8}\begin{pmatrix} C^{xxxx} + 2\,C^{xxyy} + C^{yyyy} \\ 4\,C^{xxxx} - 4\,C^{yyyy} \\ 4\,C^{xxxy} + 4\,C^{xyyy} \\ C^{xxxx} - 6\,C^{xxyy} + C^{yyyy} \\ 4\,C^{xxxy} - 4\,C^{xyyy} \end{pmatrix}$$

$$\cdots \tag{27}$$

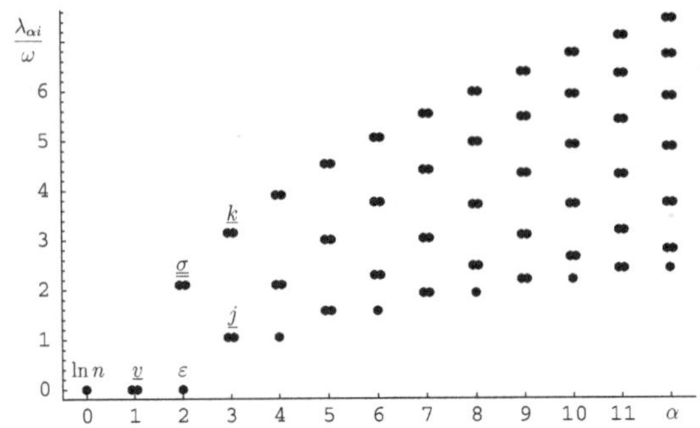

Fig. 3. Spectrum of eigenvalues for the linearized collision operator up to order $N_\alpha = 12$. For each approximation order α the eigenvalues of the corresponding submatrix $\underline{\underline{\tilde{B}_s^{\alpha\alpha}}}$ are shown

In the notation used for the indices, α enumerates the cumulant order, r enumerates the rate of relaxation in ascending order and i enumerates eigenvariables with degenerate relaxation rates and is omitted if there is only a single one. For those eigenvariables the space-homogeneous equations of motion (23) decouple into separate equations

$$\partial_t E^{\alpha r i} = -\omega_{\alpha r i} E^{\alpha r i} \tag{28}$$

with some relaxation rate $\omega_{\alpha r i}$ given by the corresponding main diagonal element of the normal form $\underline{\underline{J}}$ [5].

The spectrum of the eigenvalues is shown in figure 3 for the first approximation orders. We observe that eigenvalues appear pairwise except for the lowest eigenvalue for even α. The pairs of eigenvariables for odd α are 'symmetric' with regard to interchange of x and y and could thus be characterized as flux-like specific quantities. For even α we have one scalar and pairs of 'asymmetric' eigenvariables we could characterize as energy- and stress-like specific quantities. The terminology chosen for characterization is derived from the relation of the first few eigenvariables to classical thermodynamic quantities. It appears that the first eight eigenvariables can be related [5, 6] one-to-one to well-known thermodynamic quantities which have the three appearing symmetry-properties observed for the eigenvariables of the linear approximation of the production terms:

energy-like energy ε, log-density $\ln n$

flux-like velocity $\underline{v}$, flux of specific energy $\underline{j}$

stress-like (shear/normal) stress $\underline{\underline{\sigma}}$

Thus the relaxation behavior of the cumulants for the space homogeneous BOLTZMANN equation provides the key to match the cumulants with the macroscopic thermodynamic variables density n, specific energy ε, velocity $\underline{v}$, shear stress $\sigma_{\rightleftharpoons}$ and normal stress $\sigma_\circ$ as well as heat flux $\underline{j}$:

$$
\begin{aligned}
n &= e^{C^0} & \varepsilon &= \tfrac{1}{2}(C^{xx} + C^{yy}) \\
\underline{v} &= \begin{pmatrix} C^x \\ C^y \end{pmatrix} & \underline{j} &= \tfrac{1}{2}\begin{pmatrix} C^{xxx} + C^{xyy} \\ C^{yyy} + C^{xxy} \end{pmatrix} \\
\underline{\underline{\sigma}} &= \tfrac{1}{2}\begin{pmatrix} C^{xx} - C^{yy} & 2\,C^{xy} \\ 2\,C^{xy} & C^{yy} - C^{xx} \end{pmatrix} = \begin{pmatrix} \sigma_\circ & \sigma_{\rightleftharpoons} \\ \sigma_{\rightleftharpoons} & -\sigma_\circ \end{pmatrix}
\end{aligned}
\tag{29}
$$

The motivating assumption (that high-order cumulants decay more quickly than low-order cumulants), seems to hold in principle, at least for the MAXWELL interaction model. But there is considerable overlap between the spectra for various approximation orders. This poses the question whether a simple truncation as in (21) is a proper ansatz for the characteristic function φ_s. It is reasonable to expect a one-to-one correspondence of the eigensystem (27) to the eigenfunctions discussed by WALDMANN [25]. Thus it should be possible to construct a "consistent order of magnitude" closure for the cumulants similar to [24]. This could be achieved by rewriting (23) in terms of the eigenvariables and performing a MAXWELL iteration in order to assign the orders of magnitude to the various terms. The first step of this procedure has already been used to clarify the relation of (23) to the NAVIER-STOKES equations.

4.4 Application to the space-homogeneous Boltzmann equation

So how does the cumulant method apply to a single component gas in two dimensions ($d = 2$) with only one species ($N_s = 1$)? Let us assume the space-homogeneous case and the species parameters be given by specific energy $\varepsilon_1 = \varepsilon$, density $n_1 = n$, particle mass $m_1 = 2\,\mu$ and interaction strength $\kappa_{11} = \kappa$; all chosen unity for the numerical calculations. From equation (18) we find ($\Delta_{11} = 0$)

$$
\lambda = n\sqrt{\frac{2\,\kappa}{\mu}\,\frac{\omega_2}{8}}\,. \tag{30}
$$

Using the solution (15) we can calculate an analytic solution for the time-development of the cumulants and eigenvariables. Due to simplicity and high symmetry of the solution we find that only a few eigenvariables are non-zero, namely

$$
\begin{aligned}
E^{00} &= C^0 \\
E^{20} &= \tfrac{1}{2}(C^{xx} + C^{yy}) \\
E^{40} &= \tfrac{1}{8}(C^{xxxx} + 2\,C^{xxyy} + C^{yyyy}) \\
E^{60} &= \tfrac{1}{32}(C^{xxxxxx} + 3\,C^{xxxxyy} + 3\,C^{xxyyyy} + C^{yyyyyy}) \\
E^{80} &= \tfrac{1}{64}(C^{xxxxxxxx} + 4\,C^{xxxxxxyy} + 6\,C^{xxxxyyyy} + 4\,C^{xxyyyyyy} + C^{yyyyyyyy}) \\
&\ \,\vdots
\end{aligned}
\tag{31}
$$

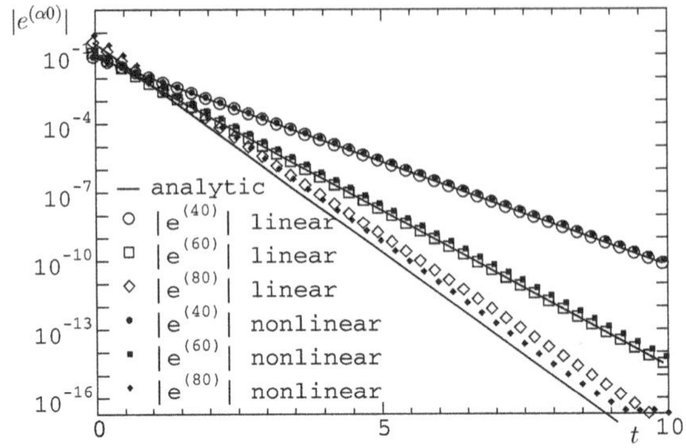

Fig. 4. Evolution of the eigenvalues for the pure Maxwell gas. We observe excellent agreement between analytic solutions and the numerical results for calculation with the non-linear production terms. However, the high-order eigenvariable E^{80} relaxes slower if linearized production terms are used

for which the time evolution is determined by solution (15) as

$$\begin{aligned} E^{00} &= \ln(n) \\ E^{20} &= \varepsilon \\ E^{40} &= -\left[\varepsilon\,\theta\,\exp\left(-\lambda\,t\right)\right]^2 \\ E^{60} &= 3\left[\varepsilon\,\theta\,\exp\left(-\lambda\,t\right)\right]^3 \\ E^{80} &= -18\left[\varepsilon\,\theta\,\exp\left(-\lambda\,t\right)\right]^4 . \end{aligned} \tag{32}$$

Thus, for the particular (isotropic) BOBYLEV/KROOK-WU solution, the only nonvanishing eigenvariables are those of even order α with the slowest relaxation rates, which are not degenerate.

Fig. 4 shows the time evolution of the eigenvalues for the single component MAXWELL gas. We observe an excellent agreement for the numerical results and the analytic solution. This is the case for all three non-trivial eigenvalues if the nonlinear production terms are used in the numerical calculation. If the linearized productions are used, we observe a slower relaxation for E^{80} than predicted by equation (32). The other two non-trivial eigenvalues are as predicted by the solution. The reason for this becomes obvious when we write the equations of motion for the eigenvariables:

$$\begin{aligned} \partial_t E^{00} &= 0 \\ \partial_t E^{20} &= 0 \\ \partial_t E^{40} &= \tfrac{\omega_2}{2}\left(2\,(E^{211})^2 + 2\,(E^{212})^2 - E^{40}\right) \\ \partial_t E^{60} &= \tfrac{3\omega_2}{4}\left(3\left((E^{311})^2 + (E^{312})^2 + (E^{321})^2 + (E^{322})^2\right) + \right. \\ &\qquad \left. 2\,E^{211}\,E^{411} + 4\,E^{212}\,E^{412} - E^{60}\right) . \end{aligned} \tag{33}$$

Even though the productions are linear in the cumulants only up to order $\alpha = 3$, the production term for E^{40} is linear in this particular case because E^{211} and E^{212} vanish for the particular initial conditions. However, E^{80} is 'truly' nonlinear even for the particular case of the BOBYLEV/KROOK-WU solution, in which we have

$$\partial_t E^{80} = \frac{1}{32}\left(36\,(4\,\omega_2 - \omega_4)\,(E^{40})^2 - (28\,\omega_2 + \omega_4)\,E^{80}\right) . \tag{34}$$

Thus we expect such deviations to occur for eigenvalues for larger α too when linearized productions are used in the simulation.

4.5 Eigenstructure of the advection tensor

The convection tensor $\underline{\underline{A_s}}$ is sparsely populated and has a particularly simple structure, as can be seen from the x-component of $\underline{\underline{A_s}}$. For the 2D case we have

$$\left[\underline{\underline{A_s}}\right]_x = \left(\begin{array}{c|cc|ccc|cccc|c}
C_s^{x} & 1 & 0 & 0 & 0 & 0 & 0 & 0 & 0 & 0 & \\
\hline
C_s^{xx} & C_s^{x} & 0 & 1 & 0 & 0 & 0 & 0 & 0 & 0 & \\
C_s^{xy} & 0 & C_s^{x} & 0 & 1 & 0 & 0 & 0 & 0 & 0 & \\
\hline
C_s^{xxx} & 2C_s^{xx} & 0 & C_s^{x} & 0 & 0 & 1 & 0 & 0 & 0 & \\
C_s^{xxy} & C_s^{xy} & C_s^{xx} & 0 & C_s^{x} & 0 & 0 & 1 & 0 & 0 & \\
C_s^{xyy} & 0 & 2C_s^{xy} & 0 & 0 & C_s^{x} & 0 & 0 & 1 & 0 & \\
\hline
C_s^{xxxx} & 3C_s^{xxx} & 0 & 3C_s^{xx} & 0 & 0 & C_s^{x} & 0 & 0 & 0 & \ddots \\
C_s^{xxxy} & 2C_s^{xxy} & C_s^{xxx} & C_s^{xy} & 2C_s^{xx} & 0 & 0 & C_s^{x} & 0 & 0 & \\
C_s^{xxyy} & C_s^{xyy} & 2C_s^{xxy} & 0 & 2C_s^{xy} & C_s^{xx} & 0 & 0 & C_s^{x} & 0 & \\
C_s^{xyyy} & 0 & 3C_s^{xyy} & 0 & 0 & 3C_s^{xy} & 0 & 0 & 0 & C_s^{x} & \\
\hline
 & \ddots & & \ddots & & & \ddots & & & & \ddots
\end{array}\right) . \tag{35}$$

Remember that multi-index superscripts refer to the actual cumulant tensor components. By taking the limit $N_\alpha \to \infty$ equation (23) may be considered an infinite system just as (20). For a truncation of finite order N_α, however, the bottom left block in (35) vanishes.

Fig. 5 shows the spectrum of $\underline{\underline{A_s}}(\tilde{C})$ as a function of the approximation order N_α. This spectrum has been obtained by inserting the equilibrium values [4] for the cumulants in (35) and determining the eigenvalues of the resulting matrix numerically using MATHEMATICA [28]. As for EIT [2], the eigenvalue spectrum for a given order N_α contains all eigenvalue spectra for approximations of lower order. Further we observe finite but monotonically growing maximum and minimum eigenvalues. These advection tensor eigenvalues can be related to the (finite) speeds of propagation of weak discontinuities and should be real (so that (23) is hyperbolic). Consistent with EIT, growth of the maximal magnitude of the eigenvalues slows down with increasing truncation order. These eigenvalues corresponding to the highest characteristic

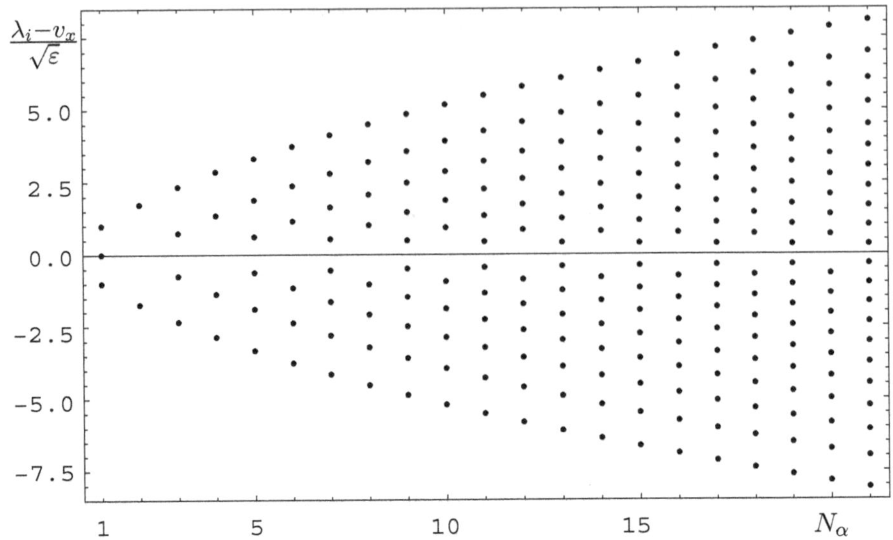

Fig. 5. Eigenspectrum of $\underline{\underline{A}}_s(\tilde{C})$ in equilibrium as it depends on the order of approximation order N_α. For each order, only the newly appearing spectral values are plotted. We observe a finite maximum magnitude, growing monotonically with N_α

speeds evaluated in equilibrium play an important role in modeling shock structures, as for shock speeds beyond that value unphysical sub-shocks appear in the solutions of the approximate equations [31, 32]. This implies that many moments or cumulants have to be considered for fast shocks and might be considered a drawback of the cumulant method and moment methods in general.

Solving (29) for the (low-order) cumulants, and inserting these into (35), we can calculate [33] the dependence of the convection tensor eigen-spectrum on the classical variables. For this, one variable has been varied and for the others equilibrium values have been assumed. We find that demanding hyperbolicity of the cumulant equations, which requires real eigenvalues for all components of $\underline{\underline{A}}_s$, imposes different possible constraints on the domain of allowed eigenvariable values. The first case is that no constraints are imposed, as is the case for v_x and v_y. An arbitrary mean velocity just produces a shift of the eigenvalue spectrum of $\underline{\underline{A}}_s$, as we would expect from a set of equations with GALILEIAN invariance. The same situation is observed for the shear stress $\sigma_{\rightleftharpoons}$, which does not influence the spectrum at all. The second case is observed for the energy density ε, where the spectra for both the x and y component impose a lower bound on ε, independent of N_α. With the third case, boundedness of the normal stress $\sigma_{\circ}$, j_x and j_y is imposed but for two different reasons: for $\sigma_{\circ}$, the x and y component of $\underline{\underline{A}}_s$ impose either an upper or a lower bound which are the same for any N_α. For $\underline{j}$, however, one $\underline{\underline{A}}_s$ component imposes both upper and lower bounds and the other component of $\underline{\underline{A_s}}$ does not impose any bounds. But we also observe the paradoxical situation that the interval of allowed j_x-values for (23) to be hyperbolic becomes smaller with increasing N_α. Whether the allowed interval remains finite or converges to the empty set for $N_\alpha \to \infty$ remains an open question. We note that this might not be the case in real flow situations, where other cumulants may have non-equilibrium values, thereby possibly compensating this effect. Thus, 'not too far' from equilibrium, (23) might be characterized as a hyperbolic system of partial differential equations.

4.6 Numerical scheme

There are several modern, robust numerical methods for solving a time-dependent hyperbolic system of partial differential equations (see, for instance [34–38]). However, most of them rely on a conservation or balance law formulation of the governing equations. Unfortunately we are not able to find a balance law formulation for the cumulants directly, as the convection tensor components are not JACOBIan matrices. The moment equations are equations in balance form, so this might be a possible set of equations to use with these methods. But, this would ultimately require (re-)construction of the fluxes from moment values: We would have to calculate the cumulant values from moments and then the fluxes from the cumulants. Though theoretically possible (from (22) and the inverse relation) this could lead to numerical difficulties due to introduction of round-off errors in each iteration step [26].

If we would like to use finite element methods for discretization and numerical solution the problem appears that the required variational forms are usually not symmetric and therefore require stabilization [39]. On the other hand it is well known [40] that for systems where an analytic form of the entropy density is exists, the equations of motion become symmetric in a particular set of 'entropic' variables. Applying finite element discretizations to these symmetric hyperbolic equations it can be shown that the numerical solution will have the same (thermodynamic) stability properties as the continuous equations [41]. Except for necessary conditions, existence and construction of entropy functionals operating on the characteristic function has not been discussed in the literature so far.

We therefore choose an explicit finite difference approximation of (23), but should be aware that these methods are known *not* to be suited for problems with discontinuous solutions, such as shocks. We start by choosing a regular, orthogonal grid of spacing $\mathrm{d}t$ in time and $\mathrm{d}r$ in space to discretize the space and time domain. Let $(t, \underline{x})$ denote a grid point and $\underline{e}_i$ the unit vector in direction i. Then we may approximate the derivatives using the following finite difference ratios of first and second order. Thus we approximate the differentials in (23) by

$$\begin{aligned} \partial_t \tilde{C}_s &\approx \tfrac{1}{\mathrm{d}t}\,\delta_t^{(1)}\tilde{C}_s = \tfrac{1}{\mathrm{d}t}\Big(\tilde{C}_s(t+\mathrm{d}t,\underline{x}) - \tilde{C}_s(t,\underline{x})\Big) \\ \partial_{x_i} \tilde{C}_s &\approx \tfrac{1}{2\,\mathrm{d}r}\,\delta_{x_i}^{(2)}\tilde{C}_s = \tfrac{1}{2\,\mathrm{d}r}\Big(\tilde{C}_s(t,\underline{x}+\underline{e}_i\,\mathrm{d}r) - \tilde{C}_s(t,\underline{x}-\underline{e}_i\,\mathrm{d}r)\Big)\,. \end{aligned} \tag{36}$$

Solving for $\tilde{C}_s(t+\mathrm{d}t,\underline{x})$ we obtain the following simple, explicit iteration scheme with a typical finite-difference stencil

$$\begin{aligned} \tilde{C}_s(t+\mathrm{d}t,\underline{x}) = {} & \tilde{C}_s \\ & +\mathrm{d}t\Big(\tilde{E}_s + \textstyle\sum_r \tilde{B}_{rs}\Big) \\ & -\tfrac{\mathrm{d}t}{2\,\mathrm{d}r}\,\underline{\underline{A_s}}(\tilde{C}_s)\cdot\delta_{\underline{x}}^{(2)}\tilde{C}_s \end{aligned} \tag{37}$$

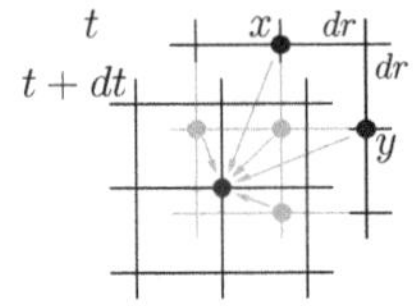

where the time and position arguments $(t,\underline{x})$ have been omitted on the right hand side. The scheme (37) is known to be unconditionally unstable. It can be stabilized by using the average over the values used for approximation of the derivatives in space instead of $\tilde{C}_s(t,\underline{x})$ [42].

4.7 Boundary conditions

As important as the details of the modeling, however, are boundary conditions. In order to formulate well-posed problems we need to give conditions for the cumulants at the boundaries $\partial\Omega$ of the flow domain Ω, e.g. if we want to describe flows past solid bodies or between solid walls. In macroscopic models these conditions need to be formulated in terms of gradients normal to the wall or in terms of the values at the wall. In kinetic theory, however, we need to give conditions in terms of the phase space density. These conditions reflect a model of interaction of the gas particles with the walls. It is due to this interaction that forces are exerted on and heat may be transferred across boundary surfaces. In order to give physically correct boundary conditions, detailed knowledge about the processes taking place at boundaries is required. As we do not possess this knowledge, difficulties in theoretical modeling arise; mainly due to lack of knowledge concerning an effective interaction potential of the gas with the surface. For a more detailed introduction to the subject, the reader is referred to [7,43–46].

In theory, arbitrarily complex boundary conditions may be derived, if they can be formulated as conditions for the distribution function f_s according to (21). In practice we might run into difficulties as there may be conditions where the required class of functions cannot be approximated well by the class of ansatz functions: Assume particles moving to the boundary are immediately re-emitted 'equilibrated' (e.g. particles leaving the wall have distribution f_s^{eq} with boundary velocity, temperature and density such that the density is conserved). These distribution functions would be discontinuous in velocity along a plane orthogonal to the wall orientation. With a continuous ansatz this could only be approximated by steep gradients, possibly requiring high order approximations.

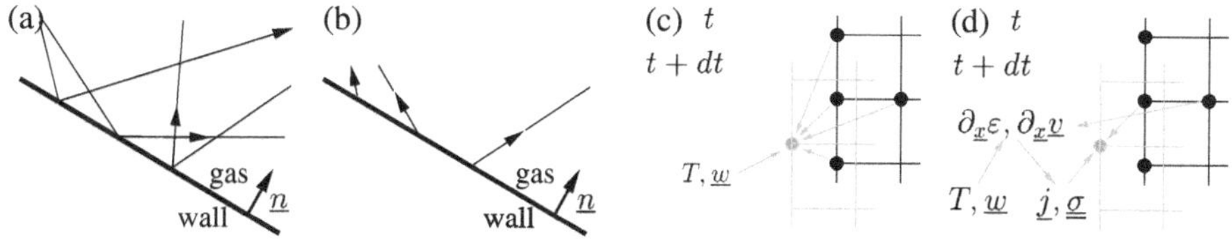

Fig. 6. The four types of boundary conditions discussed in the text: (a) adiabatic slip conditions, (b) with adiabatic no-slip conditions, (c) thermal no-slip conditions and (d) Navier-Stokes conditions

Adiabatic boundary conditions

The simplest conditions are adiabatic slip conditions, which can be modeled by an ideally reflecting boundary. In [4] we have discussed these boundary conditions already in detail, however, we give a short summary here. An ideally reflecting boundary (Fig. 6a) with orientation $\underline{n}$ at rest interacts with gas particles such that for particles moving with velocity $\underline{c}$ the normal component $\underline{n} \cdot \underline{c}$ of the particle velocity relative to the surface is inverted and the tangential component relative to the surface remains unchanged in an interaction with the boundary. Thus with this kind of interaction, the gas cannot exert shear stress on the boundary and there will be no heat exchange across the wall. The velocity tangential to the surface is arbitrary, so this poses an idealized 'slip' condition. An ideally retro-reflecting boundary (Fig. 6b) with orientation $\underline{n}$ interacts with gas particles such that for particles moving towards the surface the normal component as well as the tangential component of the particle velocity relative to the surface are reverted. This fixes the relative tangential velocity component to that of the wall (no-slip) and allows the fluid to exert shear forces on the boundary. These microscopic fluid-wall interaction models allow to derive conditions for the cumulants and their gradients at the wall, which has been discussed in [4]. Denoting the cumulant values at the node adjacent to the boundary with C_+^α, the conditions to employ for the moving, adiabatic slip boundary read

$$
\begin{aligned}
& & \delta_x^{(2)} C^0 &= 0 & \delta_y^{(2)} C^0 &= 0 \\
C_0^x &= 0 & \delta_x^{(2)} C^1 &= \tfrac{1}{dr} C_+^1 & \delta_y^{(2)} C^1 &= 0 \\
C_0^{xy} &= 0 & \delta_x^{(2)} C^2 &= 0 & \delta_y^{(2)} C^2 &= 0 \\
C_0^{xxx} = C_0^{xyy} &= 0 & \delta_x^{(2)} C^3 &= \tfrac{1}{dr} C_+^3 & \delta_y^{(2)} C^3 &= 0
\end{aligned}
\qquad (38)
$$

and the conditions to employ for the moving, adiabatic no-slip boundary read

$$
\begin{aligned}
& & \delta_x^{(2)} C^0 &= 0 & \delta_y^{(2)} C^0 &= 0 \\
C_0^1 &= \underline{w} & \delta_x^{(2)} C^1 &= \tfrac{1}{\mathrm{d}r} C_+^1 & \delta_y^{(2)} C^1 &= 0 \\
& & \delta_x^{(2)} C^2 &= 0 & \delta_y^{(2)} C^2 &= 0 \\
C_0^3 &= \underline{0} & \delta_x^{(2)} C^3 &= \tfrac{1}{\mathrm{d}r} C_+^3 & \delta_y^{(2)} C^3 &= 0
\end{aligned} \tag{39}
$$

where $\underline{w}$ denotes the wall velocity.

Thermal no-slip conditions

An important drawback of the adiabatic boundary conditions is the fact that heat dissipated in the flow region may not be transported out of the flow region. This is a considerable deficiency as almost all steady flow regimes require some kind of transport of the heat generated by dissipation out of the flow region. The main problem is that – for the two kinds of ideally reflective wall – we cannot prescribe wall velocity and temperature and have the gas develop stress and heat flux in response to the flow conditions at the wall with the (quite academic) boundary conditions presented above. For the thermal no-slip boundary conditions (Fig. 6c) used in the simulations we treat boundary nodes just as interior fluid nodes. The gradients in $\underline{n}$-direction are approximated by (first-order) one-sided differences. In each update, the wall velocity and the wall temperature substitute the value of C^1 and the trace of C^2. Otherwise the node is updated according to (37). This prescribes wall temperature and wall velocity at the boundary node. Shear stress at the wall and heat flux across the wall develop due to the gradients in the cumulants building up.

Navier-stokes conditions

In [5] we have demonstrated calculation of the production terms for the Maxwell gas. By considering states close to equilibrium for a single-component gas we found that the Jacobian of the production terms with regard to the cumulants is block-diagonal. This allows the definition of a set of eigenvariables of the linearized production terms. It turns out that the first eigenvariables can be related one by one to well-known macroscopic quantities, namely particle density n, mean particle velocity $\underline{v}$, mean energy ε, stress $\underline{\underline{\sigma}}$ and flux of specific energy $\underline{j}$. Performing the fist step of a Maxwell iteration [47], we recover the well-known constitutive relations for a Newtonian fluid with heat conduction according to Fouriers law. This motivates the Navier-Stokes boundary conditions (Fig. 6d): First we calculate the (classical) eigenvariables for the boundary node and the fluid node next to the boundary. Then, for the boundary node, replace $\underline{v}$ and specific ε by the wall velocity w and specific energy given by the wall temperature T. Further we approximate the gradients in velocity and energy by one-sided differences and calculate $\underline{\underline{\sigma}}$ and $\underline{q}$ from their constitutive relations. Now the corresponding cumulant values for the boundary node are obtained by the relation between the cumulants and the eigenvariables as obtained from the Maxwell iteration (given in [5]). With these boundary conditions employed for the numerical scheme (37) we can simulate various flow conditions that result in a stationary, non-equilibrium regime. However, in some cases properties of flows in the Navier-Stokes regime are reproduced, in other cases qualitative features of a rather dilute gas, depending on the boundary conditions employed.

5 Summary

We gave a comprehensive overview of the theory behind the cumulant method, the main results about the resulting equations and their properties, as well as a simple numerical scheme and possible boundary conditions to apply. The main ansatz is a TAYLOR expansion of the second characteristic function. From that ansatz, a set of moment equations can be derived by symbolic calculation up to (in principle) arbitrary high orders of approximation. Applying the method of deriving equations for the cumulants for the special case of a space-homogeneous gas close to a equilibrium state we can linearize the production terms and determine an eigensystem of the production terms. The low order eigen-variables can be related one-to-one to classic thermodynamic quantities. Next we have compared a numerical solution of the space-homogeneous equations to the exact BOBLEV/KROOK-WU solution. We find that the numerical solution coincides with the exact solution if the fully non-linear production terms are used. For the linearized production terms, relaxation rates may be under-estimated. The eigensystem of the advection tensor characterizes the system as hyperbolic as long as the system is not too far from equilibrium. The application of modern numerical methods appears to be difficult as the cumulant equations are not in symmetric form. Construction of symmetric equations would be possible if an entropy density could be given as a function of the cumulants. How this can be achieved remains an open question, as so far entropy functionals that operate on the first (or second) characteristic function have not been discussed extensively in the literature. Despite that, a simple finite difference scheme and microscopically or phenomenologically motivated boundary conditions have been given that can be used to obtain numerical solutions for simple flow problems.

References

1. H. Grad. On the kinetic theory of rarefied gases. *Comm. Pure Appl. Math.*, 2:331–407, 1949.
2. I. Müller and T. Ruggeri. *Rational Extended Thermodynamics*, volume 37 of *Springer Tracts in Natural Philosophy*. Springer-Verlag, New York, Berlin, Heidelberg, 2nd edition, 1998.

3. B. C. Eu. *Kinetic Theory and Irreversible Thermodynamics*, chapter 10.7 Modified Moment Method, pages 365–386. John Wiley & Sons, New York, 1992.
4. S. Seeger and K. H. Hoffmann. The cumulant method for computational kinetic theory. *Continuum Mech. Thermodyn.*, 12:403–421, 2000.
5. S. Seeger and K. H. Hoffmann. The cumulant method applied to a mixture of Maxwell gases. *Continuum Mech. Thermodyn.*, 14(2):321–335, 2002. see also Erratum: CMT 16(5):515, 2004.
6. S. Seeger. *The Cumulant Method.* Phd thesis, Chemnitz University of Technology, Chemnitz, September 2003. `http://archiv.tu-chemnitz.de/pub/2003/0120`.
7. C. Cercignani, R. Illner, and M. Pulvirenti. *The Mathematical Theory of Dilute Gases*, volume 106 of *Applied Mathematical Sciences.* Springer-Verlag, 1994.
8. J. O. Hirschfelder, C. F. Curtiss, and R. B. Bird. *Molecular Theory of Gases and Liquids.* Structure of Matter Series. John Wiley & Sons, New York, Chinchester, Brisbane, 2nd edition, 1964.
9. L. Boltzmann. *Vorlesungen über Gastheorie*, volume I, chapter Abschnitt III., pages 153–176. J.A. Barth Verlag, Leipzig, 1896.
10. E. Lukacs. *Characteristic Functions.* Griffin's Statistical Monographs and Courses. Charles Griffin & Company, London, 2nd edition, 1970.
11. A. V. Bobylev. The Fourier transform method in the theory of the Boltzmann equation for Maxwell molecules. *Doklady Akad. Nauk SSSR*, 225:1041–1044, 1975. in Russian.
12. A. V. Bobylev. The theory of the nonlinear spatially uniform Boltzmann equation for Maxwell molecules. *Sov. Sci. Rev. C: Math. Phys.*, 7:111–233, 1988.
13. C. Cercignani. *The Boltzmann Equation and Its Applications*, volume 67 of *Applied Mathematical Sciences.* Springer, 1988.
14. A. V. Bobylev. Exact solutions of the Boltzmann equation. *Dokl. Akad. Nauk SSSR*, 225(6):1296–1299, 1975. in Russian.
15. M. Krook and T. Wu. Exact solutions of the Boltzmann equation. *Phys. Fluids*, 20(10):1589–1595, 1977.
16. M. Krook and T.T. Wu. Exact solutions of Boltzmann equations for multicomponent systems. *Phys. Rev. Lett.*, 38(18):991–993, 1977.
17. B. Wennberg. On moments and uniqueness for solutions to the space homogeneous Boltzmann equation. *Transp. Theor. Stat. Phys.*, 23:533–539, 1994.
18. L. Arkeryd, R. Esposito, and M. Pulverenti. The Boltzmann equation for weakly inhomogeneous data. *Comm. Math. Phys.*, 111:393–407, 1987.
19. W. Loose and S. Hess. Nonequilibrium velocity distribution function of gases: Kinetic theory and molecular dynamics. *Phys. Rev. A*, 37(6):2099–2111, 1988.
20. S. Hess and M. Malek Mansour. Temperature profile of a dilute gas undergoing a plane Poiseuille flow. *Physica A*, 272:481–496, 1999.
21. Y. Sone, K. Aoki, S. Takata, H. Sugimoto, and A. V. Bobylev. Inappropriateness of the heat-conduction equation for description of a temperature field of a stationary gas in the continuum limit: Examination by asymptotic analysis and numerical computation of the Boltzmann equation. *Phys. Fluids*, 8(2):628–638, 1996.
22. D. Reitebuch and W. Weiss. Application of high moment theory to the plane Couette flow. *Continuum Mech. Thermodyn.*, 4:217–225, 1999.
23. C. D. Levermore. Moment closure hierarchies for kinetic theories. *J. Stat. Phys.*, 83(5/6):1021–1065, 1996.

24. I. Müller, D. Reitebuch, and W. Weiss. Extended thermodynamics – consistent in order of magnitude. *Continuum Mech. Thermodyn.*, 15:113–146, 2002.
25. L. Waldmann. *Transporterscheinungen in Gasen von mittlerem Druck*, volume XII of *Handbuch der Physik*, pages 295–514. Springer-Verlag, Berlin, 1958.
26. M. Torrilhon, J. D. Au, and H. Struchtrup. Explicit fluxes and productions for large systems of the moment method based on extended thermodynamics. *Continuum Mech. Thermodyn.*, 15(1):97–111, 2002.
27. B. C. Eu. *Kinetic Theory and Irreversible Thermodynamics.* John Wiley & Sons, Inc., New York, Chichester, Brisbane, 1992.
28. Wolfram Research, Inc. *Mathematica, Version 5.0.* Champaign, Illinois, 2004.
29. S. Seeger and K. H. Hoffmann. On symbolic derivation of the cumulant equations. *Comp. Phys. Comm.*, 168(3):165–176, 2005.
30. S. Seeger. Cumulant method implementation for MATHEMATICA. `http://www.tu-chemnitz.de/pub/2003/0120`, March 2003.
31. W. Weiss. Continuous shock structure in extended thermodynamics. *Phys. Rev. E*, 52(6):R5760–R5763, 1995.
32. G. Boillat and T. Ruggeri. On the shock structure problem for hyperbolic system of balance laws and convex entropy. *Continuum Mech. Thermodyn.*, 10(5):285–292, 1998.
33. S. Seeger and K. H. Hoffmann. On the domain of hyperbolicity of the cumulant equations. *J. Stat. Phys.*, 121(1–2):75–90, 2005.
34. E. F. Toro. *Riemann solvers and numerical methods for fluid dynamics: a practical introduction.* Springer Verlag, Berlin, Heidelberg, New York, 2nd edition, 1999.
35. D. Kröner, M. Ohlberger, and C. Rohde, editors. *An Introduction to Recent Developments in Theory and Numerics for Conservation Laws*, volume 5 of *Lecture Notes in Computational Science and Engineering*, Berlin, Heidelberg, New York, 1999. International School on Theory and Numerics for Conservation Laws 20–24th October, 1997, Springer.
36. T. J. Barth and H. Deconinck, editors. *High-Order Methods for Computational Physics*, volume 9 of *Lecture Notes in Computational Science and Engineering.* Springer, Berlin, Heidelberg, New York, 1999.
37. E. Godlewski and P.-A. Raviart. *Numerical Approximation of Hyperbolic Systems of Conservation Laws*, volume 118 of *Applied Mathematical Sciences.* Springer, New York, Heidelberg, 1996.
38. M. Fey, editor. *Hyperbolic Problems: Theory, Numerics, Applications.* Number 129 in International Series of Numercial Mathematics. Birkhäuser Verlag, Switzerland, 1999.
39. P. Houston and E. Süli. Stabilized *hp*-finite element approximation of partial differential equations with nonnegative characteristic form. *Journal of Computing*, 66(2):99–119, 2001.
40. M. Mock. Systems of conservation laws of mixed type. *J. Differential Equations*, 37:70–88, 1980.
41. T. J. R. Hughes and L. P. Franca. A new finite element formulation for computational fluid dynamics: VII. the Stokes problem with various well-posed boundary conditions: symmetric formulations that converge for all velocity/pressure spaces. *Comput. Methods Appl. Mech. Engrg.*, 65:85–96, 1987.
42. Lax. Weak solutions of nonlinear hyperbolic equations and their numerical compuation. *Comm. Pure Appl. Math.*, VII:159–193, 1954.

43. J. C. Maxwell. On stresses in rarefied gases arising from inequalities of temperature. *Phil. Trans. Royal Soc.*, 170:231–256, 1879.
44. M. Knudsen. *The Kinetic Theory of Gases.* Methuen, London, 1950.
45. A. V. Bogdanov. *Interaction of Gases with Surfaces*, volume 25 of *Lecture Notes in Physics.* Springer, Berlin, Heidelberg, 1995.
46. I. Kuščer. Phenomenological aspects of gas-surface interaction. In E. G. D. Dohen and W. Fiszdom, editors, *Fundamental Problems in Statistical Mechanics*, volume IV, pages 441–467. Ossolineum, Warsaw, 1978.
47. E. Ikenberry and C. Truesdell. On the pressures and the flux of energy in a gas according to Maxwell's kinetic theory, I. *J. Rational Mech. Anal.*, 5(1):1–126, 1956.

Index

Editorial Policy

1. Volumes in the following three categories will be published in LNCSE:
i) Research monographs
ii) Lecture and seminar notes
iii) Conference proceedings

Those considering a book which might be suitable for the series are strongly advised to contact the publisher or the series editors at an early stage.

2. Categories i) and ii). These categories will be emphasized by Lecture Notes in Computational Science and Engineering. **Submissions by interdisciplinary teams of authors are encouraged.** The goal is to report new developments – quickly, informally, and in a way that will make them accessible to non-specialists. In the evaluation of submissions timeliness of the work is an important criterion. Texts should be well-rounded, well-written and reasonably self-contained. In most cases the work will contain results of others as well as those of the author(s). In each case the author(s) should provide sufficient motivation, examples, and applications. In this respect, Ph.D. theses will usually be deemed unsuitable for the Lecture Notes series. Proposals for volumes in these categories should be submitted either to one of the series editors or to Springer-Verlag, Heidelberg, and will be refereed. A provisional judgment on the acceptability of a project can be based on partial information about the work: a detailed outline describing the contents of each chapter, the estimated length, a bibliography, and one or two sample chapters – or a first draft. A final decision whether to accept will rest on an evaluation of the completed work which should include
- at least 100 pages of text;
- a table of contents;
- an informative introduction perhaps with some historical remarks which should be accessible to readers unfamiliar with the topic treated;
- a subject index.

3. Category iii). Conference proceedings will be considered for publication provided that they are both of exceptional interest and devoted to a single topic. One (or more) expert participants will act as the scientific editor(s) of the volume. They select the papers which are suitable for inclusion and have them individually refereed as for a journal. Papers not closely related to the central topic are to be excluded. Organizers should contact Lecture Notes in Computational Science and Engineering at the planning stage.

In exceptional cases some other multi-author-volumes may be considered in this category.

4. Format. Only works in English are considered. They should be submitted in camera-ready form according to Springer-Verlag's specifications.
Electronic material can be included if appropriate. Please contact the publisher.
Technical instructions and/or LaTeX macros are available via
http://www.springer.com/east/home/math/math+authors?SGWID=5-40017-6-71391-0.
The macros can also be sent on request.

General Remarks

Lecture Notes are printed by photo-offset from the master-copy delivered in camera-ready form by the authors. For this purpose Springer-Verlag provides technical instructions for the preparation of manuscripts. See also *Editorial Policy*.

Careful preparation of manuscripts will help keep production time short and ensure a satisfactory appearance of the finished book.

The following terms and conditions hold:

Categories i), ii), and iii):
Authors receive 50 free copies of their book. No royalty is paid. Commitment to publish is made by letter of intent rather than by signing a formal contract. Springer-Verlag secures the copyright for each volume.

For conference proceedings, editors receive a total of 50 free copies of their volume for distribution to the contributing authors.

All categories:
Authors are entitled to purchase further copies of their book and other Springer mathematics books for their personal use, at a discount of 33,3 % directly from Springer-Verlag.

Addresses:

Timothy J. Barth
NASA Ames Research Center
NAS Division
Moffett Field, CA 94035, USA
e-mail: barth@nas.nasa.gov

Michael Griebel
Institut für Numerische Simulation
der Universität Bonn
Wegelerstr. 6
53115 Bonn, Germany
e-mail: griebel@ins.uni-bonn.de

David E. Keyes
Department of Applied Physics
and Applied Mathematics
Columbia University
200 S. W. Mudd Building
500 W. 120th Street
New York, NY 10027, USA
e-mail: david.keyes@columbia.edu

Risto M. Nieminen
Laboratory of Physics
Helsinki University of Technology
02150 Espoo, Finland
e-mail: rni@fyslab.hut.fi

Dirk Roose
Department of Computer Science
Katholieke Universiteit Leuven
Celestijnenlaan 200A
3001 Leuven-Heverlee, Belgium
e-mail: dirk.roose@cs.kuleuven.ac.be

Tamar Schlick
Department of Chemistry
Courant Institute of Mathematical
Sciences
New York University
and Howard Hughes Medical Institute
251 Mercer Street
New York, NY 10012, USA
e-mail: schlick@nyu.edu

Mathematics Editor at Springer: Martin Peters
Springer-Verlag, Mathematics Editorial IV
Tiergartenstrasse 17
D-69121 Heidelberg, Germany
Tel.: *49 (6221) 487-8185
Fax: *49 (6221) 487-8355
e-mail: martin.peters@springer.com

Lecture Notes in Computational Science and Engineering

Vol. 1 D. Funaro, *Spectral Elements for Transport-Dominated Equations.* 1997. X, 211 pp. Softcover. ISBN 3-540-62649-2

Vol. 2 H. P. Langtangen, *Computational Partial Differential Equations.* Numerical Methods and Diffpack Programming. 1999. XXIII, 682 pp. Hardcover. ISBN 3-540-65274-4

Vol. 3 W. Hackbusch, G. Wittum (eds.), *Multigrid Methods V.* Proceedings of the Fifth European Multigrid Conference held in Stuttgart, Germany, October 1-4, 1996. 1998. VIII, 334 pp. Softcover. ISBN 3-540-63133-X

Vol. 4 P. Deuflhard, J. Hermans, B. Leimkuhler, A. E. Mark, S. Reich, R. D. Skeel (eds.), *Computational Molecular Dynamics: Challenges, Methods, Ideas.* Proceedings of the 2nd International Symposium on Algorithms for Macromolecular Modelling, Berlin, May 21-24, 1997. 1998. XI, 489 pp. Softcover. ISBN 3-540-63242-5

Vol. 5 D. Kröner, M. Ohlberger, C. Rohde (eds.), *An Introduction to Recent Developments in Theory and Numerics for Conservation Laws.* Proceedings of the International School on Theory and Numerics for Conservation Laws, Freiburg / Littenweiler, October 20-24, 1997. 1998. VII, 285 pp. Softcover. ISBN 3-540-65081-4

Vol. 6 S. Turek, *Efficient Solvers for Incompressible Flow Problems.* An Algorithmic and Computational Approach. 1999. XVII, 352 pp, with CD-ROM. Hardcover. ISBN 3-540-65433-X

Vol. 7 R. von Schwerin, ***Multi Body System SIM**ulation.* Numerical Methods, Algorithms, and Software. 1999. XX, 338 pp. Softcover. ISBN 3-540-65662-6

Vol. 8 H.-J. Bungartz, F. Durst, C. Zenger (eds.), *High Performance Scientific and Engineering Computing.* Proceedings of the International FORTWIHR Conference on HPSEC, Munich, March 16-18, 1998. 1999. X, 471 pp. Softcover. ISBN 3-540-65730-4

Vol. 9 T. J. Barth, H. Deconinck (eds.), *High-Order Methods for Computational Physics.* 1999. VII, 582 pp. Hardcover. ISBN 3-540-65893-9

Vol. 10 H. P. Langtangen, A. M. Bruaset, E. Quak (eds.), *Advances in Software Tools for Scientific Computing.* 2000. X, 357 pp. Softcover. ISBN 3-540-66557-9

Vol. 11 B. Cockburn, G. E. Karniadakis, C.-W. Shu (eds.), *Discontinuous Galerkin Methods.* Theory, Computation and Applications. 2000. XI, 470 pp. Hardcover. ISBN 3-540-66787-3

Vol. 12 U. van Rienen, *Numerical Methods in Computational Electrodynamics.* Linear Systems in Practical Applications. 2000. XIII, 375 pp. Softcover. ISBN 3-540-67629-5

Vol. 13 B. Engquist, L. Johnsson, M. Hammill, F. Short (eds.), *Simulation and Visualization on the Grid.* Parallelldatorcentrum Seventh Annual Conference, Stockholm, December 1999, Proceedings. 2000. XIII, 301 pp. Softcover. ISBN 3-540-67264-8

Vol. 14 E. Dick, K. Riemslagh, J. Vierendeels (eds.), *Multigrid Methods VI.* Proceedings of the Sixth European Multigrid Conference Held in Gent, Belgium, September 27-30, 1999. 2000. IX, 293 pp. Softcover. ISBN 3-540-67157-9

Vol. 15 A. Frommer, T. Lippert, B. Medeke, K. Schilling (eds.), *Numerical Challenges in Lattice Quantum Chromodynamics.* Joint Interdisciplinary Workshop of John von Neumann Institute for Computing, Jülich and Institute of Applied Computer Science, Wuppertal University, August 1999. 2000. VIII, 184 pp. Softcover. ISBN 3-540-67732-1

Vol. 16 J. Lang, *Adaptive Multilevel Solution of Nonlinear Parabolic PDE Systems.* Theory, Algorithm, and Applications. 2001. XII, 157 pp. Softcover. ISBN 3-540-67900-6

Vol. 17 B. I. Wohlmuth, *Discretization Methods and Iterative Solvers Based on Domain Decomposition.* 2001. X, 197 pp. Softcover. ISBN 3-540-41083-X

Vol. 18 U. van Rienen, M. Günther, D. Hecht (eds.), *Scientific Computing in Electrical Engineering.* Proceedings of the 3rd International Workshop, August 20-23, 2000, Warnemünde, Germany. 2001. XII, 428 pp. Softcover. ISBN 3-540-42173-4

Vol. 19 I. Babuška, P. G. Ciarlet, T. Miyoshi (eds.), *Mathematical Modeling and Numerical Simulation in Continuum Mechanics.* Proceedings of the International Symposium on Mathematical Modeling and Numerical Simulation in Continuum Mechanics, September 29 - October 3, 2000, Yamaguchi, Japan. 2002. VIII, 301 pp. Softcover. ISBN 3-540-42399-0

Vol. 20 T. J. Barth, T. Chan, R. Haimes (eds.), *Multiscale and Multiresolution Methods.* Theory and Applications. 2002. X, 389 pp. Softcover. ISBN 3-540-42420-2

Vol. 21 M. Breuer, F. Durst, C. Zenger (eds.), *High Performance Scientific and Engineering Computing.* Proceedings of the 3rd International FORTWIHR Conference on HPSEC, Erlangen, March 12-14, 2001. 2002. XIII, 408 pp. Softcover. ISBN 3-540-42946-8

Vol. 22 K. Urban, *Wavelets in Numerical Simulation.* Problem Adapted Construction and Applications. 2002. XV, 181 pp. Softcover. ISBN 3-540-43055-5

Vol. 23 L. F. Pavarino, A. Toselli (eds.), *Recent Developments in Domain Decomposition Methods.* 2002. XII, 243 pp. Softcover. ISBN 3-540-43413-5

Vol. 24 T. Schlick, H. H. Gan (eds.), *Computational Methods for Macromolecules: Challenges and Applications.* Proceedings of the 3rd International Workshop on Algorithms for Macromolecular Modeling, New York, October 12-14, 2000. 2002. IX, 504 pp. Softcover. ISBN 3-540-43756-8

Vol. 25 T. J. Barth, H. Deconinck (eds.), *Error Estimation and Adaptive Discretization Methods in Computational Fluid Dynamics.* 2003. VII, 344 pp. Hardcover. ISBN 3-540-43758-4

Vol. 26 M. Griebel, M. A. Schweitzer (eds.), *Meshfree Methods for Partial Differential Equations.* 2003. IX, 466 pp. Softcover. ISBN 3-540-43891-2

Vol. 27 S. Müller, *Adaptive Multiscale Schemes for Conservation Laws.* 2003. XIV, 181 pp. Softcover. ISBN 3-540-44325-8

Vol. 28 C. Carstensen, S. Funken, W. Hackbusch, R. H. W. Hoppe, P. Monk (eds.), *Computational Electromagnetics.* Proceedings of the GAMM Workshop on "Computational Electromagnetics", Kiel, Germany, January 26-28, 2001. 2003. X, 209 pp. Softcover. ISBN 3-540-44392-4

Vol. 29 M. A. Schweitzer, *A Parallel Multilevel Partition of Unity Method for Elliptic Partial Differential Equations.* 2003. V, 194 pp. Softcover. ISBN 3-540-00351-7

Vol. 30 T. Biegler, O. Ghattas, M. Heinkenschloss, B. van Bloemen Waanders (eds.), *Large-Scale PDE-Constrained Optimization.* 2003. VI, 349 pp. Softcover. ISBN 3-540-05045-0

Vol. 31 M. Ainsworth, P. Davies, D. Duncan, P. Martin, B. Rynne (eds.), *Topics in Computational Wave Propagation.* Direct and Inverse Problems. 2003. VIII, 399 pp. Softcover. ISBN 3-540-00744-X

Vol. 32 H. Emmerich, B. Nestler, M. Schreckenberg (eds.), *Interface and Transport Dynamics.* Computational Modelling. 2003. XV, 432 pp. Hardcover. ISBN 3-540-40367-1

Vol. 33 H. P. Langtangen, A. Tveito (eds.), *Advanced Topics in Computational Partial Differential Equations.* Numerical Methods and Diffpack Programming. 2003. XIX, 658 pp. Softcover. ISBN 3-540-01438-1

Vol. 34 V. John, *Large Eddy Simulation of Turbulent Incompressible Flows.* Analytical and Numerical Results for a Class of LES Models. 2004. XII, 261 pp. Softcover. ISBN 3-540-40643-3

Vol. 35 E. Bänsch (ed.), *Challenges in Scientific Computing - CISC 2002.* Proceedings of the Conference *Challenges in Scientific Computing*, Berlin, October 2-5, 2002. 2003. VIII, 287 pp. Hardcover. ISBN 3-540-40887-8

Vol. 36 B. N. Khoromskij, G. Wittum, *Numerical Solution of Elliptic Differential Equations by Reduction to the Interface.* 2004. XI, 293 pp. Softcover. ISBN 3-540-20406-7

Vol. 37 A. Iske, *Multiresolution Methods in Scattered Data Modelling.* 2004. XII, 182 pp. Softcover. ISBN 3-540-20479-2

Vol. 38 S.-I. Niculescu, K. Gu (eds.), *Advances in Time-Delay Systems.* 2004. XIV, 446 pp. Softcover. ISBN 3-540-20890-9

Vol. 39 S. Attinger, P. Koumoutsakos (eds.), *Multiscale Modelling and Simulation.* 2004. VIII, 277 pp. Softcover. ISBN 3-540-21180-2

Vol. 40 R. Kornhuber, R. Hoppe, J. Périaux, O. Pironneau, O. Wildlund, J. Xu (eds.), *Domain Decomposition Methods in Science and Engineering.* 2005. XVIII, 690 pp. Softcover. ISBN 3-540-22523-4

Vol. 41 T. Plewa, T. Linde, V.G. Weirs (eds.), *Adaptive Mesh Refinement – Theory and Applications.* 2005. XIV, 552 pp. Softcover. ISBN 3-540-21147-0

Vol. 42 A. Schmidt, K.G. Siebert, *Design of Adaptive Finite Element Software.* The Finite Element Toolbox ALBERTA. 2005. XII, 322 pp. Hardcover. ISBN 3-540-22842-X

Vol. 43 M. Griebel, M.A. Schweitzer (eds.), *Meshfree Methods for Partial Differential Equations II.* 2005. XIII, 303 pp. Softcover. ISBN 3-540-23026-2

Vol. 44 B. Engquist, P. Lötstedt, O. Runborg (eds.), *Multiscale Methods in Science and Engineering.* 2005. XII, 291 pp. Softcover. ISBN 3-540-25335-1

Vol. 45 P. Benner, V. Mehrmann, D.C. Sorensen (eds.), *Dimension Reduction of Large-Scale Systems.* 2005. XII, 402 pp. Softcover. ISBN 3-540-24545-6

Vol. 46 D. Kressner (ed.), *Numerical Methods for General and Structured Eigenvalue Problems.* 2005. XIV, 258 pp. Softcover. ISBN 3-540-24546-4

Vol. 47 A. Boriçi, A. Frommer, B. Joó, A. Kennedy, B. Pendleton (eds.), *QCD and Numerical Analysis III.* 2005. XIII, 201 pp. Softcover. ISBN 3-540-21257-4

Vol. 48 F. Graziani (ed.), *Computational Methods in Transport.* 2006. VIII, 524 pp. Softcover. ISBN 3-540-28122-3

Vol. 49 B. Leimkuhler, C. Chipot, R. Elber, A. Laaksonen, A. Mark, T. Schlick, C. Schütte, R. Skeel (eds.), *New Algorithms for Macromolecular Simulation.* 2006. XVI, 376 pp. Softcover. ISBN 3-540-25542-7

Vol. 50 M. Bücker, G. Corliss, P. Hovland, U. Naumann, B. Norris (eds.), *Automatic Differentiation: Applications, Theory, and Implementations.* 2006. XVIII, 362 pp. Softcover. ISBN 3-540-28403-6

Vol. 51 A.M. Bruaset, A. Tveito (eds.), *Numerical Solution of Partial Differential Equations on Parallel Computers* 2006. XII, 482 pp. Softcover. ISBN 3-540-29076-1

Vol. 52 K.H. Hoffmann, A. Meyer (eds.), *Parallel Algorithms and Cluster Computing.* 2006. X, 374 pp. Softcover. ISBN 3-540-33539-0

For further information on these books please have a look at our mathematics catalogue at the following URL: `www.springer.com/series/3527`

Monographs in Computational Science and Engineering

Vol. 1 J. Sundnes, G.T. Lines, X. Cai, B.F. Nielsen, K.-A. Mardal, A. Tveito, *Computing the Electrical Activity in the Heart.* 2006. XI, 318 pp. Hardcover. ISBN 3-540-33432-7

For further information on this book, please have a look at our mathematics catalogue at the following URL: `www.springer.com/series/7417`

Texts in Computational Science and Engineering

Vol. 1 H. P. Langtangen, *Computational Partial Differential Equations.* Numerical Methods and Diffpack Programming. 2nd Edition 2003. XXVI, 855 pp. Hardcover. ISBN 3-540-43416-X

Vol. 2 A. Quarteroni, F. Saleri, *Scientific Computing with MATLAB and Octave.* 2nd Edition 2006. XIV, 318 pp. Hardcover. ISBN 3-540-32612-X

Vol. 3 H. P. Langtangen, *Python Scripting for Computational Science.* 2nd Edition 2006. XXIV, 736 pp. Hardcover. ISBN 3-540-29415-5

For further information on these books please have a look at our mathematics catalogue at the following URL: www.springer.com/series/5151

FSC
www.fsc.org
MIX
Papier | Fördert
gute Waldnutzung
FSC® C083411

Kolibri 360 GmbH
Ferdinand-Jühlke-Straße 7
99095 Erfurt, Deutschland
produktsicherheit@kolibri360.de